TERRAIN ANALYSIS

TERRAIN ANALYSIS
PRINCIPLES AND APPLICATIONS

Edited by

John P. Wilson
University of Southern California
Los Angeles

John C. Gallant
CSIRO Land and Water
Canberra, Australia

JOHN WILEY & SONS, INC.

New York • Chichester • Weinheim • Brisbane • Singapore • Toronto

This book is printed on acid-free paper. ∞

This publication is designed to provide accurate and authoritative information in regard to the subject matter covered. It is sold with the understanding that the publisher is not engaged in rendering professional services. If professional advice or other expert assistance is required, the services of a competent professional person should be sought.

Library of Congress Cataloging-in-Publication Data:
Terrain analysis : principles and applications / edited by John P. Wilson, John C. Gallant.
    p. cm.
   Includes bibliographical references.
   ISBN 0-471-32188-5 (alk. paper)
   1. Landforms—Measurement.   2. Geomorphology—Technique.   I. Wilson, John P. (John Peter), 1955-   II. Gallant, John C.

GB400.4 .T37 2000
551.41—dc21

                99-089635

Printed in the United States of America.

10 9 8 7 6 5 4 3

This book celebrates the life and scientific accomplishments of Ian Moore
(1951–1993) who served as a friend and mentor at key times in both of our lives.

CONTENTS

3. Primary Topographic Attributes   **51**
by John C. Gallant and John P. Wilson

16. Toward a Spatial Model of Boreal Forest Ecosystems: The Role of Digital Terrain Analysis   **391**

by Brendan G. Mackey, Ian C. Mullen, Kenneth A. Baldwin, John C. Gallant, Richard A. Sims, and Daniel W. McKenney

17. Future Directions for Terrain Analysis   **423**

by John C. Gallant, Michael F. Hutchinson, and John P. Wilson

PREFACE

The origins of this book can be traced to the pioneering work of Professor Ian D. Moore and our own efforts to continue the development and support of his terrain-analysis tools following his death in September 1993. Ian Moore saw terrain analysis as a robust method for modeling over large areas the complex spatial patterns of environmental systems, a vision that is borne out by the large variety of hydrologic, geomorphic, and biological applications demonstrated in this book. The book itself is the result of a collaborative effort involving thirteen research groups based in four countries. Special thanks are owed to the authors who wrote and revised the chapters that form the core of this book, and to a series of anonymous reviewers who provided numerous ideas and suggestions for improving earlier drafts of individual chapters. Our contributions have been the five introductory chapters that are the foundations for the remainder of the book and that provide essential supporting material and references for the reader.

We are particularly grateful to Ian Moore, our friend and mentor, who started us down the path exemplified by this book. The book is based on his terrain-analysis tools, which he called TAPES: Terrain Analysis Programs for the Environmental Sciences. Ian's professional life was split between the United States and Australia, and numerous organizations in both countries contributed support for the development of these terrain analysis tools. The Australian contributors included the Australian National University, Australian Research Council, Commonwealth Scientific and Industrial Research Organization, Department of Industry, Science and Technology, Land and Water Resources Research and Development Corporation, Murray–Darling Basin Commission, National Soil Conservation Program, Water Research Advisory Council, and the Water Research Foundation of Australia. The U.S. contributors included the Department of Agriculture–Agricultural Research Service's Great Plains Systems Unit, Department of Agriculture–Cooperative States Research Service, M.J. Murdock Charitable Trust, Montana Agricultural Experiment Station, Montana State University, Minnesota Agricultural Experiment Station, Minnesota Water Resources Research Center, National Science Foundation, and the University of Minnesota. Our own work with Ian covered the period 1988–1993 and focused on the development, testing, and application of selected terrain-analysis algorithms. Professor Moore was a source of inspiration as well as new ideas, and many of his ideas are still embedded in our ongoing work.

This book is an example of such work. It illustrates some of the applications of terrain analysis, it explains the algorithms used by current terrain-analysis software, and deals with details such as the quirks of particular algorithms, interpretation of terrain attributes, use of terrain attributes in predictive models, and the effects of scale and resolution. We hope it will inspire others to work on new digital terrain-analysis algorithms and applications.

John P. Wilson
John C. Gallant

ACKNOWLEDGMENTS

Jay Bell, David Grigal, and Peter Bates acknowledge the support of the United States Department of Agriculture–Forest Service Northern Station's Global Change Program and the Minnesota Agricultural Experiment Station. Special thanks are due to Jim Engstrom for playing a major role in the development of the initial digital geographic database and to Charlie Butler for assisting with parts of the statistical analysis in Chapter 12.

Jinfan Duan and Gordon Grant (Chapter 13) acknowledge the support of CLAMS and the National Science Foundation-sponsored Andrew Forest LTER program. Special thanks are due to Fred Swanson for contributing ideas and feedback on numerous occasions, and to Hazel Hammond, George Lienkaemper, and Beverly Wemple for providing GIS support. The observed landslide inventory, soil type and depth, vegetation zones, and land use history were obtained from the Forest Science Data Bank maintained by the Forestry Science Laboratory in Corvallis, Oregon.

Janet Franklin, Paul McCullough, and Curtis Gray thank Ellen Bauder, John Gallant, Art Getis, Hazel Gordon, William Hetrich, Brian Lees, David McKinsey, John O'Leary, Brian Ripley, Joseph Shandley, P. A. Walker, Tom White, John Wilson, Paul Zedler, and an anonymous reviewer for their assistance and advice on earlier drafts of Chapter 14.

Jeremy Fried, Daniel Brown, Mark Zweifler, and Michael Gold acknowledge the support of the Michigan State University Office of Outreach (the research reported in Chapter 7 was supported in part through a competitive Statewide Issues Response Grant) and the Michigan Agricultural Experiment Station. Special thanks are due to Jerry Fulcher and Rick Popp of the Michigan Department of Environmental Quality for the map interpretation work on which part of the model validation is based. John Gallant, John Wilson, Demetrios Gatziolis, the Michigan Department of Environmental Quality, and Michigan State University's Institute for Water Research and Center for Remote Sensing provided data, advice, and technical assistance.

Michael Hutchinson, John Gallant, and John Wilson thank Ed DeYoung for preparing the digital contour coverage of the Cottonwood Creek catchment used in Chapters 2, 3, and 4, and Chenyin Zhong for preparing the saturated soil coverage of the same area used in Chapter 4.

Valentina Krysanova, Dirk-Ingmar Müller-Wohlfeil, Wolfgang Cramer, and Alfred Becker acknowledge the support of the German Federal Ministry for Research and Technology. Special thanks are due to John Gallant and Keith Beven for providing the TAPES and TOPMODEL code, and to John Gallant, Keith Beven, and Paul Quinn for providing useful advice at key stages of model implementation. The research reported in Chapter 6 also benefited from and contributes to research being carried out as part of the core project on Biospheric Aspects of the Hydrologic Cycle of the International Geosphere–Biosphere Program.

Brendan Mackey, Ian Mullen, Ken Baldwin, John Gallant, Richard Sims, and Daniel McKenney thank Gerry Racey and the Ontario Ministry of Natural Resources who contributed logistical support and use of the Rinker Lake field station facilities. Special thanks are also owed Janice McKee, Kevin Lawrence, Norm Szclyrek, and Brian Zavitz of the Canadian Forest Service in Sault Ste. Marie for help with various aspects of database preparation. The Rinker Lake field survey activities described in Chapter 16 were funded by a grant from the Northern Forestry Program of the joint Federal/Provincial Northern Ontario Development Agreement. A digital version of the Forest Resources Inventory for the Rinker Lake Research Area was provided by Abitibi-Consolidated, and the *Landsat* TM classification was developed in conjunction with the Institute of Space and Terrestrial Studies at the University of Waterloo and Geomatics International.

Neil McKenzie, Paul Gessler, Philip Ryan, and Deborah O'Connell thank Linda Ashton and Andrew Loughhead for computing and technical support. Funds for the research reported in Chapter 10 were provided by the Commonwealth Scientific and Industrial Research Organization, Murray Darling Basin Commission, State Forests New South Wales, Australian Geological Survey Organization, and the Forest and Wood Products Research and Development Corporation.

Gregory Pohl and John Warwick acknowledge the support of the United States Department of Energy (the research reported on in Chapter 9 was supported in part by Grant FC08-93NV11359). Special thanks are due to J. Simunek and M. Th. Van Genuchten for providing the SWMS_2D code, to A. F. Tompson for providing the Turning Bands code, and to Rodger Grayson and Gunter Blöschl for providing helpful advice on the use of the TAPES-C and THALES code.

Stephen Ventura and Barbara Irvin acknowledge the collaboration of Kevin McSweeney and Brian Slater on research underlying Chapter 11. This research was supported in part through grants from the National Science Foundation (SBR 9210093) and the United States Department of Agriculture (Hatch WISO 3866).

Jonathan Wheatley, John Wilson, Roland Redmond, Zhenkui Ma, and Jeff DiBenedetto (Chapter 15) acknowledge the funding and logistical support provided by the Custer National Forest and United States Department of Agriculture–Forest Service Northern Region Office. Special thanks are also owed to Damian Spangrud for help with the development of Arc Macro Language programs and selected map products.

John Wilson and John Gallant thank Ed DeYoung for preparing the final figures reproduced in Chapters 1 and 4. Yan Xu also helped with the preparation of the maps reproduced in Figure 1.3.

John Wilson, Philip Repetto, and Robert Snyder thank Ed DeYoung for helping with the preparation of the figures in Chapter 5. Special thanks are owed to Allan Busacca for providing the Idaho cesium-137 data as well.

Alan Yeakley, George Hornberger, Wayne Swank, Paul Bolstad, and James Vose (Chapter 8) acknowledge the support of the University of Virginia, United States Department of Agriculture–Forest Service, and the National Science Foundation (Grant DEB 9596191). Ian Moore and Keith Beven provided code for TAPES-C and IHDM4, and field assistance was provided by J. S. Hill, A. G. Uhlenhopp, U. C. Tomlinson, G. A. Hamilton, and S. M. Steiner. Helpful comments at various stages in the preparation of this manuscript were received from John Gallant, B. P. Hayden, H. H. Stugart, L. W. Swift, D. L. Urban, and S. L. Yu.

CONTRIBUTORS

Kenneth A. Baldwin, Canadian Forest Service, Great Lakes Forestry Centre, 1219 Queen Street East, P.O. Box 490, Sault Ste. Marie, Ontario P6A 5M7, Canada. E-mail: *kbaldwin@nrcan.gc.ca*

Peter C. Bates, Department of Natural Resources, Western Carolina University, Cullawee, NC 28723, USA. E-mail: *bates@wpoff.wcu.edu*

Alfred Becker, Global Change and Natural Systems Department, Potsdam Institute for Climate Impact Research, P.O. Box 60 12 03, D-14412 Potsdam, Germany. E-mail: *becker@pik-potsdam.de*

Jay C. Bell, Department of Soil, Water, and Climate, University of Minnesota, Borlaug Building, 1991 Upper Buford Circle, St Paul, MN 55108, USA. E-mail: *jbell@soils.umn.edu*

Paul V. Bolstad, Department of Forest Resources, University of Minnesota, St. Paul, MN 55108, USA. E-mail: *pbolstad@forestry.umn.edu*

Daniel G. Brown, School of Natural Resources and Environment, University of Michigan, Ann Arbor, MI 48103, USA. E-mail: *danbrown@umich.edu*

Wolfgang Cramer, Global Change and Natural Systems Department, Potsdam Institute for Climate Impact Research, P.O. Box 60 12 03, D-14412 Potsdam, Germany

Jeff DiBenedetto, Custer National Forest, P.O. Box 50760, Billings, MT 59760, USA. E-mail: *dibennedetto_jeff/rl_custer@fs.fed.us*

Jinfan Duan, Pacific Northwest Research Station, United States Forest Service, Corvallis, OR 97331, USA.

Janet Franklin, Department of Geography, San Diego State University, San Diego, CA 92182-4493, USA. E-mail: *janet.franklin@sdsu.edu*

Jeremy S. Fried, Pacific Northwest Forest Inventory and Monitoring Program, Portland Forestry Sciences Laboratory, P.O. Box 3890, Portland, OR 97208, USA. E-mail: *jfried/r6pnw_portland@fs.fed.us*

John C. Gallant, Division of Land and Water, Commonwealth Scientific and Industrial Research Organization, G.P.O. Box 1666, Canberra, ACT 2601, Australia. E-mail: *john.gallant@cbr.clw.csiro.au*

Paul E. Gessler, Department of Forest Resources, University of Idaho, Moscow, ID 83844-1133, USA. E-mail: *paulg@uidaho.edu*

Michael A. Gold, Center for Agroforestry, University of Missouri, 203 ABNR Building, Columbia, MO 65211, USA. E-mail: *goldm@missouri.edu*

Gordon E. Grant, Pacific Northwest Research Station, United States Forest Service, Corvallis, OR 97331, USA. E-mail: *grantg@ccmail.orst.edu*

Curtis Gray, Pacific Meridian Resources, Remote Sensing Laboratory, Forest Service, United States Department of Agriculture, 1920 20th Street, Sacramento, CA 95814, USA. E-mail: *curtis@cdf.ca.gov*

David F. Grigal, Department of Soil, Water, and Climate, University of Minnesota, Borlaug Building, 1991 Upper Buford Circle, St. Paul, MN 55108, USA. E-mail: *dgrigal@soils.umn.edu*

George M. Hornberger, Department of Environmental Sciences, University of Virginia, Charlottesville, VA 22903, USA. E-mail: *gmh3k@virginia.edu*

Michael F. Hutchinson, Centre for Resource and Environmental Studies, Australian National University, Canberra, ACT 0200, Australia. E-mail: *hutch@cres20.anu.edu.au*

Barbara J. Irvin, GeoAnalytics, Inc., 1716 Fordem Avenue, Madison, WI 53704-4604, USA. E-mail: *irvinb@geoanalytics.com*

Valentina Krysanova, Global Change and Natural Systems Department, Potsdam Institute for Climate Impact Research, P.O. Box 60 12 03, D-14412 Potsdam, Germany. E-mail: *valen@pik-potsdam.de*

Zhenkui Ma, Weyerhauser Corporation, WWC-2E2, 33405 8th Avenue South, Federal Way, WA 98003, USA. E-mail: *maz@wdni.com*

Brendan G. Mackey, Department of Geography, Australian National University, Canberra, ACT 0200, Australia. E-mail: *brendan@geography.anu.edu.au*

Paul McCullough, San Diego Data Processing Corporation, 5975 Santa Fe Street, San Diego, CA 92109, USA. E-mail: *hemizonia@earthlink.com*

Daniel W. McKenney, Canadian Forest Service, Great Lakes Forestry Centre, 1219 Queen Street East, P.O. Box 490, Sault Ste. Marie, Ontario P6A 5M7, Canada. E-mail: *dmckenne@nrcan.gc.ca*

Neil J. McKenzie, Division of Land and Water, Commonwealth Scientific and Industrial Research Organization, G.P.O. Box 1666, Canberra, ACT 2601, Australia. E-mail: *neil.mckenzie@cbr.clw.csiro.au*

Ian C. Mullen, Land and Water Sciences Division, Bureau of Rural Sciences, P.O. Box E11, ACT 2604, Australia. E-mail: *ian.mullen@brs.gov.au*

Dirk-Ingmar Müller-Wohlfeil, National Environmental Research Institute, Vejlsøevej 25, DK-8600 Sikeborg, Denmark. E-mail: *dmw@dmu.dk*

Deborah O'Connell, Division of Land and Water, Commonwealth Scientific and Industrial Research Organization, G.P.O. Box 1666, Canberra, ACT 2601, Australia. E-mail: *deb.jamie@rainbowis.com.au*

Gregory M. Pohll, Water Resources Center, Desert Research Institute, University of Nevada System, P.O. Box 60220, Reno, NV 89506, USA. E-mail: *pohll@maxey.dri.edu*

Roland L. Redmond, Wildlife Spatial Analysis Laboratory, University of Montana, Missoula, MT 59812, USA. E-mail: *red@selway.umt.edu*

Philip L. Repetto, Baker GeoResearch, Inc., 115 North Broadway Avenue, Billings, MT 59102, USA. E-mail: *skip_repetto@hotmail.com*

Philip J. Ryan, Division of Forestry and Forest Products, Commonwealth Scientific and Industrial Research Organization, P.O. Box E4008, Kingston, ACT 2604, Australia. E-mail: *philip.ryan@ffp.csiro.au*

Richard A. Sims, EBA Engineering Consultants Ltd., 550-1100 Melville Street, Vancouver, British Columbia V6E 4A6, Canada. E-mail: *rsims@eba.ca*

Robert D. Snyder, Geographic Information and Analysis Center, Montana State University, Bozeman, MT 59717-0348, USA. E-mail: *bob@papaya.giac.montana.edu*

Wayne T. Swank, Coweeta Hydrologic Laboratory, Forest Service, United States Department of Agriculture, Otto, NC 28763, USA. E-mail: *wswank@sparc.ecology.uga.edu*

Stephen J. Ventura, Department of Soil Science, University of Wisconsin–Madison, Madison, WI 53706, USA. E-mail: *sventura@facstaff.wisc.edu*

James M. Vose, Coweeta Hydrologic Laboratory, Forest Service, United States Department of Agriculture, Otto, NC 28763, USA. E-mail: *jvose@sparc.ecology.uga.edu*

John J. Warwick, Department of Environmental Engineering Sciences, University of Florida, 217 Black Hall, P.O. Box 116450, Gainesville, FL 32611-6450, USA. E-mail: *warwick@eng.ufl.edu*

Jonathan M. Wheatley, Department of Earth Resources, Colorado State University, CO 80523, USA. E-mail: *apres_ski@hotmail.com*

John P. Wilson, Department of Geography, University of Southern California, Los Angeles, CA 90089-0255, USA. E-mail: *jpwilson@usc.edu*

J. Alan Yeakley, Department of Environmental Science and Resources, Portland State University, Portland, OR 97207-0751, USA. E-mail: *yeakley@pdx.edu*

Mark O. Zweifler, Pacific Meridian Resources, Forest Management Division, Michigan Department of Natural Resources, P.O. Box 30452, Lansing, MI 48909-7952, USA.

TERRAIN ANALYSIS

■■■■■■ CHAPTER 1

Digital Terrain Analysis

John P. Wilson and John C. Gallant

1.1 PRINCIPLES AND APPLICATIONS

The development and application of the *TAPES: Terrain Analysis Programs for the Environmental Sciences* software tools described in this book was motivated by our view of the world as a stage on which a series of hierarchically scaled biophysical processes are played out (Figure 1.1). This approach is useful because it can handle the complexity of individual landscape processes and patterns as well as some of the difficulties that are encountered in delineating the appropriate spatial and temporal scales (O'Neill et al. 1986, Mackey 1996, Malanson and Armstrong 1997). Many of the important biophysical processes operating at or near the earth's surface are influenced by both past events and contemporary controls, interactions, and thresholds (Dietrich et al. 1992, Grayson et al. 1993, Montgomery and Dietrich 1995). These interrelationships are complicated and may be best understood using a dynamic systems modeling approach (Kirkby et al. 1996). The boundaries separating different spatial and temporal scales are not very clear and they may vary with individual processes and/or landscapes (cf. Sivapalan and Wood 1986, Mackey 1996, Malanson and Armstrong 1997).

This state of affairs suggests that additional work is required to identify the important spatial and temporal scales and the factors that influence or control the processes and patterns operating at particular scales. The potential benefits may be substantial. Schaffer (1981), working with interacting systems of populations in community ecology, and Phillips (1986), working on examples in fluvial geomorphology, have demonstrated that the key processes operating over different timescales can be considered independently of each other. Phillips (1988) has also shown how the key processes operating at different spatial scales and affecting the hydraulic gradient of a desert stream in Arizona can be considered independently of each other. Band et al. (1991) generated landscape units with low internal variance and high between-unit

Terrain Analysis: Principles and Applications, Edited by John P. Wilson and John C. Gallant.
ISBN 0-471-32188-5   © 2000 John Wiley & Sons, Inc.

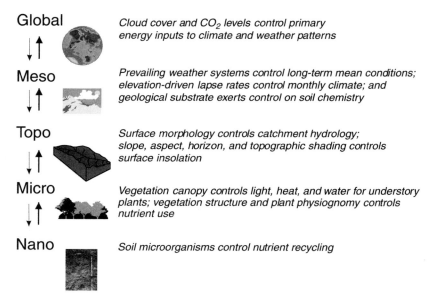

Global Cloud cover and CO_2 levels control primary energy inputs to climate and weather patterns

Meso Prevailing weather systems control long-term mean conditions; elevation-driven lapse rates control monthly climate; and geological substrate exerts control on soil chemistry

Topo Surface morphology controls catchment hydrology; slope, aspect, horizon, and topographic shading controls surface insolation

Micro Vegetation canopy controls light, heat, and water for understory plants; vegetation structure and plant physiognomy controls nutrient use

Nano Soil microorganisms control nutrient recycling

Figure 1.1. Scales at which various biophysical processes dominate calculation of primary environmental regimes. Reprinted with permission from Mackey (1996) The role of GIS and environmental modeling in the conservation of biodiversity. In *Proceedings of the Third International Conference on Integrating GIS and Environmental Modeling, Santa Fe, New Mexico, 21–25 January, 1996,* edited by NCGIA. Copyright © 1996 by National Center for Geographic Information and Analysis, University of California, Santa Barbara.

variance for the important parameters in a nonlinear, deterministic model designed to simulate carbon, water, and nitrogen cycles in a forest ecosystem using a series of hillslope and watershed templates. However, this result may not be universally applicable. Phillips (1988) warned that the key differences in spatial scales cannot be related to fundamental landscape units in numerous instances. Grayson et al. (1993) argued that we should avoid implementing at one scale models developed at a different scale because the simplifying assumptions will often undermine the validity of the original models. Kirkby et al. (1996) concluded that different processes and interactions are likely to emerge as dominant as we move from the plot scale to catchment and regional scales in soil erosion modeling applications. This state of affairs is true of other hydrological, geomorphological, and biological settings as well.

Most of the hydrological, geomorphological, and ecological research of the past century has been conducted at the global and nano- or microscales identified in Figure 1.1 (Mackey 1996). The meso- and toposcales have received much less attention, and yet these scales are important because many of the solutions to environmental problems, such as accelerated soil erosion and non-point-source pollution, will require changes in management strategies at these landscape scales (Moore and Hutchinson 1991). The influence of geologic substrate on soil chemistry (e.g., Likens et al. 1977) and impact of prevailing weather systems and elevation-driven lapse rates on long-term average monthly climate (e.g., Daly et al. 1994, Hutchinson 1995) exemplify some of the controls operating at the mesoscale. The influence of surface morphology on catchment hydrology and the impact of slope, aspect, and horizon shading on insolation probably represent the most important controls operating at toposcales. Numerous studies have shown how the shape of the land surface can affect the lateral migration and accumulation of water, sediments, and other constituents (e.g., Moore et al. 1988a). These variables, in turn, influence soil development (e.g., Kreznor et al. 1989) and exert a strong influence on the spatial and temporal distributions of the light, heat, water, and mineral nutrients required by photosynthesizing plants (Mackey 1996). The increased popularity of work at these two intermediate scales during the past decade has capitalized on the increasing availability of high-resolution, continuous, digital elevation data and the development of new computerized terrain-analysis tools (Wilson 1996, Burrough and McDonnell 1998, Wilson and Burrough 1999).

1.1.1 Digital Elevation Data Sources and Structures

Most of the currently available digital elevation data sets are the product of photogrammetric data capture (I. D. Moore et al. 1991). These sources rely on the stereoscopic interpretation of aerial photographs or satellite imagery using manual or automatic stereoplotters (Carter 1988, Weibel and Heller 1991). Additional elevation data sets can be acquired by digitizing the contour lines on topographic maps and conducting ground surveys. The advent and widespread use of Global Positioning Systems (GPS) in agriculture and other settings provides many new and affordable opportunities for the collection of large numbers of special-purpose, one-of-a-kind elevation data sets (Fix and Burt 1995, Twigg 1998, Wilson 1999a).

These digital elevation data are usually organized into one of three data structures—(1) regular grids, (2) triangulated irregular networks, and (3) contours—depending on the source and/or preferred method of analysis (Figure 1.2). Square-grid digital elevation models (DEMs) have emerged as the most widely used data structure during the past decade because of their simplicity (i.e., simple elevation matrices that record topological relations between data points implicitly) and ease of computer implementation (I. D. Moore et al. 1991, 1993f, Wise 1998). These advantages offset at least three disadvantages. First, the size of the grid mesh will often affect the storage requirements, computational efficiency, and the quality of the results (Collins and Moon 1981, I. D. Moore et al. 1991). Second, square grids cannot handle abrupt changes in elevation easily and they will often skip important details of the land surface in flat areas (Carter 1988). However, it is worth noting that many of the problems in flat areas occur because the U.S. Geological Survey (USGS) and others persist in recording elevations in whole meters. Third, the computed upslope flow paths will tend to zigzag across the landscape and increase the difficulty of calculating specific catchment areas accurately (Zevenbergen and Thorne 1987, I. D. Moore et al. 1991). Several of these obstacles have been overcome in recent years. For example, there is no generic reason why regular DEMs cannot represent shape well in flat areas, so long as the terrain attributes are calculated by a method that respects surface drainage. ANUDEM (Hutchinson 1988, 1989b) is one such method and is described in more detail in Chapter 2. Similarly, the advent of several new compression techniques have reduced the storage requirements and improved computational efficiency in recent years (e.g., Kidner and Smith 1992, Smith and Lewis 1994). DEMs with grid sizes of 500, 100, 30, 10, and even 1 m are increasingly available for different parts of the globe (see U.S. Geological Survey 1993, Ordinance Survey 1993, and Hutchinson et al. 1996 for examples).

Triangulated irregular networks (TINs) have also found widespread use (e.g., Tajchman 1981, Jones et al. 1990, Yu et al. 1997). TINs are based on triangular elements (facets) with vertices at the sample points (I. D. Moore et al. 1991). These facets consist of planes joining the three adjacent points in the network and are usually constructed using Delauney triangulation (Weibel and Heller 1991). Lee (1991) compared several methods for building TINs from gridded DEMs. However, the best TINs sample surface-specific points, such as peaks, ridges, and breaks in slope, and form an irregular network of points stored as a set of x, y, and z values together with pointers to their neighbors in the net (I. D. Moore et al. 1991). TINs can easily incorporate discontinuities and may constitute efficient data structures because the density of the triangles can be varied to match the roughness of the terrain (I. D. Moore et al. 1991). This arrangement may cancel out the additional storage that is incurred when the topological relations are computed and recorded explicitly (Kumler 1994).

The third structure incorporates the stream tube concept first proposed by Onstad and Brakensiek (1968) and divides landscapes into small, irregularly shaped polygons (elements) based on contour lines and their orthogonals (Figure 1.2) (O'Loughlin 1986, I. D. Moore et al. 1988a). This structure is used most frequently in hydrological applications because it can reduce complex three-dimensional flow equations into a series of coupled one-dimensional equations in areas of complex terrain (e.g., Moore and Foster 1990, Moore and Grayson 1991, Grayson et al. 1994). Excellent reviews of digital elevation data sources and data structures are presented by Carter (1988), Weibel and Heller (1991), and I. D. Moore et al. (1991).

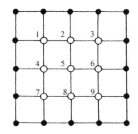  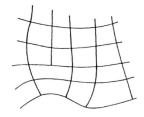

Figure 1.2. Methods of structuring an elevation data network: (a) square-grid network showing a moving 3 by 3 submatrix centered on node 5; (b) triangulated irregular network; and (c) contour-based network. Reprinted with permission from Moore, Grayson, and Ladson (1991) Digital terrain modeling: A review of hydrological, geomorphological, and ecological applications. *Hydrological Processes* 5: 3–30. Copyright © 1991 by John Wiley and Sons Ltd.

The proliferation of digital elevation sources and preprocessing tools means that the initial choice of data structure is not as critical as it once was (Kemp 1997a, b). Numerous methods have been proposed to convert digital elevation data from one structure to another, although care must be exercised with each of these methods to minimize unwanted artifacts (e.g., Krajewski and Gibbs 1994). In addition, larger quantities of data do not necessarily produce better results: Eklundh and Martensson (1995), for example, used ANUDEM (Hutchinson 1988, 1989b) to derive square grids from contours and demonstrated that point sampling produces faster and more accurate square-grid DEMs than the digitizing of contours. Similarly, Wilson et al. (1998) used ANUDEM to derive square grids from irregular point samples and showed that many of the *x, y, z* data points acquired with a truck-mounted GPS were not required to produce satisfactory square-grid DEMs. ANUDEM calculates ridge and streamlines from points of maximum local curvature on contour lines and incorporates a drainage enforcement algorithm that automatically removes spurious sinks or pits in the fitted elevation surface (Hutchinson 1988, 1989b). ANUDEM is one of several programs of this type and an early version has been implemented in the ARC/INFO (Environmental Systems Research Institute, Redlands, CA) geographical information system (GIS) with the TOPOGRID command. Qian et al. (1990) describe an alternative approach that utilizes local operators and global reasoning to automatically extract drainage networks and ridge lines from digital elevation data. Similarly, Smith et al. (1990) proposed a two-step, knowledge-based procedure for extracting channel networks from noisy DEM data. Kumler (1994) described the method used by the U.S. Geological Survey to generate square-grid DEMs from digital contour lines.

Carrara et al. (1997) compared several methods for generating DEMs from contour lines; however, the range of terrain types, sample structures, and modeling routines is so great that attempts to make generalizations about "best" models is tremendously difficult (Burrough and McDonnell 1998, Dixon et al. 1998, Wilson 1999b). In addition, some of the interpolation methods that have been proposed are difficult to use and Eklundh and Martensson (1995) recommended that less experienced users focus on the quality of the input data instead of learning sophisticated interpolation methods. Simpler interpolation methods will give satisfactory results so long as the input data are well sampled and sophisticated algorithms are likely to produce unsatisfactory results if applied to poor data (e.g., Wilson et al. 1998).

1.1.2 Calculation and Use of Topographic Attributes in Hydrological, Geomorphological, and Biological Applications

Many of the most popular topographic attributes, such as slope, specific catchment area, aspect, and plan and profile curvature, can be derived from all three types of elevation data for each and every element as a function of its surroundings (I. D. Moore et al. 1991, 1993f). Individual terrain-analysis tools have been classified in various ways based on the characteristics of the computed attributes and/or their spatial extent. Some authors distinguish tools that perform operations on local neighborhoods (i.e., 3 by 3 moving windows) from those that perform operations on extended neighborhoods (calculation of upslope drainage areas, viewsheds, etc.) (e.g., Burrough and McDonnell 1998). We usually distinguish primary attributes that are computed directly from the DEM and secondary or compound attributes that involve combinations of primary attributes and constitute physically based or empirically derived indices that can characterize the spatial variability of specific processes occurring in the landscape (I. D. Moore et al. 1991, 1993f). This same logic is adopted here.

Primary attributes include slope, aspect, plan and profile curvature, flow-path length, and upslope contributing area (see Table 1.1 for a more complete list). Most of these topographic attributes are calculated from the directional derivatives of a topographic surface. They can be computed directly with a second-order finite difference scheme or by fitting a bivariate interpolation function $z = f(x, y)$ to the DEM and then calculating the derivatives of the function (Moore et al. 1993d, Mitasova et al. 1996, Florinsky 1998). We may or may not want to calculate a depressionless DEM first and we must specify one or more rules to determine drainage directions and the connectivity of individual elements in order to calculate flow-path lengths and upslope contributing areas (e.g., Jenson and Domingue 1988, Martz and De Jong 1988). The overall aim is to be able to use the computed attributes to describe the morphometry, catchment position, and surface attributes of hillslopes and stream channels comprising drainage basins (e.g., Speight 1974, 1980, Band 1986, 1993a, b, Jenson and Domingue 1988, Montgomery and Foufoula-Georgiou 1993, Moore et al. 1993a). Dikau (1989), Dymond et al. (1995), Brabyn (1997), Giles (1998), and Burrough et al. (2000a, b) have all used computed topographic attributes to generate formal landform classifications.

The secondary attributes that are computed from two or more primary attributes are important because they offer an opportunity to describe pattern as a function of process (Table 1.2). Those attributes that quantify the role played by topography in redistributing water in the landscape and in modifying the amount of solar radiation received at the surface have important hydrological, geomorphological, and ecological consequences in many landscapes. These attributes may affect soil characteristics (because the pedogenesis of the soil catena is affected by the way water moves through the environment in many landscapes), distribution and abundance of soil water, susceptibility of landscapes to erosion by water, and the distribution and abundance of flora and fauna. Three sets of compound topographic indices are discussed below to illustrate how these attributes are constructed and used in hydrological, geomorphological, and ecological applications.

Two topographic wetness indices have been used extensively to describe the effects of topography on the location and size of saturated source areas of runoff generation as follows:

$$W_T = \ln \left(\frac{A_s}{T \tan \beta} \right) \tag{1.1}$$

$$W = \ln \left(\frac{A_s}{\tan \beta} \right) \tag{1.2}$$

where A_s is the specific catchment area ($m^2 m^{-1}$), T is the soil transmissivity when the soil profile is saturated, and β is the slope gradient (in degrees) (I. D. Moore et al. 1991,

TABLE 1.1   Primary Topographic Attributes That Can Be Computed by Terrain Analysis from DEM Data

Attribute	Definition	Significance
Altitude	Elevation	Climate, vegetation, potential energy
Upslope height	Mean height of upslope area	Potential energy
Aspect	Slope azimuth	Solar insolation, evapotranspiration, flora and fauna distribution and abundance
Slope	Gradient	Overland and subsurface flow velocity and runoff rate, precipitation, vegetation, geomorphology, soil water content, land capability class
Upslope slope	Mean slope of upslope area	Runoff velocity
Dispersal slope	Mean slope of dispersal area	Rate of soil drainage
Catchment slope	Average slope over the catchment	Time of concentration
Upslope area	Catchment area above a short length of contour	Runoff volume, steady-state runoff rate
Dispersal area	Area downslope from a short length of contour	Soil drainage rate
Catchment area	Area draining to catchment outlet	Runoff volume
Specific catchment area	Upslope area per unit width of contour	Runoff volume, steady-state runoff rate, soil characteristics, soil-water content, geomorphology
Flow path length	Maximum distance of water flow to a point in the catchment	Erosion rates, sediment yield, time of concentration
Upslope length	Mean length of flow paths to a point in the catchment	Flow acceleration, erosion rates
Dispersal length	Distance from a point in the catchment to the outlet	Impedance of soil drainage
Catchment length	Distance from highest point to outlet	Overland flow attenuation
Profile curvature	Slope profile curvature	Flow acceleration, erosion/deposition rate, geomorphology
Plan curvature	Contour curvature	Converging/diverging flow, soil-water content, soil characteristics
Tangential curvature	Plan curvature multiplied by slope	Provides alternative measure of local flow convergence and divergence
Elevation percentile	Proportion of cells in a user-defined circle lower than the center cell	Relative landscape position, flora and fauna distribution and abundance

Source. Adapted with permission from Moore, Grayson, and Ladson (1991) Digital terrain modeling: A review of hydrological, geomorphological, and ecological applications. *Hydrological Processes* 5: 3–30. Copyright © 1991 by John Wiley and Sons Ltd.

TABLE 1.2 Secondary Topographic Attributes That Can Be Computed by Terrain Analysis from DEM Data

Attribute	Definition	Significance
Topographic wetness indices	$W_T = \ln\left(\dfrac{A_s}{T \tan \beta}\right)$	This equation assumes steady-state conditions and describes the spatial distribution and extent of zones of saturation (i.e., variable source areas) for runoff generation as a function of upslope contributing area, soil transmissivity, and slope gradient.
	$W = \ln\left(\dfrac{A_s}{\tan \beta}\right)$	This particular equation assumes steady-state conditions and uniform soil properties (i.e., transmissivity is constant throughout the catchment and equal to unity). This pair of equations predicts zones of saturation where A_s is large (typically in converging segments of landscapes), β is small (at base of concave slopes where slope gradient is reduced), and T_i is small (on shallow soils). These conditions are usually encountered along drainage paths and in zones of water concentration in landscapes.
	$W = \ln\left(\dfrac{A_e}{\tan \beta}\right)$	This quasi-dynamic index substitutes effective drainage area for upslope contributing area and thereby overcomes the limitations of the steady-state assumption used in the first pair of equations.
Stream-power indices	$SPI = A_s \tan \beta$	Measure of erosive power of flowing water based on assumption that discharge (q) is proportional to specific catchment area (A_s). Predicts net erosion in areas of profile convexity and tangential concavity (flow acceleration and convergence zones) and net deposition in areas of profile concavity (zones of decreasing flow velocity).
	$LS = (m + 1)\left(\dfrac{A_s}{22.13}\right)^m \left(\dfrac{\sin \beta}{0.0896}\right)^n$	This sediment transport capacity index was derived from unit stream power theory and is equivalent to the length–slope factor in the Revised Universal Soil Loss Equation in certain circumstances. Another form of this equation is sometimes used to predict locations of net erosion and net deposition areas.
	$CIT = A_s \, (\tan \beta)^2$	Variation of stream-power index sometimes used to predict the locations of headwaters of first-order streams (i.e., channel initiation).

Radiation indices	$R_t = (R_{th} - R_{dh})\, F + R_{dh}\, v + R_{th}\, (1 - v)\, \alpha$	This equation estimates the total short-wave irradiance incident at the earth's surface for some user-defined period ranging in length from 1 day to 1 year. The three main terms account for direct-beam, diffuse, and reflected irradiance. A variety of methods are used by different authors to calculate these individual components. The methods vary tremendously in terms of sophistication, input data, and accuracy.
	$L_{in} = \varepsilon_a \sigma T_a^4\, v + (1 - v)\, L_{out}$	This equation estimates the incoming or atmospheric long-wave irradiance.
	$L_{out} = \varepsilon_s\, \sigma\, T_s^4$	This equation estimates the outgoing long-wave irradiance.
	$R_n = (1 - \alpha)\, R_t + \varepsilon_s L_{in} - L_{out}$	This equation estimates the net radiation or surface energy budget at the earth's surface for some user-defined period. May or may not account for the effects of clouds depending on the methods and data sources used to estimate individual short-wave radiation components (see Chapter 4 for additional details).
Temperature indices	$T = T_b - \dfrac{T_{lapse}(Z - Z_b)}{1000} + CS \left(1 - \dfrac{LAI}{LAI_{max}}\right)$	This equation is used to extrapolate minimum air, maximum air, and surface temperatures for a nearby climate station to other parts of the landscape. This equation corrects for elevation via a lapse rate, slope–aspect effects via the short-wave radiation ratio, and vegetation effects via a leaf area index.

1993d). The second equation contains one less term because it assumes uniform soil properties (i.e., that the soil transmissivity is constant throughout the landscape). Wood et al. (1990) have shown that the variation in the topographic component is often far greater than the local variability in soil transmissivity and that Equation 1.2 can be used in place of Equation 1.1 in many landscapes. Both of these indices predict that points lower in the catchment, and particularly those points near the outlets of the main channels, are the wettest points in the catchment, and the soil-water content decreases as the flow lines are retraced upslope to the catchment divide (Wilson and Gallant 1998).

These indices are used in the TOPMODEL (Beven and Kirkby 1979) hydrologic model to characterize the spatial distribution and extent of zones of saturation and variable source areas for runoff generation. O'Loughlin (1986) also used these indices to identify surface saturation zones in landscapes. Burt and Butcher (1986), Jones (1986), and Moore et al. (1988a) used variants of these compound topographic wetness indices to describe the spatial distribution of soil-water content. Moore et al. (1986) showed how the wetness index versus percent saturated source area relationship can be combined with observed stream-flow data and used to estimate the effective transmissivity of a small forested catchment. Sivapalan et al. (1987) used this index to characterize hydrologic similarity, and Phillips (1990) used it to delineate wetlands in a coastal plain drainage basin. Moore et al. (1993b, c) used slope and topographic wetness index to characterize the spatial variability of soil properties for a toposequence in Colorado. Montgomery and Dietrich (1995) used the TOPOG (O'Loughlin 1986) hydrologic model to predict the degree of soil saturation in response to a steady-state rainfall for topographic elements defined by the intersection of contours and stream-tube boundaries. This measure of relative saturation was then used to analyze the stability of each topographic element for the case of cohesionless soils of spatially constant thickness and saturated conductivity in three California, Oregon, and Washington study areas.

These types of static indices must be used carefully to predict the distribution of dynamic phenomena like soil-water content because surface saturation is a threshold process and because of hysteretic effects (Burt and Butcher 1986, I. D. Moore et al. 1991). In addition, there are several important and implicit assumptions in the derivation of the two wetness indices described above. Most notably, the gradient of the piezometric head, which dictates the direction of subsurface flow, is assumed to be parallel to the land surface and steady-state conditions are assumed to apply (Moore et al. 1993d). Several authors have described the pitfalls of using these indices in inappropriate ways. Jones (1986, 1987), for example, discussed the advantages and limitations of wetness indices as indicators of spatial patterns of soil-water content and drainage. Quinn et al. (1995) summarized various problems and described how steady-state topographic wetness indices can be calculated and used effectively in the TOPMODEL hydrologic modeling framework.

In an attempt to overcome the limitations of the steady-state assumption, Barling (1992) proposed a quasi-dynamic topographic wetness index of the form

$$W = \ln \left(\frac{A_e}{\tan \beta} \right) \qquad (1.3)$$

where A_e is the effective specific catchment area. Barling et al. (1994) calculated steady-state and quasi-dynamic indices for a catchment near Wagga Wagga in Australia and found that only the quasi-dynamic index correctly predicted that the topographic hollows and not the drainage channels themselves determined the hydrologic response of the catchment. Wood et al. (1997) proposed an alternative index of saturated zone thickness incorporating both spatial and temporal variation in recharge. However, the suitability of these methods as generally applicable tools has yet to be demonstrated.

Several terrain-based stream-power and sediment transport capacity indices have also been proposed (Table 1.2). Stream power is the time rate of energy expenditure and has been used extensively in studies of erosion, sediment transport, and geomorphology as a measure of the erosive power of flowing water (I. D. Moore et al. 1991). It is usually computed as

$$\Omega = \rho g q \tan \beta \qquad (1.4)$$

where ρg is the unit weight of water, q is the discharge per unit width, and β is the slope gradient (in degrees). The compound topographic index $A_s \tan \beta$ is, therefore, a measure of stream power, since ρg is essentially constant and q is often assumed to be proportional to A_s. Several researchers have used variations of this index to predict the locations of ephemeral gullies. Thorne et al. (1986) multiplied this index by plan curvature and predicted both the locations of ephemeral gullies and the cross-sectional areas of the gullies after 1 year of development with variants of this new index. Moore et al. (1988a) showed that ephemeral gullies formed where $W > 6.8$ and $A_s \tan \beta > 18$ for a small semiarid catchment in Australia; and Srivastava and Moore (1989) found that ephemeral gullies formed where $W > 8.3$ and $A_s \tan \beta > 18$ on a small catchment in Antigua. I. D. Moore et al. (1991) concluded that threshold values of these indices are likely to vary from place to place because of differences in soil properties. Moore and Nieber (1989) used the stream-power index to identify places where soil conservation measures that reduce the erosive effects of concentrated flow, such as grassed waterways, should be installed. Montgomery and Dietrich (1989, 1992) and Montgomery and Foufoula-Georgiou (1993) used a variation of this index, $A_s (\tan \beta)^2$, to predict the headwaters of first-order streams (i.e., the locations of channel initiation).

A second compound index was derived by Moore and Burch (1986a–c) from unit stream-power theory and a variant used in place of the length–slope factor in the Revised Universal Soil Loss Equation (RUSLE) for slope lengths < 100 m and slopes < 14° as follows:

$$LS = (m + 1) \left(\frac{A_s}{22.13} \right)^m (\sin \beta / 0.0896)^n \qquad (1.5)$$

where $m = 0.4$ and $n = 1.3$ (Moore and Wilson 1992, 1994). Both this and the next equation are nonlinear functions of slope and specific discharge. This new index calculates a spatially distributed sediment transport capacity and may be better suited to landscape assessments of erosion than the original empirical equation because it explicitly accounts for flow convergence and divergence (Moore and Wilson 1992, Desmet and Govers 1996b).

Another terrain-based sediment transport capacity index can be used to differentiate net erosion and net deposition areas:

$$\Delta T_{cj} = \phi \, A_{sj-}^{m} \, (\sin \beta_{j-})^{n} - A_{sj-}^{m} \, (\sin \beta_{j})^{n} \qquad (1.6)$$

where ϕ is a constant, subscript j signifies the outlet of cell j, and subscript $j-$ signifies the inlet to cell j (Moore and Wilson 1992, 1994). This index will predict erosion in areas experiencing an increase in sediment transport capacity and deposition in areas experiencing a decrease in sediment transport capacity. Mitasova et al. (1996) implemented variants of these equations in the GRASS (U.S. Army Corps Engineers 1987) GIS. Net erosion areas coincided with areas of profile convexity and tangential concavity (flow acceleration and convergence), and net deposition areas coincided with areas of profile concavity (decreasing flow velocity). These patterns match those observed by Martz and De Jong (1987), Foster (1990), Sutherland (1991), and Busacca et al. (1993) in a variety of landscapes.

Several authors have criticized the use of Equation 1.5 in place of the original slope gradient and length terms in the Revised Universal Soil Loss Equation (Renard et al. 1991) and its predecessors. Interested readers should consult Moore and Wilson (1992, 1994), Foster (1994), Mitasova et al. (1996, 1997), and Desmet and Govers (1996b, 1997) for additional details. Wilson and Lorang (1999) recently summarized the key elements of this debate, and why the terrain-based approach of Mitas et al. (1996) probably represents a superior approach for simulating the impact of complex terrain and various soil and land cover changes on the spatial distribution of soil erosion and deposition.

The third and final set of compound indices is used to estimate the spatial and temporal distribution of solar radiation at the earth's surface. Topography may exert a large impact on the amount of solar energy incident at a location on the earth's surface (Moore et al. 1993f, Dubayah and Rich 1995). Variations in elevation, slope, aspect, and local topographic horizon can cause substantial differences in solar radiation and thereby affect such biophysical processes as air and soil heating, evapotranspiration, and primary production (Gates 1980, Linacre 1992, Dubayah 1992, 1994, Dubayah and Rich 1995). These processes may, in turn, affect the distribution and abundance of flora and fauna. Moore et al. (1993e), for example, used computed radiation and temperature indices to characterize the fine-scale environmental heterogeneity and environmental domains of the five major subalpine forest types for a 20-km^2 study area in the Brindabella Range in southeastern Australia. Hutchins et al. (1976), Kirkpatrick and Numez (1980), Tajchman and Lacey (1986), Austin et al. (1983, 1984), and Noguchi (1992a, b) have also shown that the distributions of solar radiation and vegetation are highly correlated.

Numerous approaches have been proposed to calculate the radiation fluxes and temperature indices used in these types of applications. Most of the radiation models incorporate one or more of the following equations. The net radiation, R_n, received by an inclined surface can be written as

$$R_n = (1 - \alpha)\,(R_{dir} + R_{dif} + R_{ref}) + \varepsilon_s\,L_{in} - L_{out} = (1 - \alpha)\,R_t + L_n \qquad (1.7)$$

where α is the surface albedo, ε_s is the surface emissivity, R_{dir}, R_{dif}, and R_{ref} are the direct, diffuse, and reflected short-wave irradiance, respectively, for which $R_t = R_{dir} + R_{dif} + R_{ref}$, the global short-wave irradiance, L_{in} is the incoming or atmospheric long-wave irradiance, L_{out} is the outgoing or surface long-wave irradiance, for which $\varepsilon_s\,L_{in} - L_{out} = L_n$, the net long-wave irradiance.

The total short-wave irradiance is estimated by

$$R_t = (R_{th} - R_{dh})\,F + R_{dh}\,v + R_{th}\,(1 - v)\alpha \qquad (1.8)$$

where R_{th} and R_{dh} are the total and diffuse radiation on a horizontal surface, and F is the potential solar radiation ratio ($= R_o/R_{oh}$), which is the ratio of the potential solar radiation (R_o) on a sloping surface to that on a horizontal surface (R_{oh}), and v is the skyview factor, which is the fraction of the sky that can be seen by the sloping surface. The total and diffuse short-wave irradiances on a horizontal surface are often expressed as functions of the total and diffuse transmittances of the atmosphere and the potential solar radiation on a horizontal surface. These transmittances are functions of the thickness and composition of the atmosphere, such as the water vapor, dust, and aerosol content (Lee 1978, Gates 1980).

The long-wave irradiance components are approximated on a cell-by-cell basis using

$$L_{out} = \varepsilon_s\,\sigma T_s^{\,4} \qquad (1.9)$$

$$L_{in} = \varepsilon_a \sigma T_a^{\,4}\,v + (1 - v)\,L_{out} \qquad (1.10)$$

where ε_a is the atmospheric emissivity (a function of air temperature, vapor pressure, and cloudiness), σ is the Stefan–Boltzman constant, T_s is the mean surface temperature, and T_a is the mean air temperature. An equation that utilizes modifications of a simple approach proposed by Running et al. (1987), Hungerford et al. (1989), and Running (1991) for estimating the spatial distribution of minimum, maximum, and average air temperature is summarized in Table 1.2.

The different indices that have been proposed vary in terms of the methods, data sources, and assumptions used to estimate individual components. Moore et al. (1993e) developed an approximate method for estimating each of the above fluxes at any location in a topographically heterogeneous landscape in one of these applications. The variation in the potential solar radiation was estimated over a catchment as a function of slope, aspect, topographic shading, and time of year, and then adjusted for cloud, atmospheric, and land cover effects. The variables that serve as model inputs, such as albedo, cloudiness, emissivity, sunshine fraction, mean air and surface temperatures, and clear-sky transmittances, can be varied on a monthly or annual basis (Wilson and Gallant 1999). Hetrick et al. (1993a, b) developed the SOLARFLUX model in the GRID module of the ARC/INFO GIS and used it with latitude, atmospheric transmissivity, slope, aspect, topographic shading, and time of year to estimate direct and diffuse irradiance at each grid point. The effects of cloud cover, which are likely to be substantial in many humid environments, were not accounted for in this model. Kumar et al. (1997) chose a simpler approach and used latitude and a series of topographic attributes derived from a square-grid DEM to estimate clear-sky, direct-beam, short-wave radiation. These relationships are generally straightforward and numerous authors have summarized the appropriate equations for both horizontal and sloping sites (see Lee 1978, Gates 1980, Iqbal 1983, Linacre 1992 for additional details). Most of the challenges (problems) are encountered when atmospheric effects (precipitable water, dust, etc.), cloud cover, and land surface characteristics (albedo) are considered. Dubayah and Rich (1995) have reviewed many of the important computational challenges and errors that are likely to be encountered in building accurate, physically based topographic solar radiation models. Many of their insights are derived from their work combining the Atmospheric and Topographic (ATM) model (Dubayah 1992), ground measurements, and satellite imagery in the Konza Prairie and Rio Grande Basin (Dubayah 1992, 1994; Dubayah and van Katwijk 1992).

Although it is not always apparent to users of terrain analysis, the three sets of indices described above are simplified process models and are not applicable in all situations (Wilson and Gallant 1998). The topographic wetness index, for example, is based on the assumption that the soil hydraulic conductivity decreases exponentially with depth so that subsurface flow is confined to a shallow layer. If this is not the case, the steady-state and quasi-dynamic topographic wetness indices will be poor predictors of the spatial distribution of soil water. An alternative index might be developed to better represent the topographic effect on water distribution, perhaps based on groundwater potential expressed as a simple elevation difference above a local mean or minimum (e.g., Hinton et al. 1993).

The topographic indices introduced on the preceding pages account for the component of the spatial variability of processes that is due to topographic effects. Other spatially variable factors are usually involved, such as soil hydraulic properties and vegetation in the case of soil water. In some instances, the spatial variations in these other attributes are themselves linked to the topographic indices. The spatial variability of soil properties is one case where significant links have been established (e.g., Moore et al. 1993f, Wilson et al. 1994). There are other properties though where explicit incorporation of the spatial variation of other important components of process models would substantially improve the predictive accuracy of topographic indices particularly when working at a broad landscape scale as opposed to the small catchment scale. Surficial geology and, in some cases, climate are likely candidates for inclusion in these types of applications. Some of the applications reported in this book make use of such additional information.

Additional problems may be encountered by the terrain analyst or user because the spatial and statistical distributions of the computed primary and secondary topographic attributes may be affected by the presence of errors in the source data and the choice of computer algorithm and/or element size (i.e., grid spacing). These aspects are often interconnected, although the review that follows treats each problem separately.

1.1.3   Identification and Treatment of Error and Uncertainty

Nonsystematic and systematic errors in DEMs may confound the expected relationships between computed terrain attributes and terrain-controlled site conditions. These problems may be amplified when first- and second-order derivatives, such as slope and convexity, are calculated (e.g., Bolstad and Stowe 1994). The most serious problems are usually encountered when secondary attributes are derived: The topographic wetness and sediment transport capacity indices are very sensitive to the presence of errors in source (elevation) data in flat areas and to the choice of flow routing algorithm (Moore et al. 1993f).

Many studies have examined the causes, detection, visualization, and correction of DEM errors. Carter (1988), Weibel and Heller (1991) and Kumler (1994), for example, describe the causes of errors in DEMs compiled by different methods. Several methods have been proposed for the detection of errors and estimation of the magnitude and/or spatial distribution of errors (e.g., Polidori et al. 1991, Brown and Bara 1994, Felicisimo 1994, Fryer et al. 1994, Li 1994, Garbrecht and Starks 1995, Lopez 1997). Most of the quantitative estimates have used topographic map elements (e.g., Evans 1980, Skidmore 1989), field measurements (e.g., Bolstad and Stowe 1994, Hammer et al. 1995, Giles and Franklin 1996), or hypothetical (imaginary) DEMs (e.g., Chang and Tsai 1991, Carter 1992, Hodgson 1995) as reference values for these assessments. Most assessments have also examined specific DEM products (e.g., Sasowsky et al. 1992, Brown and Bara 1994), although a few have compared two or more products (e.g., Bolstad and Stowe 1994, Hammer et al. 1995). Other researchers have focused on the development of methods for the visualization (e.g., Kraus 1994, Hunter and Goodchild 1995, 1996, McCullagh 1988) and correction of errors (e.g., Hannah 1981, O'Callaghan and Mark 1984, Jenson and Domingue 1988, Brown and Bara 1994). Several recent studies are discussed in more detail below to illustrate the key issues that have been addressed.

The horizontal and vertical resolution of most square-grid DEMs is such that flow lines become trapped in pits and depressions in key parts of the landscape. Guercio and Soccodato (1996), Jenson and Domingue (1988), Hutchinson (1988, 1989b), Martz and Garbrecht (1998), and Reiger (1998) have all proposed methods for correcting DEMs and/or avoiding these problems. Topographic attributes are computed for depressionless DEMs in many (most) applications that rely on published DEM data sets (as noted earlier).

Brown and Bara (1994) used semivariograms and fractals to detect the presence of errors in 7.5' USGS 30-m DEMs and evaluated several types of filters for reducing the magnitude of these errors. Their method does not require reference values. It identified the anisotropic conditions (i.e., where the variation in one direction is different from the variation in another direction) that are consistent with the "banding" or "striping" remnants produced by the aerial photograph scanning procedures used in the production of many USGS DEMs (U.S. Geological Survey 1993). Brown and Bara (1994) also showed how the semivariance and fractal dimensions provided a quantitative basis for applying corrections to mitigate the severity of these problems for a high-relief study area in Glacier National Park, Montana. The anisotropic conditions were greater for slope and curvature than for elevation.

The use of these types of methods for correcting DEMs may cause additional problems. Garbrecht and Starks (1995), for example, reported similar errors for a low relief drainage basin in Nebraska and concluded that the 7.5' USGS 30-m DEMs could not be used for wetland drainage analysis in these types of landscapes. However, these authors rejected subsequent manipulation of the original USGS DEMs to remove striping because of the additional data degradation that would have occurred. There is no easy way to distinguish real terrain features and errors when applying correction methods like those of Brown and Bara (1994) in many types of landscapes. Garbrecht and Starks (1995) therefore generated a completely new square-grid DEM based on interpolation from digital contours in order to conduct the drainage and wetland analysis tasks at hand.

Hammer et al. (1995) compared field measured and computer-generated slopes. They divided two 16-ha sites in Atchison County, Missouri into 10-m grid cells and field measured slope for each grid cell. The field-measured slope class maps served as templates for cell-by-cell comparisons with computer-generated slope class maps derived from 10- and 30-m DEMs and a standard 1:24,000-scale USDA soil survey. More than 50% of the areas were classified into correct slope classes with 10-m DEM maps. Two iterations of low-pass filters increased the accuracy of these particular maps. The 30-m DEM maps were 30 and 21% correctly classified in the two areas, and the soil survey maps correctly classified >30% of each area but did not capture the fine-scale landscape heterogeneity. The DEM-derived maps underestimated slopes on convexities and overestimated slopes on concavities at both sites.

There are at least three sets of problems connected with the above approach. First, the use of reference values for assessments of multiple attributes over large areas is impractical (Ruiz 1997). Second, Li (1991) and Kumler (1994) both examined the impact of the number, distribution, and accuracy of the checkpoints used for experimental tests of DEM accuracy on the resulting accuracy estimates, and found that the results were often highly sensitive to the choice of reference values. Third, some researchers have argued that the accuracy of primary and secondary topographic attributes cannot be determined by a comparison of calculated and "reference" values because the land surface is not mathematically smooth and there are no actual values of any attributes except elevation (e.g., Shary 1991).

Florinsky (1998) has argued that the accuracy of these attributes depends on the accuracy of the initial data (i.e., the DEM) and the precision of the calculation technique(s). His novel approach incorporated three steps and avoided all three sets of problems noted above. First, he showed that the Evans (1980) method was the most precise method for computing four local topographic variables (i.e., slope, aspect, and plan and profile curvature). The other methods tried were those of Zevenbergen and Thorne (1987), Moore et al. (1993b), and Shary (1995). Second, he derived formulae to calculate root mean square errors (RMSEs) based on the partial derivatives of the elevation surface for these variables (provided they are estimated with the Evans method). Third, Florinsky (1998) argued that mapping is the most convenient and practical way to implement the formulae that were derived (Figure 1.3). Overall, four general observations about errors can be deduced from this approach:

1. The values of the slope, aspect, and plan and profile curvature RMSEs are directly proportional to the elevation RMSE.
2. The values of these RMSEs increase with decreasing grid spacing.
3. The values of the plan and profile curvature RMSEs are more responsive than slope gradient and aspect to changes in grid spacing.
4. The values of all four RMSEs can become large with decreasing slope gradients (i.e., in flat areas).

These observations also highlight the important roles played by computation methods and grid spacing (grid resolution) in the identification and treatment of error.

Many researchers, including Florinsky (1998), have examined the sensitivity of computed attributes to the method of computation. Gao (1994), for example, proposed a C program for computing slope, aspect, and plan and profile curvature that included special rules for handling edge cells and special areas (summits, ridge lines, stream lines, etc.). Most terrain-analysis methods incorporate special rules for handling these cases. Srinivasen and Engel (1991) compared the performance of four slope algorithms with topographic map and field assessments of slope steepness. Moore et al. (1993d) showed that the D8 slope algorithm (described in more detail below) predicted slightly larger slopes than the finite difference method for a forested study area in southeastern Australia. Weih and Smith (1997) examined the influence of several cell slope computation algorithms on a common forest management decision.

At least six algorithms have also been proposed for routing flow and computing contributing areas from square-grid DEMs. Five of these algorithms—the D8 (deterministic eight-node) algorithm of O'Callaghan and Mark (1984), the Rho8 (random eight-node) algorithm of Fairfield and Leymarie (1991), the FD8 and FRho8 algorithms, and the DEMON algorithm of Costa-Cabral and Burges (1994)—have been implemented in TAPES-G (Moore 1992, Wilson and Gallant 1998). The sixth method uses a vector–grid approach and has been implemented as the *r.flow* routine in the GRASS GIS (Mitasova and Hofierka 1993, Mitasova et al. 1995, 1996).

The D8 algorithm allows flow to one of only eight neighbors based on the direction of steepest descent. This popular algorithm is often criticized because it tends to predict flow in parallel lines along preferred directions that will agree with aspect only when aspect is a multiple of 45° and it cannot model flow dispersion (e.g., Moore et al. 1993d). Rho8 is a stochastic version of D8 that simulates more realistic flow networks but still cannot model flow dispersion. Moore et al. (1993d) found that Rho8 breaks up the long, linear flow paths produced by the D8 method while generating more single-cell drainage areas. Both FD8 and FRho8 allow flow to be distributed to multiple nearest-neighbor nodes in upland areas above defined channels and use either the D8 or Rho8 algorithms below points of channel initiation (Moore et al.

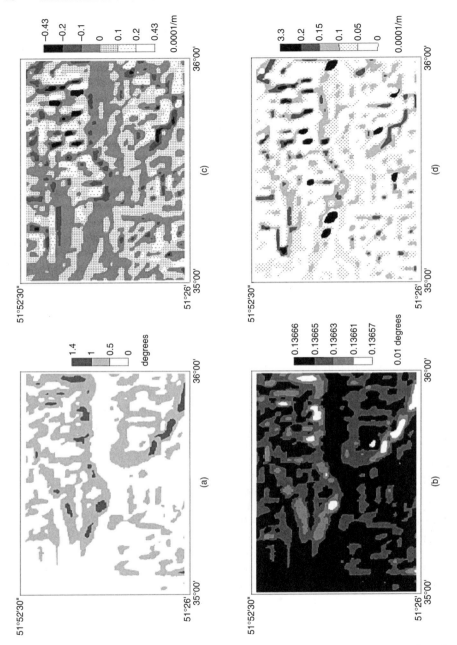

Figure 1.3. Maps of Kursk Region in Russia showing (a) slope gradient; (b) RMSE of gradient; (c) profile curvature; and (d) RMSE of profile curvature. Reprinted with permission from Florinsky (1998) Accuracy of local topographic variables derived from digital elevation models. *International Journal of Geographical Information Science* 12: 47–61 (http://www.tandf.co.uk/journals). Copyright © 1998 by Taylor and Francis.

1993d). Flow in upland areas is assigned to multiple downstream nearest neighbors with these algorithms in TAPES-G using slope-weighted methods similar to those of Freeman (1991) and Quinn et al. (1991). Moore et al. (1993d) showed that the FD8 and FRho8 algorithms implemented in TAPES-G produced almost identical catchment area frequency distributions with very few single-cell drainages. DEMON avoids these problems by representing flow in two directions as directed by aspect. This approach permits the representation of varying flow width over nonplanar topography (similar to contour-based models) (Moore 1996). The final vector–grid method constructs flow lines downhill from each grid cell until they reach a cell with a slope lower than some specified minimum, a boundary line, or some other barrier to calculate upslope contributing areas. These flow lines follow the aspect direction of flow and they are represented in vector format, avoiding the artificial nature of cell-to-cell flow routing in the previous methods. The points defining the flow lines are computed as the points of intersection of a line constructed in the flow direction given by the aspect angle and a grid cell edge. The DEMON and vector–grid algorithms share many similarities and are likely to produce similar results (Wilson and Lorang 1999).

Wolock and McCabe (1995), Moore (1996), Desmet and Govers (1996a), and the two case studies discussed in Chapter 5 provide detailed assessments of the performance of many of the existing algorithms. Wolock and McCabe (1995) compared several single- and multiple-flow-direction algorithms for calculating the topographic parameters used in TOPMODEL. Moore (1996) compared the D8, Rho8, FD8/FRho8, DEMON, and contour-based algorithms in terms of specific catchment area calculations. Desmet and Govers (1996a) compared six flow-routing algorithms in terms of contributing area calculations and the prediction of ephemeral gully locations. These comparisons showed that the single-flow (D8, Rho8) and multiple-flow (FD8, FRho8) direction algorithms will perform very differently in most types of landscapes. The identification of problems with existing algorithms and fundamental role of flowing water in controlling or explaining many key environmental processes and patterns are likely to promote further methodological innovation in this area. Holmgren (1994) and Quinn et al. (1995), for example, have recently proposed new methods for computing the weights used with the FD8 and/or FRho8 algorithms. Burrough et al. (1999a, b) introduced random errors and computed several hundred realizations of the flow network with the D8 algorithm to overcome the presence of DEM errors and/or shortcomings of this algorithm noted earlier.

An even larger group of studies have examined the sensitivity of selected attributes to the choice of data source, structure, and/or cell size. Panuska et al. (1991) and Vieux and Needham (1993) quantified the effects of data structure and cell size on Agricultural Non-Point Source (AGNPS) pollution model inputs, and showed how the computed flow-path lengths and upslope contributing areas varied with element size. Vieux (1993) examined the sensitivity of a direct surface runoff model to the effects of cell size aggregation and smoothing using different-sized windows. Moore et al. (1993d) examined the sensitivity of computed slope and steady-state topographic wetness index values across 22 grid spacings for three moderately large ($\approx$100 km^2) catchments in southeastern Australia. Hodgson (1995) demonstrated that the slopes and aspects calculated from 30-m DEMs are representative of grid spacings two or three times larger than the original DEM grid spacing. Issacson and Ripple (1991) compared 1° USGS 3-arc-second and 7.5′ USGS 30-m DEMs and Lagacherie et al. (1996) examined the effect of DEM data source and sampling pattern on computed topographic attributes and the performance of a terrain-based hydrology model. Chairat and Delleur (1993) quantified the effects of DEM resolution and contour length on the distribution of the topographic wetness index as used by TOPMODEL and the model's peak flow predictions. Wolock and Price (1994) and Zhang and Montgomery (1994) also examined the effects of DEM source scale and DEM cell spacing on the topographic wetness index and TOPMODEL watershed model predictions. Garbrecht and Martz (1994) examined the impact of DEM resolution on extracted drainage properties for an 84-km^2 study area in Oklahoma using hypothetical drainage network configurations and DEMs of increasing size. They derived various quantitative relationships and concluded that the grid spacing must be selected relative to the size of the smallest drainage features that are considered important for the work at hand. Bates et al. (1998) showed how high-frequency information is lost at progressively larger grid spacings.

1.2   THE PURPOSE OF THIS BOOK

The preceding review is instructive in at least four ways. First, it highlights the tremendous interest in digital terrain analysis that has emerged during the past decade and the key contributions of Ian D. Moore during the period 1985–1993. Second, it describes the most popular topographic attributes and the methods that have been used to calculate them. Third, it illustrates some of the ways in which the computed topographic attributes have been used to improve our understanding of hydrological, geomorphological, and ecological systems. Finally, it describes many of the subtleties and challenges that must be overcome in order to use digital elevation data and terrain-analysis tools effectively.

Most of the chapters in this book look past the problems raised at the end of the previous section and demonstrate some of the ways in which these continuously varying but gridded landform attributes can be used to quantify topographic controls on hydrological, geomorphological, and ecological systems. The individual chapters included in this book describe the TAPES terrain-analysis methods and show how computed terrain attributes can be utilized to describe key environmental patterns as a function of process. The applications confirm that the current methods and data sources are best suited to work at intermediate spatial (hillslopes and catchments) and temporal scales (measured in terms of months or years).

In all of this work, we must take care to ensure that our simplifying assumptions resolve rather than introduce computational complexity. Figure 1.4 is adapted from a similar diagram in Grayson et al. (1993) and shows how the terrain-based "hydrological information content" can be expected to vary with changing element size (i.e., spatial resolution). The relationship in the right-hand side of Figure 1.4 shows how the quantity of information declines as the element size increases beyond the scale of the measurements (topographic attributes in this instance). This is a generally accepted notion and is caused by the lumping of subgrid information (I. D. Moore et al. 1991, Bates et al. 1998). Two scenarios are captured as element size is reduced on the left-hand side of Figure 1.4. In one instance, the topography controls the lateral migration and accumulation of water and the finer resolution increases the information content. In the second case, the hydrological behavior is dominated by soil characteristics, such as preferential flow paths, that are not related to topography and the finer spatial resolution does not increase the level of hydrological information (Grayson et al. 1993). These examples suggest at least three challenges. One is the need to increase our understanding of the key processes affecting sediment and water behavior at a variety of scales (e.g., Moore and Grayson 1991, Grayson and Moore 1991, Grayson et al. 1992a, Robinson et al. 1995). The second is concerned with the development and testing of methods to measure and/or interpolate values for these variables across landscapes (Phillips 1988). The final challenge involves the identification of indicators that can be used to monitor changes in individual processes (Kirkby et al. 1996).

The individual chapters included in this book raise many of the conceptual and methodological issues that we will need to consider as we forge ahead with these

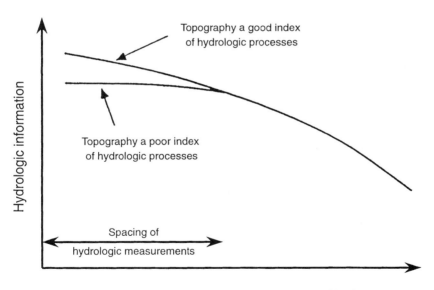

Figure 1.4. Conceptual representation of the relationship between hydrologic information and element size for topography-based interpolations. Reprinted with permission from Grayson, Blöschl, Barling, and Moore (1993) Process, scale, and constraints to hydrological modeling in GIS. p. 83–92 in *Application of Geographic Information Systems in Hydrology and Water Resources: Proceedings of the HydroGIS '93 Conference held in Vienna, April 1993,* edited by K Kovar and H P Nachtnebel. Copyright © 1993 by International Association of Hydrological Sciences, Wallingford, United Kingdom.

types of applications in the future. The applications presented in this book also demonstrate how simple spatial models can be combined with qualitative reasoning to improve our understanding and management of environmental systems. Grayson et al. (1993) and others have argued that this approach is consistent with our current ability to represent biophysical systems and the availability of data. We hope that the methods and applications described in this book will encourage others to adopt this approach and thereby help to increase the quantity and quality of the scientific concepts and geospatial information used in environmental assessments and management applications.

1.3 OVERVIEW

This book is divided into four main sections. The first section is concerned with methods and data and the remaining three sections illustrate hydrological, geomorphological, and biological applications, respectively. The final chapter offers some concluding remarks and future predictions. Our contributions consist of two chapters in addition to this one and co-authorship of five other chapters. The remainder of the authors were selected because of their utilization of the TAPES software tools and participation in a workshop celebrating the life and scientific contributions of Ian D. Moore (1951–1993) that we organized at the Third International Conference Integrating GIS and Environmental Modeling in 1996. Their contributions have evolved in various ways in the three years that have elapsed since the workshop and the following subsections summarize the contents of the individual chapters.

1.3.1 Digital Terrain Analysis Methods

This initial chapter and the four that follow describe the conceptual foundations and methods of digital terrain analysis. These chapters therefore provide important background material for the subsequent chapters on hydrological, geomorphological, and biological applications in this book, and for readers who may be interested in similar types of applications. Table 1.3 lists the individual terrain-analysis programs by book chapter for those interested in specific topographic attributes and/or terrain-analysis methods.

   Michael Hutchinson and John Gallant describe DEM data sources and interpolation methods in relation to the accurate representation of terrain shape and scales of source data and applications in Chapter 2. A contour map for the Cottonwood Creek catchment on the Red Bluff Montana Agricultural Experiment Station was digitized and is used to illustrate the application of the interpolation methods described in this chapter. The contour map and a 15-m square-grid DEM that was produced from it are used in Chapters 3 and 4 to illustrate the application of six additional terrain-analysis programs.

   We describe the methods used by TAPES-C and TAPES-G to calculate a series of primary topographic attributes in Chapter 3. TAPES-C and TAPES-G start with contours and regular grids, respectively. Both programs generate spatially variable esti-

TABLE 1.3   Utilization of Individual Terrain Analysis Programs by Book Chapter

Chapter	TAPES-C	TAPES-G	DYNWET	EROS	SRAD	WET
3	X	X				
4			X	X	X	X
5		X		X		
6		X				X
7		X	X			
8	X					
9	X					
10		X				X
11		X				
12		X				
13		X				
14		X				
15		X	X		X	
16		X			X	

mates of slope, aspect, profile and plan curvature, flow-path length, specific catchment area, and several other topographic attributes. The inputs, estimation methods, and outputs are described for each program in turn, and the 15-m Cottonwood Creek DEM produced in Chapter 2 is used to illustrate the performance of both programs.

We describe the methods used by four grid-based programs (EROS, SRAD, WET, DYNWET) to calculate several sets of secondary topographic attributes in Chapter 4. EROS estimates the spatial distribution of soil loss and erosion and deposition potential in a catchment. SRAD computes the radiation budget using incoming short-wave irradiance and incoming and outgoing long-wave irradiance for periods ranging from 1 day to a year. Slope, aspect, topographic shading, and monthly variations in cloudiness, atmospheric transmissivity, and vegetation properties are taken into account, and this program can also be used to estimate surface and minimum, maximum, and average air temperatures. WET calculates equilibrium soil moisture, evaporation, and runoff based on topographic attributes and spatial estimates of solar radiation, and DYNWET calculates topographic wetness indices based on both the steady-state and quasi-dynamic subsurface flow assumptions. The inputs, estimation methods, and outputs are described for each program in turn, and the 15-m Cottonwood Creek DEM produced in Chapter 2 is used to illustrate the performance of all four programs.

John Wilson, Philip Repetto, and Robert Snyder examined the effect of DEM data source, grid resolution, and flowing routing method on computed topographic attributes in a pair of experiments conducted in southwest Montana and northern Idaho (Chapter 5). The Montana experiment quantified the sensitivity of five attributes in the 105-km^2 Squaw Creek catchment. The agreement between maps of computed topographic attributes derived from DEMs of different size was poor when the attributes were reclassified into five classes (28–49% agreement) and slightly better when 7.5′ USGS 30-m DEMs were used with different flow-routing algorithms (49–71%

agreement). The Idaho experiment was conducted on a single farm field and examined topographic controls on soil erosion and the ability of the stream-power index used with 7.5′ USGS 30-m DEMs to distinguish field areas experiencing net erosion and net deposition. Numerous explanations are offered for the poor performance of the stream-power index in this particular instance. Overall, the results from these experiments showed why care must be exercised when choosing DEM sources, grid resolutions, and terrain-analysis methods for different types of applications and landscapes, and in that sense, they corroborate many of the arguments raised in Chapters 1 and 2.

1.3.2 Hydrological Applications

The four chapters in this section incorporate one or more topographic attributes in a variety of GIS-based hydrologic modeling frameworks. All four applications explore the topographic influences on soil moisture and runoff behavior, but each uses the computed topographic attributes with successively more complicated hydrologic models and smaller study areas. Hence, the four study areas used in these applications ranged from 96,000 km^2 to 2.5 ha in spatial extent.

In Chapter 6 Valentina Krysanova, Dirk-Ingmar Müller-Wohlfeil, Wolfgang Cramer, and Alfred Becker combined the WET model and a variant of the USLE with a GIS to explore spatial patterns of soil moisture and potential soil loss in the 96,000-km^2 German portion of the Elbe River Basin. The models are used to identify the areas that are expected to experience long-term average soil moisture deficits and accelerated soil erosion as a function of long-term average climate, topography, soil, and land-use data. The maps of wetness index generated with WET showed good agreement with maps of long-term water availability expressed as groundwater table depths and the map of soil loss potential showed good agreement with previous studies. This application illustrates how the computed topographic attributes can be used to identify areas that warrant more detailed analysis.

Jeremy Fried, Daniel Brown, Mark Zweifler, and Michael Gold (Chapter 7) construct a series of stream buffers using cumulative cost distance calculated over fuzzy set combinations of relative topographic wetness and stream-power indices for a 17-km^2 first-order Michigan drainage basin. Their investigative buffers (models) are grounded on the assumption that riparian segments receiving the greatest discharge have upslope contributing areas dominated by saturated soils and have sufficient stream power for saturated flow to reach the stream. The resultant models are evaluated using field data collected during a post-storm-event GPS field survey of ponded storm flow accumulations and concentrated storm flow discharge sites. This application is instructive because it shows how topographic indices might be combined with qualitative reasoning to guide site-specific water pollution remediation efforts in the future.

Alan Yeakley, George Hornberger, Wayne Swank, Paul Bolstad, and James Vose describe the development, calibration, and testing of a terrain-based hillslope hydrology modeling framework for simulating soil moisture distributions in forested landscapes in the southern Appalachian Mountains (Chapter 8). Their approach incorporates an above-ground interception model, contour-based topographic attributes computed with TAPES-C, and a watershed hydrology model. These models were applied to a 12.3-ha first-order drainage basin that is part of the Coweeta Hydrologic Laboratory, where previous work has demonstrated the dominance of subsurface flow and absence of overland flow due to the high infiltration capacities of the forest soils. The new modeling framework captured the mean soil moisture response during storm events and over several years (seasons) but did less well at capturing soil moisture extremes and depicting spatial variability. These shortcomings indicate why increased knowledge of the spatial variation in soil hydraulic properties and more accurate representation of soil moisture near seepage faces are needed to build more successful (complete) explanatory models.

In the final chapter in this set (Chapter 9), Greg Pohll and John Warwick describe some work to build an improved explanatory model for the hydrologic behavior of a subsidence crater at the Nevada Nuclear Test Site. An increased understanding of the linkages between surface and subsurface components is required to predict large-scale radionuclide transport and to assess the feasibility of using subsidence craters as low-level nuclear waste storage sites. The crater used in this particular study has a diameter of 180 m and drains a 2.5-ha catchment. The TAPES-C terrain analysis program was dynamically linked with an overland flow simulator (THALES) and a numerical Richards' equation solver (SWMS-2D) to represent moisture migration within the vadose zone and simulate the movement of water during overland flow, ponding, infiltration, and seepage at the study site. The outputs were compared with a simpler vadose zone model with static surface boundary conditions to determine the effectiveness of each model. The results were mixed—the simpler and more complicated models offered superior predictions of deep moisture migration and the temporal distribution of infiltration, respectively—and, like the results from the previous chapter, demonstrate the difficulty of building realistic hydrologic process models.

1.3.3 Geomorphological Applications

The four chapters in this section examine pedological and geomorphological applications of digital terrain analysis. The breadth of the applications declines from one chapter to the next. The first chapter starts with a review of theories of pedogenesis and the role of soil survey in summarizing and communicating knowledge of soil properties, and the last chapter describes a terrain-based model to delineate shallow landslide areas in steep forested watersheds. The study areas used in the four chapters vary tremendously in size and character as well.

In the first chapter in this set (Chapter 10), Neil McKenzie, Paul Gessler, Philip Ryan, and Deborah O'Connell review the role of soil survey and describe a series of soil survey applications from southeastern Australia. Digital terrain analysis is used to characterize microclimates, develop explicit statistical sampling plans, and generate spatial predictions of soil properties at resolutions unmatched by comparable conventional methods for the 500-km^2 Bago-Maragle study area in southern New South Wales. These authors argue that digital terrain analysis has created an opportunity for a more scientifically based method of soil survey and that they may (one day) be used to generate improved quantitative spatial predictions of specific soil properties. The next two chapters illustrate both of these possibilities in specific environments.

Stephen Ventura and Barbara Irvin (Chapter 11) describe the use of topographic attributes to automate the classification of landform elements for a 49-ha study area in the "Driftless Area" of Wisconsin. Continuous (fuzzy logic) and unsupervised classification techniques were used to assign DEM cells a membership of a landform element class. The unsupervised classification assigned cells to single landform classes, whereas the continuous classification allocated relative class memberships for every class in every cell. These classes were determined by the natural clustering of the data in attribute space in both instances. The clusters formed readily recognizable patches on the landscape that matched manually interpreted landform classes and soil survey map units.

Jay Bell, David Grigal, and Peter Bates (Chapter 12) developed several quantitative models that predict soil organic carbon (SOC) storage as a function of topographic attributes and vegetative cover for a 22-km^2 study area in the Cedar Creek Natural History Area of Minnesota. Different models were constructed for mineral soils and peatlands, and the final model described approximately 50% of the variation in SOC over the entire study area. Slope and several relative elevation and distance measures were included as explanatory variables in the final models.

Jinfan Duan and Gordon Grant used topographic attributes with an infinite slope model to predict shallow landform areas in the final chapter in this set (Chapter 13). Their approach incorporated a dynamic simulation of rainfall intensities and treated the spatial distribution of key soil and vegetation parameters stochastically using a Monte Carlo simulation approach. This model was tested using observed landslides for a 64-km^2 drainage basin in western Oregon. The agreement with the locations of observed slides was only fair, and the authors concluded that their model may be most useful in predicting average slide frequencies under different management regimes rather than identifying specific slide locations, as other, more deterministic models do. This application is similar to the final two hydrologic applications in Chapters 8 and 9 in that it highlights some of the difficulties that are encountered in using static topographic indices to represent dynamic landscape processes.

1.3.4   Biological Applications

The three chapters in this set examine the linkages between computed topographic attributes and vegetation patterns in three North American landscapes.

Janet Franklin, Paul McCullogh, and Curtis Gray developed classification tree models relating topographic attributes and spectral variables derived from satellite imagery to chaparral species associations and riparian vegetation types for a study area in the Laguna Mountains of the Peninsula Ranges in San Diego County, California (Chapter 14). The models were then applied to digital maps of the topographic and spectral variables to produce predictive maps of vegetation distribution with accuracy estimated in the 52–62% range. Very detailed vegetation types (classes) were identified, and the authors concluded that their approach provides a classification method that avoids the need for a priori judgments about the terrain/satellite imagery/vegetation relationships.

Jonathan Wheatley, John Wilson, Roland Redmond, Zhenkui Ma, and Jeff DiBenedetto examined whether one or more topographic attributes can be added to an existing satellite interpretation method to improve land cover classification accuracy in the Little Missouri Grassland of North Dakota (Chapter 15). Quasi-dynamic topographic wetness and incident short-wave solar radiation indices were added to the third stage of a four-step *Landsat* TM satellite interpretation method. Error matrices were developed using a "bootstrap" process that removed each plot from the training data set, one at a time, and used the remaining plots to classify each one. Accuracy remained in the 51–57% range using 13 land cover types and 173 ground-truth plots. The terrain-analysis tools did help with the identification of the channel system and the results confirmed why satellite-based land cover maps must be used with care, since different source data and levels of spatial aggregation will predict different patterns of existing vegetation.

Brendan Mackey, Ian Mullen, Kenneth Baldwin, John Gallant, Richard Sims, and Daniel McKenney examined topographic controls of boreal forest ecosystems in the Rinker Lake region of northwestern Ontario, Canada (Chapter 16). A nonparametric statistical model is used to correlate the distribution of Jack Pine with environmental field measurements and computed topographic attributes. The results showed that topographic indices derived from a 20-m DEM were better predictors of Jack Pine than in situ observations of either substrate or topography. This implies that there are strong topographic controls on the distribution of Jack Pine in the Rinker Lake area. Various explanations are offered for this relationship and they all point to the need for additional information and knowledge if topographic attributes are to be interpreted and used correctly.

CHAPTER 2

Digital Elevation Models and Representation of Terrain Shape

Michael F. Hutchinson and John C. Gallant

2.1 INTRODUCTION

Terrain plays a fundamental role in modulating earth surface and atmospheric processes. This linkage is so strong that an understanding of the nature of terrain can directly confer understanding of the nature of these processes, in both subjective and analytical terms. It is therefore natural to place representations of terrain, in the form of digital elevation models (DEMs), at the center of the flow chart shown in Figure 2.1, reproduced from Hutchinson and Gallant (1999). This flow chart shows DEMs to be at the center of interactions between source data capture and applications. These interactions are supported by DEM generation methods and a steadily increasing range of techniques for DEM interpretation and visualization. Visualization techniques are often used to support interpretations of DEMs and to assess data quality.

The issue of spatial scale arises at various points in this scheme. The scale of source data should guide the choice of resolution of generated DEMs, and the scales of DEM interpretations should match the natural scales of terrain-dependent applications. The spatial resolution of a regular-grid DEM can provide a practical index of scale, as well as a measure of information content (Hutchinson 1996). The determination of appropriate scales for hydrological modeling is an active research issue (Zhang and Montgomery 1994, Blöschl and Sivaplan 1995). Incorporation of terrain structure into considerations of spatial scale is also an emerging issue in terrain analysis (Gallant and Hutchinson 1996).

The range of spatial scales of hydro-ecological applications of DEMs and the corresponding common primary topographic data sources are indicated in Table 2.1. Here, DEM resolution is used as an index of scale. The general trend has been to move from broader continental and regional scales, closely allied to the representation of major drainage divisions (Jenson 1991, Hutchinson and Dowling 1991), to

Terrain Analysis: Principles and Applications, Edited by John P. Wilson and John C. Gallant.
ISBN 0-471-32188-5   © 2000 John Wiley & Sons, Inc.

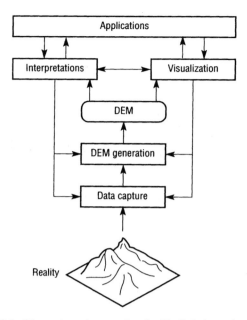

Figure 2.1. The main tasks associated with digital terrain modeling.

mesoscale representations of surface climate (Hutchinson 1995, 1998, Running and Thornton 1996, Daly et al. 1994) and associated flora and fauna (Nix 1986), to finer toposcales suited to the modeling of surface hydrology, vegetation, and soil properties (I. D. Moore et al. 1991, Quinn et al. 1991, Mackey 1996, Gessler et al. 1996, Zhang and Montgomery 1994). This has been accompanied by improvements in methods for representing fine-scale shape and structure of DEMs, supported by the steady increase in the capacity of computing platforms. These finer scale applications are the main focus of this book. At the same time, coarser scale processes, particularly mesoscale climate, can have a significant impact on the spatial distribution of hydrological and ecological processes.

There is naturally some overlap between the divisions shown in Table 2.1, but there is a genuine distinction between fine and coarse toposcale, in terms of common topographic data sources and in terms of modeling applications. Of the applications listed, only representations of surface temperature and rainfall have a direct dependence on elevation. All others depend on measures of surface shape and roughness, as exemplified by the primary and secondary terrain attributes listed in Chapter 1. This underlies the importance of DEMs providing accurate representations of surface shape and drainage structure. This is particularly so in low-relief areas where elevations must be recorded with submeter precision to accurately reflect small elevation gradients.

Though actual terrain can vary across a wide range of spatial scales, in practice, source topographic data are commonly acquired at a particular scale. This places practical limits on the range of DEM resolutions that can be truly supported by a par-

TABLE 2.1 Spatial Scales of Applications of Digital Elevation Models (DEMs) and Common Sources of Topographic Data for Generation of DEMs

Scale	DEM Resolution	Common Topographic Data Sources	Hydrological and Ecological Applications
Fine toposcale	5–50 m	Contour and stream-line data from aerial photography and existing topographic maps at scales from 1:5,000 to 1:50,000 Surface-specific point and stream-line data obtained by ground survey using GPS Remotely sensed elevation data using airborne and spaceborne radar and laser	Spatially distributed hydrological modeling Spatial analysis of soil properties Topographic aspect corrections to remotely sensed data Topographic aspect effects on solar radiation, evaporation, and vegetation patterns
Coarse toposcale	50–200 m	Contour and stream-line data from aerial photography and existing topographic maps at scales from 1:50,000 to 1:200,000 Surface-specific point and stream-line data digitized from existing topographic maps at 1:100,000 scale	Broader scale distributed parameter hydrological modeling Subcatchment analysis for lumped parameter hydrological modeling and assessment of biodiversity
Mesoscale	200 m–5 km	Surface-specific point and stream-line data digitized from existing topographic maps at scales from 1:100,000 to 1:250,000	Elevation-dependent representations of surface temperature and precipitation Topographic aspect effects on precipitation Surface roughness effects on wind Determination of continental drainage divisions
Macroscale	5–500 km	Surface-specific point data digitized from existing topographic maps at scales from 1:250,000 to 1:1,000,000. National archives of ground surveyed topographic data including trigonometric points and benchmarks	Major orographic barriers for general circulation models

Note. DEMs at coarser scales are often obtained by local averaging of finer scale DEM data.

ticular source data set. The following section describes the data sources commonly supporting generation of DEMs at each of the scales listed in Table 2.1. This is followed by an examination of issues arising in the generation of DEMs from these data, including both the quality of source data and the accuracy of generated DEMs. These issues are illustrated by applying the ANUDEM locally adaptive gridding program (Hutchinson 1988, 1989b, 1996) to interpolate digital contours and streamlines, obtained from 1:24,000 scale mapping, to derive the DEM used to calculate the terrain parameters described in Chapters 3 and 4. The choice of DEM resolution is shown to be important in minimizing errors in representation of terrain shape, as measured by various primary terrain attributes, as well as matching the true information content of the source data.

2.2   SOURCES OF TOPOGRAPHIC DATA

Three main classes of source topographic data may be recognized, for which different DEM generation techniques are applicable, as discussed below.

2.2.1   Surface-Specific Point Elevation Data

Surface-specific point elevations, including high and low points, saddle points, and points on streams and ridges, make up the skeleton of terrain (Clarke 1990). They are an ideal data source for most interpolation techniques, including triangulation methods and specially adapted gridding methods. These data may be obtained by ground survey and by manually assisted photogrammetric stereo models (Makarovic 1984). They can also be obtained from gridded DEMs to construct triangulated irregular network (TIN) models (Heller 1990, Lee 1991). The advent of the global positioning system (GPS) has enhanced the availability of accurate ground-surveyed data (Dixon 1991, Lange and Gilbert 1999). Such data are now commonly obtained for detailed surveys of relatively small experimental catchments. They are less often used for larger areas.

2.2.2   Contour and Stream-Line Data

Contour data are still the most common terrain data source for larger areas. Many of these data have been digitized from existing topographic maps, which are the only source of elevation data for some parts of the world. The conversion of contour maps to digital form is a major activity of mapping organizations worldwide (Hobbs 1995). Contours can also be generated automatically from photogrammetric stereo models (Lemmens 1988), although these methods are subject to error due to variations in surface cover. A sample contour and stream-line data set, together with some additional point data, is shown in Figure 2.2. Contours implicitly encode a number of terrain features, including points on stream lines and ridges. The main disadvantage of contour data is that they can significantly undersample the areas between contour lines, especially in areas of low relief, such as the lower right-hand portion of Figure

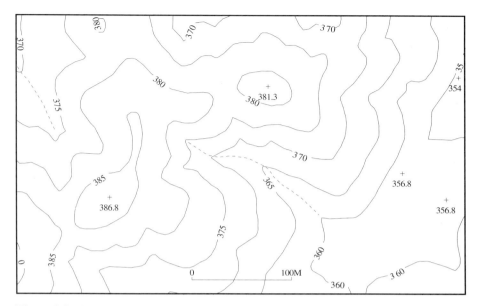

Figure 2.2. Contour, stream, and point elevation data.

2.2. This has led most investigators to prefer contour-specific algorithms over general-purpose algorithms when interpolating contour data (Clarke et al. 1982, Mark 1986).

Contour data differ from other elevation data sources in that they imply a degree of smoothness of the underlying terrain. When contours are obtained by manually assisted photogrammetric techniques, the operator can remove the effects of obstructions such as vegetation cover and buildings. Contour data, when coupled with a suitable interpolation technique, can, in fact, be a superior data source in low-relief areas (Garbrecht and Starks 1995), where moderate elevation errors in remotely sensed data can effectively preclude accurate determination of surface shape and drainage.

Streamlines are also widely available from topographic maps and provide important structural information about the landscape. However, few interpolation techniques are able to make use of stream-line data without associated elevation values. The method developed by Hutchinson (1988, 1989b) can use such streamline data, provided that the stream lines are digitized in the downhill direction. This imposes a significant editing task, which can be achieved by using a Geographic Information System (GIS) with network capabilities.

2.2.3 Remotely Sensed Elevation Data

Gridded DEMs may be calculated directly by stereoscopic interpretation of data collected by airborne and satellite sensors. The traditional source of these data is aerial photography (Kelly et al. 1977), which, in the absence of vegetation cover, can deliver elevations to submeter accuracy (Ackermann 1978, Lemmens 1988). Stereoscopic methods have been applied to SPOT imagery (Konecny et al. 1987, Day and Muller 1988), and more recently to airborne and spaceborne synthetic aperture radar (SAR). Spaceborne laser can also provide elevation data in narrow swathes (Harding et al. 1994). A major impetus for these developments is the yet unrealized goal of generating high-resolution DEMs with global coverage (Zebker et al. 1994, Dixon 1995).

Remote sensing methods can provide broad spatial coverage, but have a number of generic limitations. None of the sensors can reliably measure the ground elevations underneath vegetation cover. Even in the absence of ground cover, all methods measure elevations with significant random errors, which depend on the inherent limitations of the observing instruments, as well as surface slope and roughness (Harding et al. 1994, Dixon 1995). The methods also require accurately located ground control points to minimize systematic error. These points are not always easy to locate, especially in remote regions. Best possible standard elevation errors with spaceborne systems currently range between 1 and 10 meters, but elevation errors can be much larger, up to 100 meters, under unfavorable conditions (Sasowsky et al. 1992, Harding et al. 1994, Zebker et al. 1994, Lanari et al. 1997). Averaging of data obtained from multiple passes of the sensor can reduce these errors, but at greater cost.

Airborne SAR data are available for areas of limited extent. Standard elevation errors for DEMs derived from these data can be as small as 1 to 3 meters (Dixon 1995). Raw SAR DEMs have occasional large errors and random elevation errors across the whole DEM. They can also have systematic anomalies in the form of spurious ridges along tree-lined watercourses and missing data in areas of topographic shading. Careful filtering and interpolation of such data are required to derive useful representations of surface shape and drainage structure.

2.2.4   Scales of Source Topographic Data

The three main source data types described above may also be characterized according to their usage at different scales, as shown in Table 2.1. Remotely sensed data sources are normally used only at the finest scale. Contour data are normally used at both fine and coarse toposcales. At scales coarser than the toposcale, contour data tend to be too generalized to accurately depict surface shape and drainage structure, and their use at these scales is often supplanted by surface-specific points obtained from coarser scale maps.

Stream lines, at various levels of generalization, are used at all scales except the macroscale, where it is unrealistic to expect DEMs to accurately reflect surface drainage. At this broad scale "happenstance" data, such as benchmarks and trigonometric points recorded in national archives, can be used with profit (Hutchinson and Dowling 1991).

2.3   DEM INTERPOLATION METHODS

Interpolation is required to generate DEMs from surface-specific points and from contour and stream-line data. Since data sets are usually very large, high-quality global interpolation methods, such as thin plate splines, in which every interpolated point depends explicitly on every data point, are computationally impracticable. Such methods cannot be easily adapted to the strong anisotropy evidenced by real terrain surfaces. On the other hand, local interpolation methods, such as inverse distance weighting, local kriging, and unconstrained triangulation methods, achieve computational efficiency at the expense of somewhat arbitrary restrictions on the form of the fitted surface. Three classes of interpolation methods are in use: triangulation, local surface patches, and locally adaptive gridding. All achieve a degree of local adaptivity to anisotropic terrain structure.

2.3.1   Triangulation

Interpolation based on triangulation is achieved by constructing a triangulation of the data points, which form the vertices of the triangles, and then fitting local polynomial functions across each triangle. Linear interpolation is the simplest case, but a variety of higher order interpolants have been devised to ensure that the interpolated surface has continuous first derivatives (Akima 1978, Sibson 1981, Watson and Philip 1984, Auerbach and Schaeben 1990, Sambridge et al. 1995). Considerable attention has been directed toward methods for constructing the triangulation. The Delauney triangulation is the most popular method and several efficient algorithms have been devised (e.g., Heller 1990, Aurenhammer 1991, Tsai 1993).

Triangulation methods are attractive because they can be adapted to various terrain structures, such as ridge lines and streams, using a minimal number of data points (McCullagh 1988). However, these points are difficult to obtain as primary data. Triangulation methods are sensitive to the positions of the data points and the triangulation needs to be constrained to produce optimal results (Weibel and Heller 1991, Pries 1995). Triangulation methods have difficulties interpolating contour data, which generate many flat triangles, unless additional structural data points along streams and ridges can be provided (Clarke 1990).

2.3.2   Local Surface Patches

Interpolation by local surface patches is achieved by applying a global interpolation method to overlapping regions, usually rectangular in shape, and then smoothly blending the overlapping surfaces. Franke (1982) and Mitasova and Mitas (1993) have used bivariate spline functions in this way. These methods overcome the computational problems posed by large data sets and permit a degree of local anisotropy. They can also perform data smoothing when the data have elevation errors. There are some difficulties in defining patches when data are very irregularly spaced and anisotropy is limited to one direction across each surface patch. Nevertheless, Mitasova and Mitas (1993) have obtained good performance on contour data. An advantage of this method is that topographic parameters such as slope and curvature, as well as flow lines and catchment areas, can be calculated directly from the fitted surface patches, which have continuous first and second derivatives (Mitasova et al. 1996). Local surface patches can also be readily converted into regular grids.

2.3.3   Locally Adaptive Gridding

Direct gridding or finite difference methods can provide a computationally efficient means of applying high-quality interpolation methods to large elevation data sets. Iterative methods that fit discretized splines in tension have been described by Hutchinson (1989b) and Smith and Wessel (1990). Both methods have their origin in the method developed by Briggs (1974).

Computational efficiency is achieved by using a simple multigrid strategy that optimizes computational time in the sense that it is proportional to the number of interpolated DEM points (Hutchinson 1989b). The use of splines in tension is indicated by the statistical nature of actual terrain surfaces (Frederiksen et al. 1985, Goodchild and Mark 1987). It overcomes the tendency of minimum curvature splines to generate spurious surface oscillations in complex areas.

Former limitations in the ability of general gridding methods to adapt to strong anisotropic structure in actual terrain surfaces, as noted by Ebner et al. (1988), have been largely overcome by applying a series of locally adaptive constraints to the basic gridding procedure. These constraints can be applied between each pair of adjacent grid points, allowing maximum flexibility. Constraints that have direct relevance for hydrological applications are those imposed by the drainage enforcement algorithm devised by Hutchinson (1989b). This algorithm removes spurious depressions in the fitted DEM, in recognition of the fact that sinks are usually quite rare in nature (Band 1986, Goodchild and Mark 1987). This can significantly improve the drainage quality and overall structure of the fitted DEM, especially in data sparse areas.

A related locally adaptive feature is an algorithm that automatically calculates curvilinear ridge and stream lines from points of locally maximum curvature on contour lines (Hutchinson 1988). This permits interpolation of the fine structure in contours across the area between the contour lines in a more reliable fashion than methods that use linear or cubic interpolation along straight lines in a limited number of directions (Clarke et al. 1982, Oswald and Raetzsch 1984, Legates and Willmott 1986, Cole et al. 1990). An analogous approach combining triangulation and grid structures has been described by Aumann et al. (1992).

The result of applying the ANUDEM program (Hutchinson 1997) to the contour, streamline and point data in Figure 2.2 is shown Figure 2.3. The inferred stream and ridge lines are particularly curvilinear in the data sparse, low-relief portion of the map, and there are no spurious depressions. The derived contours also closely match the data contours. This locally adaptive gridding method has overcome problems formerly encountered by gridding methods in accurately representing drainage structure in low-relief areas (Douglas 1986, Carter 1988).

The procedure also yields a generic classification of the landscape into simple, connected, approximately planar, terrain elements, bounded by contour segments and flow line segments. These are similar to the elements calculated by Moore et al. (1988b) and described in Chapter 3, but they are determined in a more stable manner that incorporates uphill searches on ridges and downhill searches in valleys.

Recent developments in this locally adaptive gridding method include a locally adaptive data smoothing algorithm, which allows for the local slope-dependent

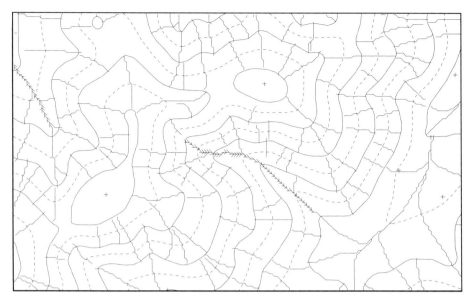

Figure 2.3. Locally adaptive gridding by ANUDEM of the contour, stream, and point data shown in Figure 2.2. Structure lines (ridges and stream lines) are generated automatically by ANUDEM. All contours are derived from the interpolated DEM. Dashed contour lines are shown at elevations midway between the data contour elevations.

errors naturally associated with the finite difference representation of terrain, and a locally adaptive surface roughness penalty, which minimizes profile curvature (Hutchinson 1996). The smoothing method has yielded useful error estimates for gridded DEMs and a criterion for matching grid resolution to the information content of source data.

2.4   FILTERING OF REMOTELY SENSED GRIDDED DEMS

Filtering of remotely sensed gridded DEMs is required to remove surface noise, which can have both random and systematic components. This is usually associated with a coarsening of the DEM resolution. Methods include simple nearest-neighbor subsampling techniques and standard filtering techniques, including median and moving average filtering in the spatial domain and low-pass filtering in the frequency domain. Several authors have recognized the desirability of filtering remotely sensed DEMs to improve the representation of surface shape.

Sasowsky et al. (1992) and Bolstad and Stowe (1994) used the nearest-neighbor method to subsample SPOT DEMs, with a spatial resolution of 10 m, to DEMs with spatial resolutions ranging from 20 to 70 m. This generally enhanced the representation of surface shape, although significant errors remained. Giles and Franklin

(1996) applied median and moving average filtering methods to a 20-m-resolution SPOT DEM. This similarly improved representation of slope and solar incidence angles, although elevation errors were as large as 80 m and no effective representation of profile curvature could be obtained.

Hutchinson et al. (1997) removed large outliers from airborne SAR data in an area of low relief and then applied moving average smoothing to generate a 50-m-resolution DEM. This provided an accurate representation of surface aspect, except in those areas affected by vegetation cover. Lanari et al. (1997) have applied a Kalman filter to spaceborne SAR data obtained on three different wavelengths. Standard elevation errors ranged between about 5 and 80 m, depending on land surface conditions.

2.5 QUALITY ASSESSMENT OF DEMS

The quality of a derived DEM can vary greatly depending on the source data and the interpolation technique. The desired quality depends on the application for which the DEM is to be used, but a DEM created for one application is often used for other purposes. Any DEM should therefore be created with care, using the best available data sources and processing techniques. As indicated in Figure 2.1, efficient detection of spurious features in DEMs can lead to improvements in DEM generation techniques, as well as detection of errors in source data.

Since most applications of DEMs depend on representations of surface shape and drainage structure, absolute measures of elevation error do not provide a complete assessment of DEM quality. A number of graphical techniques for assessing data quality have been developed. These are nonclassical measures of data quality that offer means of confirmatory data analysis without the use of accurate reference data. Assessment of DEMs in terms of their representation of surface aspect has also been examined by Wise (1998).

2.5.1 Spurious Sinks and Drainage Analysis

Spurious sinks or local depressions in DEMs are frequently encountered and are a significant source of problems in hydrological applications. Sinks may be caused by incorrect or insufficient data, or by an interpolation technique that does not enforce surface drainage. They are easily detected by comparing elevations with surrounding neighbors. Hutchinson and Dowling (1991) noted the sensitivity of this method in detecting elevation errors as small as 20 m in source data used to interpolate a continentwide DEM with a horizontal resolution of 2.5 km. More subtle drainage artifacts in a DEM can be detected by performing a full drainage analysis to derive catchment boundaries and streamline networks, using the technique of Jenson and Domingue (1988).

2.5.2   Views of Shaded Relief and Other Terrain Attributes

Computing shaded relief allows a rapid visual inspection of the DEM for local anomalies that show up as bright or dark spots. It can indicate both random and systematic errors. It can identify problems with insufficient vertical resolution, since low-relief areas will show as highly visible steps between flat areas. It can also detect edge-matching problems (Hunter and Goodchild 1995). Shaded relief is a graphical way of checking the representation of slope and aspect in the DEM. Views of other primary terrain attributes, particularly profile curvature, can provide a sensitive assessment of the accuracy of the DEM in representing terrain shape, as discussed in the example in Section 2.8.

2.5.3   Derived Elevation Contours

Contours derived from a DEM provide a sensitive check on terrain structure since their position, aspect, and curvature depend directly on the elevation, aspect, and plan curvature, respectively, of the DEM. Derived contours are a particularly useful diagnostic tool because of their sensitivity to elevation errors in source data. Subtle errors in labeling source data contours digitized from topographic maps are common, particularly for small contour isolations, which may have no label on the printed map. A simple example of derived contours indicating a single-point elevation data error is shown in Figure 2.4. It would be difficult to detect this error from only a shaded relief view of the DEM.

2.5.4   Frequency Histograms of Primary Terrain Attributes

Other deficiencies in the quality of a DEM can be detected by examining frequency histograms of elevation and aspect. DEMs derived from contour data usually show an increased frequency at the data contour elevations in the elevation histogram. The severity of this bias depends on the interpolation algorithm. Its impact is minimal for applications that depend primarily on drainage analyses that are defined primarily by topographic aspect. The frequency histogram of aspect can be biased toward multiples of $45°$ and $90°$ by simpler interpolation algorithms that restrict searching to a few specific directions between pairs of data points.

2.6   OPTIMIZATION OF DEM RESOLUTION

Determination of the appropriate resolution of an interpolated or filtered DEM is usually a compromise between achieving fidelity to the true surface and respecting practical limits related to the density and accuracy of the source data. Determination of the DEM resolution that matches the information content of the source data is desirable for several reasons. It directly facilitates efficient data inventory, since DEM storage requirements are quite sensitive to resolution. It also permits interpre-

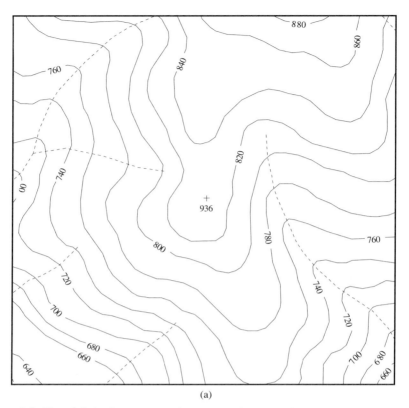

(a)

Figure 2.4. Use of derived contours to detect errors in source data: (a) contour, stream, and point elevation data with one erroneous elevation value; (b) contours derived from a DEM fitted to the erroneous data in (a).

tation of the horizontal resolution of the DEM as an index of information content. This is an important consideration when linking DEMs to other gridded data sets and when filtering remotely sensed DEMs. Moreover, it can facilitate assessment of the scale dependence of terrain-dependent applications, such as the determination of the spatial distributions of soil properties (Gessler et al. 1996).

A simple method for matching DEM resolution to source data information content has been developed by Hutchinson (1996). The method monitors the root mean square slope of all DEM points associated with elevation data as a function of DEM resolution. The optimum resolution is determined by refining the DEM resolution until further refinements produce no significant increase in the root mean square DEM slope. The method is particularly appropriate when source data have been obtained in a spatially uniform manner, such as elevation contours from topographic maps at a fixed scale, or from remotely sensed gridded elevation data. This procedure is used to aid optimization of DEM resolution in the next section.

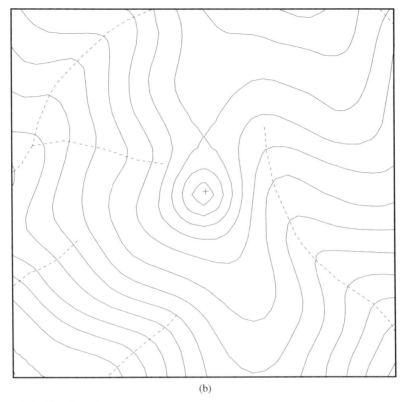

(b)

Figure 2.4. *(Continued)*

2.7   INTERPOLATION OF THE COTTONWOOD DEM USING ANUDEM

Digital elevation contours and stream-line data were obtained by digitizing 1:24,000 scale topographic maps of the Cottonwood catchment, as shown in Figure 2.5. The area spans approximately 1.4 km west to east, and nearly 3 km north to south. The vertical relief of the catchment, more than 1000 ft, is relatively large. Terrain shape is thus well determined by the elevation contours, which are spaced every 20 ft in the vertical direction. Practical limits on digitizing accuracy are indicated by slight irregularities in position of the closely spaced digitized contours in steeper areas. The lowest slopes are indicated by the more widely spaced contours along stream lines and along ridges defining the catchment boundary.

The ANUDEM program was applied to these data to produce a series of gridded DEMs at horizontal resolutions of 30, 15, and 7.5 m. The root mean square slope criterion, together with views of selected primary terrain attributes derived from the interpolated DEMs, was used to gauge the optimal DEM resolution, which was eventually chosen to be 15 m. A shaded relief view of the chosen 15-m DEM is

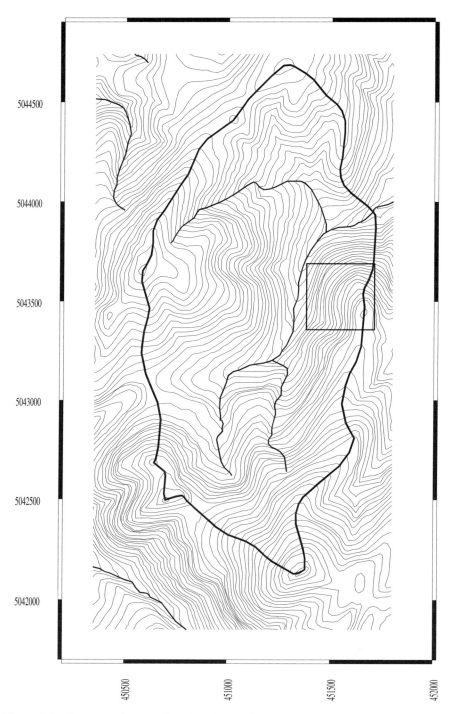

Figure 2.5. Contour and stream-line data for the Cottonwood catchment. The small square indicates the 330 × 330-m subarea used in later analysis.

42

shown in Figure 2.6 (see color insert). The aim of the procedure was to optimize DEM resolution by matching the true information content of the source data and maximizing surface accuracy. The range of candidate DEM resolutions was specified for consistency with common practice in the United States of calculating 30-m-resolution DEMs from 1:24,000 scale source data (USGS 1999). Practical issues arising in generating the DEM are now described. Optimization of DEM resolution is discussed in Section 2.8.

2.7.1 Specification of ANUDEM Options

Two ANUDEM options merit comment. First, drainage enforcement should be enacted to remove spurious sinks, since it is clear from the data in Figure 2.5 that the entire catchment should drain to the catchment outlet. Second, automatic determination of ridge and stream lines from corners in contour data should also be enacted, since these interpretations are plainly valid for the fine-scale contour data provided here. The effect of this option is that interpolation is defined by minimum curvature in the relatively planar areas away from ridge and stream lines, but the interpolation is constrained by approximate linear descent down each ridge and stream line. This ensures interpolation of curvilinear contour structure between data contour lines, as indicated in the example in Figure 2.4.

2.7.2 Elevation Units and Vertical Precision

Though a DEM with vertical units in meters was eventually required, this was not achieved by first converting the contour elevations in feet to elevations in meters. This would have introduced systematic errors if the converted elevations were stored in integer form, as is common practice. Instead, contour elevations in integer feet were retained and the derived DEM was then simply scaled to have elevations in meters. Provided that the DEM elevations are stored in real (floating point) form, this introduces no significant error.

Storage of DEM elevations in real form is generally recommended, particularly in low-relief areas where elevations in integer meters (or feet) are insufficient to adequately represent drainage direction and other primary terrain attributes. This has been a common problem with some generally available DEMs. In fact, even DEM elevations in higher relief areas, such as the one examined here, should be stored in real form. This maximizes the accuracy of terrain attributes derived from the DEM, particularly slope, aspect, and curvature, which depend on first- and second-order derivatives.

The practice of storing DEM elevations in integer meters (or feet) has arisen from a confusion between precision and absolute accuracy. The precision required for accurate determination of first and second derivatives is much higher than that required to satisfy basic vertical accuracy requirements. This is best indicated by observing that an arbitrary constant systematic error in DEM elevations would have no impact on the accuracy of most derived terrain attributes, even though absolute elevation errors could be arbitrarily large. It is difficult to smooth DEMs with integer vertical precision to recover sensible terrain shape and drainage attributes, especially in low-relief areas.

2.8   ASSESSMENT OF RESOLUTION AND QUALITY OF THE COTTONWOOD DEM

Close inspection of Figure 2.5 reveals probable slight systematic positional errors in the contour data, with some neighboring contour lines distinctly closer than adjacent pairs of contour lines. Such systematic errors are common. They indicate that some data smoothing is required. This is achieved in this case by carefully choosing the DEM resolution using several criteria, all of which relate to accuracy of representation of terrain shape.

2.8.1   Optimization of Resolution Using the Root Mean Square Slope Criterion

The standard approach recommended by Hutchinson (1996) is to monitor root mean square slope of the DEM as a function of DEM resolution. This is shown in Figure 2.7, where the resolutions decrease from left to right by successive halving from 240 to 7.5 m. The last three steps are the candidate DEM resolutions of 30, 15, and 7.5 m with respective root mean square slopes of 27.6, 31.9, and 34.4%. The curve begins to flatten at the last step from 15-m-resolution to 7.5-m-resolution. This suggests that the optimum resolution is approximately 15 m.

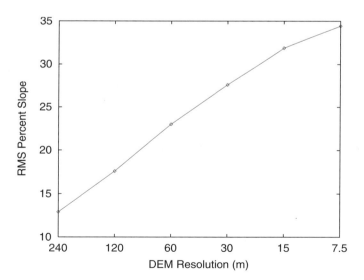

Figure 2.7. Root mean square slope of interpolated DEMs as a function of DEM resolutions successively halved from 240 to 7.5 m.

The flattening in Figure 2.7 is less marked than in Figure 2 of Hutchinson (1996), but this can be attributed to the positional error in the contour data. As the DEM resolution is refined, the fitted DEM comes closer to honoring the data contours. But if the data contours were honored exactly, then the DEM would have spurious variability, which would be reflected in over large values of the root mean square slope.

Additional assessments of DEM quality were made to confirm this choice of resolution. As indicated in Section 2.5, spurious sinks can provide an efficient way of detecting source data errors and assessing the general quality of the drainage structure of the DEM. However, in this case no sinks remained in the DEMs at all resolutions. This indicates the overall quality of the contour data, and its consistency with the stream-line data, as plotted in Figure 2.5. It also indicates the success of the drainage enforcement algorithm associated with the ANUDEM program.

2.8.2   Comparison of Data Contours with Derived Contours

A 330×330-m subgrid of each DEM, as indicated in Figure 2.5, was further examined. Figure 2.8 shows the source data contours for this area, and the corresponding contours derived from each of the three candidate DEMs. Both random and systematic positional errors are evident in the data contours. All three sets of derived contours generalize the data contours by varying amounts. The 30-m DEM plainly removes the peak in the lower right-hand corner and is therefore too coarse. This is consistent with the root mean square slope analysis above and the 30-m DEM was omitted from further consideration. The 15- and 7.5-m DEMs, on the other hand, both retain the peak. They appear to vary only in the degree to which they override the systematic positional error evident in the data. It is difficult to determine from these views which of these two resolutions is superior.

2.8.3   Views of Slope and Profile Curvature

Given the above assessment of the contour views in Figure 2.8, it would be difficult to separate the 15- and 7.5-m DEMs in terms of either shaded relief or plan curvature, the latter being directly represented by contour curvature. Views of slope and profile curvature for the two resolutions are therefore shown in Figures 2.9 and 2.10 (see color insert). Slope for the 7.5-m DEM shows only slightly more detail than the 15-m DEM. Thus, as for the contour views in Figure 2.10, it is difficult to separate the two resolutions in terms of their representation of slope.

On the other hand, differences in profile curvature between the two resolutions are quite marked. The linear features in Figure 2.10b are plainly closely associated with the systematic positional error in the data contours. These linear features are completely absent from Figure 2.10a, which shows broader scale variation in profile curvature not tied to any particular data contour. On the grounds that the linear features in the profile curvature in Figure 2.10b are spurious, the optimal DEM resolution was confirmed to be 15 meters. This was consistent with the initial indication given by the root mean square slope analysis shown in Figure 2.7.

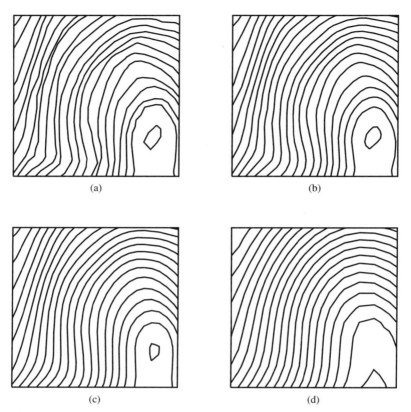

Figure 2.8. Data elevation contours and derived contours for the 330 × 330-m subarea indicated in Figure 2.5: (a) data contours; (b) contours derived from the 7.5-m DEM; (c) contours derived from the 15-m DEM; (d) contours derived from the 30-m DEM.

2.8.4   Histograms of Elevation and Aspect

To complete the assessments of DEM quality recommended in Section 2.5, histogram plots of elevation and aspect for the whole Cottonwood catchment were prepared for the 15- and 7.5-m-resolution DEMs. These are shown in Figures 2.11 and 2.12. The elevation histograms in Figure 2.11 show the expected small bias toward the data contour elevations, with little difference between the two DEM resolutions.

The aspect histogram for the 15-m-resolution DEM shows no marked bias toward multiples of 45°, with the overall distribution of aspect consistent with the predominant eastern and northwestern downslope orientations of the contour data shown in Figure 2.5. The aspect histogram for the 7.5-m-resolution DEM is consistent with this distribution, but has sharper peaks at multiples of 45°, again indicating some deficiencies with this resolution. However, overall the aspect histograms indicate a relative insensitivity of aspect to DEM resolution. This observation has been made by

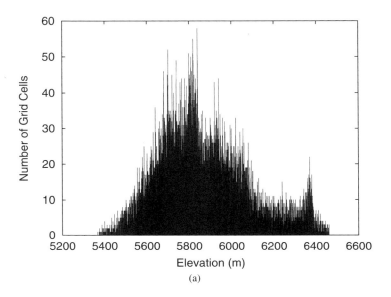

(a)

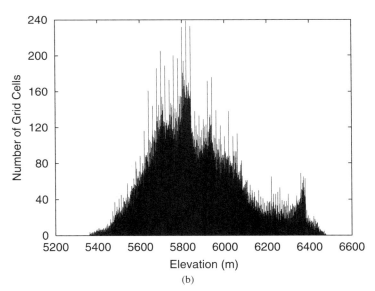

(b)

Figure 2.11. Elevation histograms for the Cottonwood catchment from (a) the 15-m DEM and (b) the 7.5-m DEM.

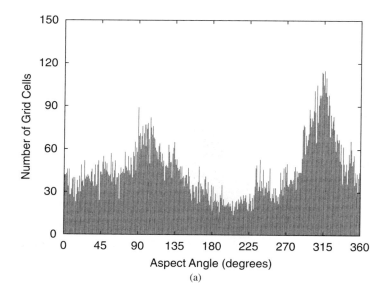

(a)

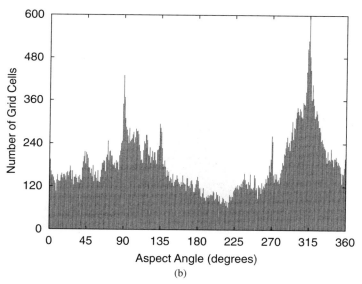

(b)

Figure 2.12. Aspect histograms for the Cottonwood catchment from (a) the 15-m DEM and (b) the 7.5-m DEM.

several authors. It has implications for the development of iterative terrain interpolation procedures defined in terms of aspect (Hutchinson 1996).

2.8.5   Summary Recommendation

The root mean square slope criterion appears to be a reliable, nongraphical, but nevertheless shape-based, way of matching DEM resolution to the information content of the source contour and stream-line data. This criterion can be refined, especially when source data have positional errors, by examining plots of derived contours and profile curvature. Examination of derived contours can prevent selection of a DEM resolution that is too *coarse* to adequately represent terrain structure. Examination of profile curvature can prevent selection of a resolution that is too *fine,* the latter leading to systematic errors in derived primary terrain attributes. Slope and plan curvature were less sensitive to DEM resolution than profile curvature. The above process of optimizing DEM resolution also illustrates the interactions between source data capture, DEM generation, and applications, as shown in Figure 2.1.

2.9   CONCLUSIONS

An important theme for providers of source topographic data and DEM interpolation methods is the need by most applications for accurate representations of terrain shape and drainage structure. The locally adaptive ANUDEM gridding procedure is shown to be able to produce such representations from contour, stream-line, and point data. These data are particularly appropriate for producing DEMs at the toposcale, at resolutions ranging from 5 to 200 m, where source contour data tend to accurately reflect terrain shape and drainage structure. The toposcale is also the scale at which applications are governed primarily by terrain shape, as can be directly measured by various primary and secondary terrain attributes that can be derived from the interpolated DEMs.

This success of locally adaptive gridding methods has prompted renewed interest in contour and stream-line data sources, which have wide global coverage. Remotely sensed elevation data sources also hold the promise of providing DEMs with global coverage, but appropriate filtering and interpolation methods that respect surface structure and drainage are required to reduce the inherent errors in these data, particularly in areas with low relief.

The process of producing a DEM from source data requires careful attention to the accuracy of the source data and the quality of the interpolated DEM. Several shape-based measures of DEM quality, which are readily plotted, can greatly assist in assessing DEM quality and in detecting data errors. These measures do not require the existence of separate reference elevation data.

DEM resolution can be optimized to match the true information content of source data as well as to filter positional errors in source data. The root mean square slope criterion, together with plots of derived contours and derived profile curvature, all assisted in determining an optimal resolution of around 15 m for a DEM derived from 1:24,000 scale contour and stream-line data. The optimization procedure clarified the importance of selecting DEM resolution carefully, especially when applications of the DEM depend on sensitive measures of terrain shape, such as profile curvature.

To respect the shape-based needs of applications, elevations of DEMs should be stored in real (floating point) form. DEMs with elevations in integer meters or integer feet have serious deficiencies in representing terrain shape and drainage structure, particularly in areas with low relief.

CHAPTER 3

Primary Topographic Attributes

John C. Gallant and John P. Wilson

3.1 TAPES-G: TERRAIN ANALYSIS ON GRIDDED DEMS

A gridded digital elevation model (DEM) represents the terrain surface as a regular lattice of point elevations. A grid can also be thought of as a tessellation using square tiles with the point elevation at the center of the square, as shown in Figure 3.1; the tiles should not be thought of as being flat with a constant elevation. The shape of the surface is represented by the change in elevations between adjacent lattice points.

The DEM lattice is nearly always aligned along geographic axes (N-S and E-W) and usually has equal spacing in both directions. When the DEM coordinates are length units, such as meters or feet, the spacing between DEM points is the same over the entire DEM, making analysis quite straightforward. Some DEMs (particularly at broad scales) use geographic coordinates (degrees, minutes, and seconds) and the actual distance between points varies across the DEM. Analysis of these DEMs must recognize this variable spacing. Most terrain-analysis programs, TAPES-G included, do not recognize the variable spacing of DEMs in geographic coordinates, so this type of DEM is difficult to work with unless it is projected to a regular grid in real coordinates before analysis. Unfortunately this type of transformation almost invariably results in the disturbance of surface drainage pathways, resulting in sinks (local minima). A better alternative is to project the source data and regenerate the DEM, but this is not always possible. This last strategy was used with ANUDEM (Hutchinson 1989b) to derive the 100- and 200-m DEMs from 1° to 2° USGS 3-arc-second DEMs used in Chapter 5, for example.

3.1.1 Surface Derivatives

Most of the primary topographic attributes listed in Table 1.1 can be determined locally from the derivatives of the topographic surface. These derivatives measure the

Terrain Analysis: Principles and Applications, Edited by John P. Wilson and John C. Gallant.
ISBN 0-471-32188-5 © 2000 John Wiley & Sons, Inc.

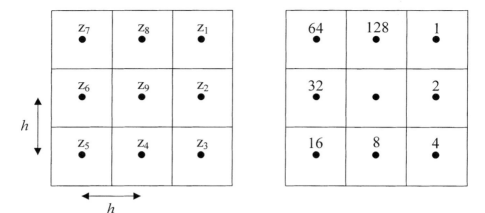

Figure 3.1. 3×3 subgrid for gridded DEM showing (a) node numbering convention and (b) flow direction numbering convention.

rate at which elevation changes in response to changes in location (x and y). The derivatives are estimated using centered finite differences:

$$z_x = \frac{\partial z}{\partial x} \approx \frac{z_2 - z_6}{2h} \tag{3.1}$$

$$z_y = \frac{\partial z}{\partial y} \approx \frac{z_8 - z_4}{2h} \tag{3.2}$$

$$z_{xx} = \frac{\partial^2 z}{\partial x^2} \approx \frac{z_2 - 2z_9 + z_6}{h^2} \tag{3.3}$$

$$z_{yy} = \frac{\partial^2 z}{\partial y^2} \approx \frac{z_8 - 2z_9 + z_4}{h^2} \tag{3.4}$$

$$z_{xy} = \frac{\partial^2 z}{\partial x \, \partial y} \approx \frac{-z_7 + z_1 + z_5 - z_3}{4h^2} \tag{3.5}$$

$$p = z_x^2 + z_y^2 \tag{3.6}$$

$$q = p + 1 \tag{3.7}$$

Equations 3.1 and 3.2 are the first-order derivatives that describe the rate of change of elevation with distance along the x and y axes, or the slope in those directions; these slope values can be positive or negative. Equations 3.3 and 3.4 are the second derivatives that describe the rate of change of the first derivative in the x and y directions, or the curvature in those directions. Equation 3.5 is a mixed second derivative that describes the rate of change of the x derivative in the y direction, or the twisting of the surface. Equations 3.6 and 3.7 are combinations of terms that are used in several places in subsequent equations.

Figure 3.1a shows the arrangement and numbering of the nine grid points that enter into the finite-difference equations, and h is the grid spacing of the DEM. The y axis, which points north, is up in Figure 3.1: Note that this is the opposite sense to the notation sometimes used in image processing where the origin is at the top left (northwest) corner and y points south. The conventions for numbering the nodes vary through the literature, and the notation for the finite difference formulae varies accordingly.

At the edges of the DEM or adjacent to no-data areas where some of the nine points are undefined, forward and backward finite differences are used to avoid reference to nonexistent data. For example, at the left edge where z_6 is not defined the x derivative is calculated as

$$z_x = \frac{z_2 - z_9}{h} \tag{3.8}$$

3.1.2 Slope

Slope S measures the rate of change of elevation in the direction of steepest descent. Slope is the means by which gravity induces flow of water and other materials, so it is of great significance in hydrology and geomorphology. It affects the velocity of both surface and subsurface flow and hence soil water content, erosion potential, soil formation, and many other important processes.

Slope is normally calculated using finite differences:

$$S_{\mathrm{FD}} = \sqrt{p} \tag{3.9}$$

This formula depends only on the elevations in the four cardinal directions. Some other finite difference formulae for slope are also used: For example, the ARC/INFO GIS uses a formula that depends on all eight adjacent cells. There is little difference between the results from these two formulae, although there is some evidence (Jones 1996) that Equation 3.9 is slightly more accurate.

Slope can also be computed using the steepest downhill slope to one of the eight nearest neighbors, the D8 method:

$$S_{\mathrm{D8}} = \max_{i=1,8} \frac{z_9 - z_i}{h\phi(i)} \tag{3.10}$$

where $\phi(i) = 1$ for cardinal (north, south, east, and west) neighbors ($i = 2, 4, 6$, and 8) and $\phi(i) = \sqrt{2}$ for diagonal neighbors to account for the extra distance to those cells (Figure 3.1a).

The D8 method gives slightly smaller average slopes than the finite difference method, because the direction in which the elevation difference is calculated is not always that of the steepest descent. An earlier report in Moore et al. (1993d) that the D8 approach predicted slightly larger average slopes was due to the use of a different formula that gave the steepest slope in any direction, not the downslope direction.

The finite difference method (Equation 3.9) is preferred because of its greater accuracy. The D8 slope estimate is useful when the slope of channels is required, because the finite difference estimate of channel cell slope may be affected by steep slopes adjacent to the channel. Using the D8 slope option guarantees that the slope calculated at a cell corresponds to the slope in the primary flow direction.

Regardless of which estimation method is used, TAPES-G reports slope in percent, or $S \times 100$. Note that slopes greater than 100% are possible in steep areas and signify slope angles greater than 45°. Slope angle β can be calculated from slope using

$$\beta = \arctan(S) \tag{3.11}$$

after dividing the TAPES-G slope value by 100. Figure 3.2 shows slope in percent computed using the finite difference method for the Cottonwood Creek 15-m DEM introduced in Chapter 2. The low slopes along valley bottoms and ridge tops are clearly visible, as are the steep slopes on the hillsides.

3.1.3 Aspect and Primary Flow Direction

Aspect ψ is the orientation of the line of steepest descent and is usually measured in degrees clockwise from north. Aspect is frequently recorded as a site attribute in eco-

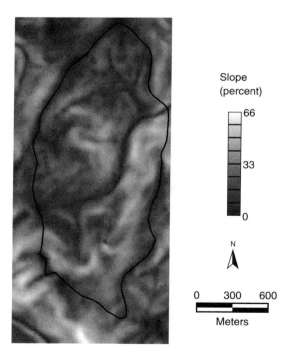

Figure 3.2. Percent slope derived from the Cottonwood Creek DEM using the finite difference formula, with the catchment boundary overlaid.

logical surveys and together with slope can be used to estimate solar radiation (although more sophisticated tools are available for this, such as SRAD described in Chapter 4). Aspect is useful for visualizing landscapes, especially when using a well-chosen color scale (Kimerling and Moellering 1989), although proper hill shading works better. The frequency distribution of aspect can also be used as a diagnostic tool to detect poor quality interpolation with a bias toward multiples of 45° or 90°. Aspect in degrees is calculated using finite differences as

$$\psi_{FD} = 180 - \arctan\left(\frac{z_y}{z_x}\right) + 90\left(\frac{z_x}{|z_x|}\right) \tag{3.12}$$

which is the angle determined by the x and y derivatives via arctan, modified to give degrees clockwise from north.

Aspect becomes rather meaningless when the slope is very small (and mathematically undefined when the slope is zero) so cells with slope less than some minimum value S_{min} may be considered to have undefined aspect (Mitasova and Hofierka 1993). TAPES-G does not currently do this, but it does give undefined aspect (NODATA value) when the slope is exactly zero.

The primary flow direction, FLOWD, is an approximate surrogate for aspect. This is essentially the primary flow direction for water moving over the land surface and identifies the direction to the nearest neighbor with maximum gradient, exactly as in Equation 3.10:

$$FLOWD = 2^{j-1} \quad \text{where } j = \arg\max_{i=1,8} \frac{z_9 - z_i}{h\phi(i)} \tag{3.13}$$

or, in words, j is the i that gives the largest slope value and hence is the direction of steepest descent. The approximate aspect corresponding to this flow direction is

$$\psi_{D8} = 45j \tag{3.14}$$

FLOWD is encoded using a binary notation (Figure 3.1b), which permits identification of flow to multiple nearest neighbors (although this capability is not actually used in TAPES-G). Note that this encoding is arbitrary and different software packages use different encoding schemes: ARC/INFO, for example, uses the same power-of-two encoding but with 1 representing flow to the east rather than northeast as in TAPES-G.

If none of the surrounding elevations z_1 to z_8 is higher than the central node z_9 then the node is a sink or in a flat area and has undefined aspect and flow direction. Section 3.1.5.1 describes how depressions are filled and flow directions assigned in flat areas using Jenson and Domingue's (1988) algorithm.

Figure 3.3 shows aspect computed from the Cottonwood Creek 15-m DEM using the finite difference method. The image is dominated by the abrupt changes where aspect switches from near maximum (360°) to near minimum (0°); this is an artifact of the linear gray scale and can be avoided with a scale having the same color for the minimum and maximum values. Some of the abrupt changes occur along the catch-

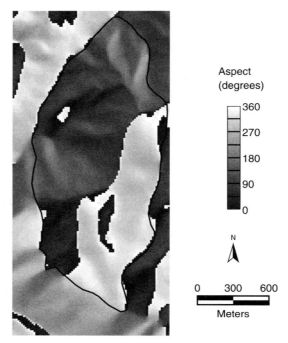

Figure 3.3. Aspect in degrees from north derived from the Cottonwood Creek DEM using the finite difference formula, with the catchment boundary overlaid.

ment boundary. The more subtle variations within the hillslopes show the changes in orientation across the surface.

3.1.4 Curvature

Curvature attributes are based on second derivatives: the rate of change of a first derivative such as slope or aspect, usually in a particular direction. The two curvatures most frequently computed are plan (or contour) curvature, K_c, the rate of change of aspect along a contour, and profile curvature, K_p, the rate of change of slope down a flow line. Profile curvature measures the rate of change of potential gradient so is important for characterizing changes in flow velocity and sediment transport processes. Plan curvature measures topographic convergence and divergence and hence the propensity of water to converge as it flows across the land. Mitasova and Hofierka (1993) have suggested that tangential curvature, K_t (plan curvature multiplied by the sine of the slope angle), is more appropriate than plan curvature for studying flow convergence and divergence because it does not take on extremely large values when slope is small. The distribution of convex and concave areas is the same for both plan and tangential curvature.

Surface curvatures can be thought of as the curvature of a line formed by the intersection of a plane and the topographic surface. The curvature of a line is the reciprocal of the radius of curvature, so a gentle curve has a small curvature value and a tight curve has a large curvature value. Plan curvature is the curvature in the horizontal plane of a contour line, while profile curvature is the curvature in the vertical plane of a flow line. Tangential curvature is curvature in an inclined plane perpendicular to both the direction of flow and the surface.

The units of curvature are radians per meter, the change in orientation resulting from traveling 1 meter along the respective line. Curvature values are typically small (nearly always less than 1), so TAPES-G reports curvature values multiplied by 100 for easier interpretation. The formulae used in TAPES-G are

$$K_p = \frac{z_{xx}z_x^2 + 2z_{xy}z_xz_y + z_{yy}z_y^2}{pq^{3/2}} \qquad (3.15)$$

$$K_c = \frac{z_{xx}z_y^2 - 2z_{xy}z_xz_y + z_{yy}z_x^2}{p^{3/2}} \qquad (3.16)$$

$$K_t = \frac{z_{xx}z_y^2 - 2z_{xy}z_xz_y + z_{yy}z_x^2}{pq^{1/2}} \qquad (3.17)$$

Using these formulae, profile curvature is negative for slope increasing downhill (convex flow profile, typically on upper slopes) and positive for slope decreasing downhill (concave, typically on lower slopes). Plan curvature is negative for diverging flow (on ridges) and positive for converging flow (in valleys). TAPES-G can optionally reverse these sign conventions to give positive values for convex curvatures, which is the more frequently used sign convention.

Total curvature is sometimes used as a measure of surface curvature. This curvature measures the curvature of the surface itself, not the curvature of a line across the surface in some direction. It can be positive or negative, with zero curvature indicating that the surface is either flat or the convexity in one direction is balanced by the concavity in another direction, as at a saddle. Total curvature is computed as

$$K = z_{xx}^2 + 2z_{xy}^2 + z_{yy}^2 \qquad (3.18)$$

Apart from their use in modeling flow characteristics, curvatures can be used to delineate geomorphic units (Dikau 1989). Plan curvature can be used to differentiate between ridges, valleys, and hillslopes, whereas profile curvature can differentiate between upper (convex) slopes and lower (concave) slopes.

Any errors in a DEM will be enhanced in the computed values of the curvature parameters because these parameters are based on second derivatives, which accentuate local irregularities. Furthermore, DEMs that have been interpolated primarily from contour data often display biases toward the elevations of the contour data, as discussed in Chapter 2, and profile curvature can alternate in sign systematically down a hillslope giving a "wavy" appearance. This result occurs because the density of data is high along the contours, but there are no data between contours. The interpolation methods give more variation in shape (curvature) where the elevation data are densest. In some landscapes this "waviness" is a real phenomenon (Moore et al. 1993d).

Figures 3.4 and 3.5 show plan and profile curvature for the Cottonwood Creek 15-m DEM using the normal sign convention of positive values for convex curves. Plan curvature clearly shows the locations of valleys and ridges with their large negative and positive plan curvatures, and the more gently curved hillslope regions between. Saddles are also visible in some places along the catchment boundary where strongly convex cells are immediately adjacent to strongly concave cells. Profile curvature clearly identifies the convex areas around hilltops and the concave areas in valley bottoms, but there is a considerable amount of pattern in the hillslope areas that is due to irregularities in the contour positions in the source data.

3.1.5   Upslope Contributing Area and Specific Catchment Area

Upslope contributing area, A, is the area above a certain length of contour that contributes flow across the contour, as shown in Figure 3.6. Specific catchment area, A_s (or a), is the ratio of the contributing area to the contour length, A/l. When calculating these areas from gridded DEMs, the contour length is approximately the size of

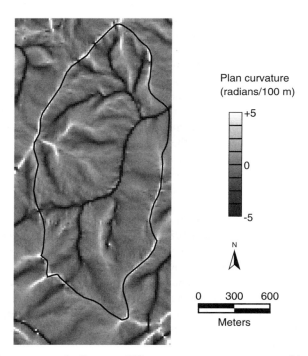

Figure 3.4. Plan curvature (radians per 100 m; convex curvatures are positive) for the Cottonwood Creek DEM, with the catchment boundary overlaid.

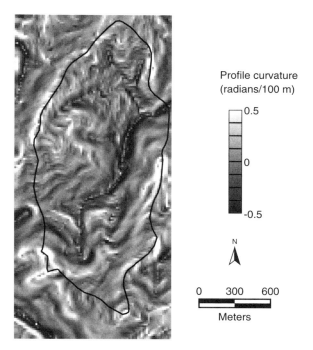

Figure 3.5. Profile curvature (radians per 100 m; convex curvatures are positive) for the Cottonwood Creek DEM, with the catchment boundary overlaid.

a grid cell, and in the simplest case the contributing area is determined by the number of cells contributing flow to a cell. The determination of contour length (or flow width) is described in the following subsection.

The estimation of both upslope contributing area and specific catchment area (i.e., the drainage area per unit width orthogonal to a flow line) is dependent on the estimation of flow direction(s) from a given node (node 9 in Figure 3.1a). Upslope contributing area is often referred to as drainage area or catchment area. TAPES-G reports contributing area in either number of cells or meters squared (m^2). Four different approaches are available in TAPES-G for calculating contributing areas: D8, Rho8, FD8/FRho8, and DEMON. Before any of these methods are applied, depressions and flat areas must be dealt with.

3.1.5.1 Depressions and Flat Areas
All contributing area algorithms follow paths of descending elevation to route flow across the surface. Any grid cells that do not have a lower neighbor represent a barrier to these algorithms and in most cases some special handling is required to allow flow accumulation to continue across the surface. There are some cases where this is not appropriate and the flow should terminate, for example, at sink holes in karst landscapes. In depressional landscapes water accumulates as ponds and lakes in depressions but routing of flow out of these

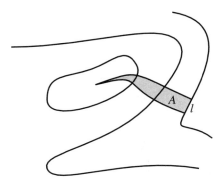

Figure 3.6. Upslope contributing area A is the area of land upslope of a length of contour l. Specific catchment area A_s is A/l.

areas is usually still required to properly represent the source areas of the streams flowing out of the lakes.

TAPES-G uses the method described in Jenson and Domingue (1988) to handle depressions and flat areas. First, depressions are filled by increasing elevations within depressions to their lowest outflow point. Care is taken to ensure that linked depressions do not drain into each other. Flow is then routed across flat areas (including filled depressions) by working backward from the drain points around the edge of each flat area. Where there is a lower point adjacent to the flat area, the flow direction is set to point to this outflow cell. Adjacent cells in the flat area then have their flow directions set to point to this drained cell and this continues across the flat area until flow directions have been assigned to all grid cells. These flow directions are used directly by the D8/Rho8 single-flow-direction, flow-accumulation algorithms and are converted to aspect for use by the DEMON algorithm in flat areas and depressions. The FD8/FRho8 multiple-flow-direction algorithm reverts to single flow directions in flat areas.

Note that in TAPES-G, slope, aspect, and curvatures are calculated from the original DEM and not the filled DEM. The filled DEM can also be saved to avoid refilling the depressions in subsequent runs. TAPES-G also reports the elevation residual between the original elevation and the elevation as modified to fill depressions, which is useful for checking whether the depressions found are substantial enough to warrant a revision of the DEM.

This depression-filling procedure completely ignores the shape of the landscape within a depression, which can be undesirable when the depressions cover many cells. Most depressions occur in valley bottoms, where the interpolation has caused a "dam" to appear across the valley. Filling the depression and then routing across the flat area backward from the outlet can cause the path of the inferred stream to deviate significantly from the lowest part of the topography in the depression. Apart from locational errors, these deviations can cause problems in secondary attributes where, for example, contributing area and slope are combined: The slope at the location of the inferred stream can be much higher than the stream gradient in these instances.

If these artifacts occur and undesirably influence the results, a revision of the DEM should be considered to remove the depressions and restore the connectivity of the drainage network. If the original source data are not available this may require manual modification of the DEM, which is only feasible when there are few sinks requiring correction. One method of restoring drainage to a DEM is "burning in" the streams: reducing by a fixed amount the elevations of cells along mapped stream lines so that the streams are at a lower elevation than the surrounding landscape. When depressions in the stream are filled the raised elevations are still below the surrounding areas so the flow accumulation remains confined to the imposed stream line. This method of restoring drainage is not recommended, because it dramatically alters the slopes around the stream line and can produce multiple stream lines. If the drainage line implicit in the DEM is more than one cell away from the imposed stream it will continue to collect flow from upslope and there will be two streams flowing across the landscape in close proximity: the imposed stream line and the DEM's drainage line. Proper incorporation of stream lines into the construction of the DEM as described in Chapter 2 gives much better results.

3.1.5.2 Single-Flow-Direction D8 Method The D8 (deterministic eight-node) algorithm developed by O'Callaghan and Mark (1984) allows flow from a cell to only one of eight nearest neighbors based on the primary flow direction. Because flow can accumulate into a cell from several upslope cells but only flow out into a single cell, this method can model flow convergence in valleys but not flow divergence in ridge areas. Another defect of this algorithm is that it tends to produce flow in parallel lines along preferred directions that will agree with the aspect only when the aspect is a multiple of $45°$. For example, on a surface with aspects ranging from 0 to $22.5°$, the D8 algorithm will predict a constant flow direction FLOWD = 128 (due north) (Figure 3.1b). Even with these significant limitations, the D8 algorithm is still frequently used for determining contributing areas, primarily because of its simplicity. In spite of its inability to model flow divergence, it is adequate for delineating catchment boundaries. For calculating the distribution of contributing area and specific catchment area across hillslopes, the more sophisticated FD8 or DEMON methods are recommended.

Figure 3.7 shows upslope contributing area computed using the D8 method on the 15-m resolution Cottonwood Creek DEM. The valley bottoms are clearly defined as a line of white high contributing area cells, and the ridges both within the catchment and at the divide are visible as dark cells with low contributing area. Many obvious straight-line artifacts are visible, due to the absence of dispersion and the inability to respond to subtle changes in surface orientation.

Note that cells of lowest contributing area do not necessarily correspond to the catchment boundary, which is determined by the connections of the flow lines. For example, at several places around the catchment divide, but particularly in the middle of the western edge, the cells of lowest contributing area do not coincide with the catchment boundary determined from the contours. Figure 3.8 shows details of the

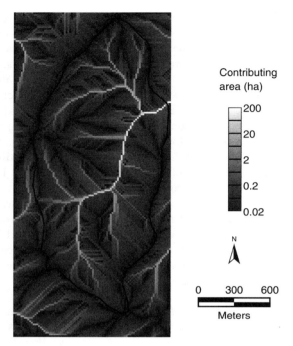

Figure 3.7. Contributing area (ha) derived from the Cottonwood Creek DEM using the D8 single flow direction algorithm, with the catchment boundary overlaid.

flow directions in this area. By following the flow lines downstream it is apparent that the line of dark cells (lowest contributing area) is not the catchment divide, since the flow lines immediately to the west of the dark cells drain eastward into the catchment. The catchment divide as determined by the D8 flow lines is within one cell of the catchment divide line derived from the contours.

The dark, low contributing area cells mark the line where the aspect shifts sufficiently east from south to change the flow direction from south to southeast: The line of dark cells is an artifact of the single-flow-direction method rather than a significant topographic feature. Another artifact of the D8 method can be seen in the valley head in the southeast corner of Figure 3.8. Three flow lines run in parallel along a valley bottom because the convergence of the valley is insufficient to force the lines together.

The effect of these artifacts is to significantly distort the spatial pattern of contributing area. The weakly divergent ridge along the catchment divide is assigned a contributing area of 20 cells or 4500 m^2 while the center of the valley bottom area immediately before the convergence of the three lines is assigned a contributing area of 32 cells or 7200 m^2. These two locations should have contributing areas differing by much more than a factor of 1.6.

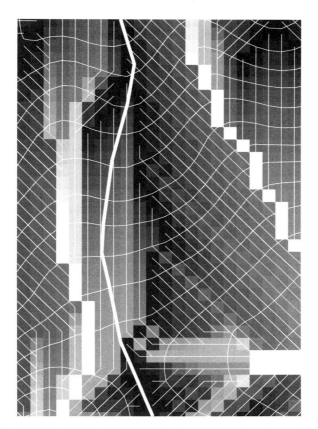

Figure 3.8. Detail of the D8 flow lines, contributing area and catchment boundary on the western edge of the Cottonwood Creek catchment. The narrow white lines are flow lines and contours; the thick white line is the mapped catchment boundary.

3.1.5.3 Randomized Single-Flow-Direction (Rho8) Method

The Rho8 (random eight-node) algorithm developed by Fairfield and Leymarie (1991) is a stochastic version of the D8 algorithm in which a degree of randomness is introduced into the flow-direction computations. This algorithm aims to break up parallel flow paths and produces a mean flow direction equal to the aspect. This is achieved by replacing $\phi(i)$, the distance factor of 1 or $\sqrt{2}$ in Equation 3.10, by $2 - r$ for the diagonal neighbors ($i = 1, 3, 5,$ or 7 in Figure 3.1a), where r is a uniformly distributed random variable between 0 and 1. The effect of this change will be most noticeable where the slopes are similar, such as on hillslopes, and will have little effect where one direction is clearly the steepest, such as in confined valley bottoms. Like the D8 algorithm, the Rho8 algorithm cannot model flow dispersion, but it does simulate more realistic-looking flow networks. The breakup of long, parallel flow paths comes at the cost of introducing many more cells that have no upslope connection. The randomizing of flow directions also results in different flow networks each time the program is run (as illustrated in Chapter 5). These characteristics are undesirable, and the Rho8 algorithm is no longer considered a useful alternative to D8. The FD8 and DEMON algorithms provide better solutions to the problem Rho8 was designed to address.

Figure 3.9 shows contributing area computed on the same 15-m Cottonwood Creek DEM using the Rho8 algorithm. Note that the straight-line artifacts produced by D8 have largely disappeared and are replaced by a more dendritic structure. However, the Rho8 surface has many more cells on the hillslopes with no upslope contributing area.

3.1.5.4 Multiple-Flow-Direction Methods (FD8 and FRho8)
The FD8 and FRho8 algorithms are modifications of D8 and Rho8 that allow flow divergence to be represented (Moore et al. 1993d). This pair of algorithms allows flow to be distributed to multiple nearest-neighbor nodes in upland areas above defined channels and uses the D8 or Rho8 algorithms below points of presumed channel initiation. On the hillslopes above channels the proportion of flow or upslope contributing area assigned to each downslope neighbor is determined on a slope-weighted basis as pro-

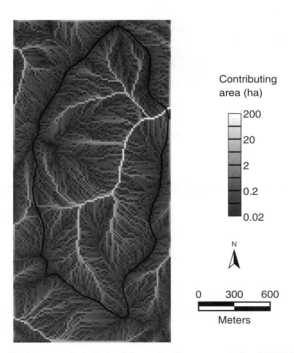

Figure 3.9. Contributing area (ha) derived from the Cottonwood Creek DEM using the Rho8 randomized single flow direction algorithm, with the catchment boundary overlaid.

posed by Freeman (1991) and Quinn et al. (1991). The fraction of contributing area passed from a cell to neighbor i is given by

$$F_i = \frac{\max(0, S_i^v)}{\sum_{i=1}^{8} \max(0, S_i^v)} \tag{3.19}$$

where S_i is the slope from the central node to neighbor i (the same term used in Equation 3.10) and v is a positive constant. Freeman (1991) found that $v = 1.1$ produced the most accurate results for artificial conical surfaces, and TAPES-G uses that value. Larger values of v concentrate the flow more to the direction of steepest slope, giving results more similar to the single-flow-direction (D8) algorithm. Holmgren (1994) reported that much higher values of v in the range 6–8 might be more appropriate. The algorithm of Quinn et al. (1991) is similar to Equation 3.19 but uses a contour-length (flow-width) term in addition to slope to derive the flow fractions.

The FD8 and FRho8 algorithms give more realistic distributions of contributing area in upslope areas, while also eliminating D8's parallel flow paths. This algorithm tends to cause considerable dispersion of flow in valleys, which is considered undesirable because stream lines usually are well defined in valleys. To overcome this, the flow-dispersion algorithm is disabled and replaced by D8 (or Rho8) wherever the contributing area exceeds a user-specified threshold, called the "maximum cross-grading area" in TAPES-G. The dispersion in valleys can be considered as representing a floodplain or riparian area, which might be desirable in some applications; in this case the maximum cross-grading area can be set to a very large value. The transition from multiple- to single-flow-direction algorithm can cause an irregularity in the frequency distribution of the contributing area, particularly if the maximum cross-grading area is set to too small a value. Depending on the landscape, a maximum cross-grading area of about 10 ha or 100,000 m^2 is a reasonable starting point. The optimum value depends on the valley density as represented in the DEM: Finer textured landscapes with high valley density would require a smaller value of cross-grading area to reflect the smaller hillslopes, and coarser landscapes with larger hillslopes would require a higher value.

Quinn et al. (1995) described a variant of the multiple-flow-direction algorithm where the exponent v was changed continuously from 1 (full dispersion) to a large value (single flow direction) as contributing area increased, giving a smooth change from multiple flow direction to single flow direction. This modification has not been implemented in TAPES-G.

The TAPES-G implementation of FD8/FRho8 takes considerably longer to run than D8 or Rho8 because of the greater density of flow connections. Figure 3.10 shows the contributing area for the Cottonwood DEM computed using the FD8 algorithm with a large maximum cross-grading area so flow dispersion is applied everywhere. Note the smooth variation of contributing area in the upland areas in contrast to the D8 and Rho8 results, and the considerable flow dispersion in the valley areas.

3.1.5.5 DEMON Stream-Tube Method
The fourth and final algorithm provides a completely different approach for modeling flow accumulation and disper-

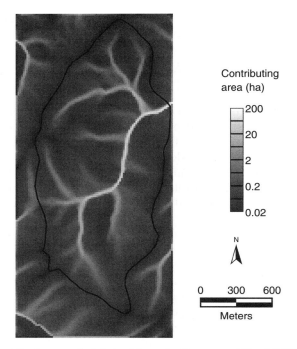

Contributing
area (ha)

200

20

2

0.2

0.02

N

0   300   600

Meters

Figure 3.10. Contributing area (ha) derived from the Cottonwood Creek DEM using the FD8 multiple flow direction algorithm with a large maximum cross-grading area, with the catchment boundary overlaid.

sion. This method, which is similar conceptually to the stream-tube approach used with contour-based DEMs by Moore and Grayson (1991), was proposed first by Lea (1992) and Costa-Cabral and Burges (1994). Costa-Cabral and Burges's program is called DEMON (digital elevation model network extraction). In DEMON, flow is generated at each pixel (source pixel) and is routed down a stream tube until the edge of the DEM or a pit is encountered. The stream tubes are not constrained to coincide with the edges of cells and can expand and contract as they traverse divergent and convergent regions of the DEM surface. The stream tubes are constructed from the points of intersections of a line drawn in the gradient direction (aspect) and a grid cell edge. The amount of flow, expressed as a fraction of the area of the source pixel, entering each pixel downstream of the source pixel is added to the flow accumulation value of that pixel. After flow has been generated on all pixels and its impact on each of the pixels has been added, the final flow accumulation value is the total upslope area contributing runoff to each pixel.

The DEMON algorithm has been incorporated into TAPES-G with three modifications. First, DEMON can be applied to either the original DEM or a derived depressionless DEM. Second, the nodes of the DEM define the centroid of the pixels rather than the vertices. Finally, the flow direction of a stream tube for each pixel is defined by the aspect (as calculated using Equation 3.12). DEMON takes a similar amount of processing time to FD8/FRho8 and gives results of similar or better quality. DEMON produces few artifacts, although a slight preference for flow in the cardinal north, south, east, and west directions is sometimes apparent.

Figure 3.11 shows the contributing area computed using the DEMON algorithm. Note the smooth variation of contributing area in the upland areas, similar to FD8, contrasted with well-defined channels in the lowlands that are more similar to D8. Minor artifacts of north–south linear features are just noticeable in this image.

3.1.5.6 Frequency Distributions of Contributing Area

The cumulative frequency distributions of contributing area calculated by the four different methods are shown in Figure 3.12. The most obvious feature of the plot is the difference at small contributing areas between D8 and Rho8 on the one hand and FD8 and DEMON on the other. The nondispersive single-flow-direction algorithms produce a much higher proportion of low contributing areas because there are many more cells that do not have any inflow. Compared to the DEMON method, which is considered the most accurate of the four, D8 has too high a proportion of low contributing areas by a factor of at least 2 up to about 0.1 ha (4 grid cells), and the effect persists up to contributing areas of about 2.2 ha (about 100 cells). Note that the Rho8 distribution is

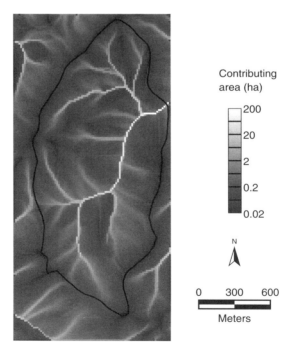

Figure 3.11. Contributing area (ha) derived from the Cottonwood Creek DEM using the DEMON stream tube algorithm, with the catchment boundary overlaid.

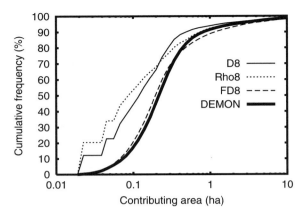

Figure 3.12. Cumulative frequency distributions of contributing area for the Cottonwood DEM computed by the four contributing area algorithms.

better than D8 for contributing areas larger than about 0.2 ha (10 cells), due to the breakup of linear flow paths and improved flow concentration on mildly convergent hillslopes, which is the effect the algorithm was designed to achieve. However, this improvement comes at the cost of a substantial increase in the number of low contributing area cells, significantly worsening the errors in this part of the distribution.

The other obvious defect of the D8 and Rho8 algorithms is the stepped nature of the distribution that is a consequence of using single flow directions. Without dispersion of flow, the contributing area can only be an integer multiple of the grid cell area, so there are steps at 0.0225, 0.045, and 0.0675 ha, and so on.

The DEMON and FD8 algorithms produce very similar results for small contributing areas, with an absence of steps due to their ability to distribute fractions of cell areas to downslope cells. The FD8 algorithm produces a slightly higher proportion of contributing areas up to about 0.4 ha (18 cells) but a lower proportion of contributing areas above that. Both these effects are due to the higher dispersion of the FD8 algorithm: The extra dispersion in strongly divergent areas produces a higher frequency of small contributing areas, while the dispersion in the valley bottoms results in a lower frequency of high contributing areas.

The frequency distributions are very similar at high contributing areas, due to the convergence of flow into the small number of valley bottom cells. The FD8 algorithm has the lowest proportion of high contributing area cells due to its dispersion of flow across valley bottom areas.

3.1.5.7 Rate of Change of Specific Catchment Area   TAPES-G computes the rate of change of specific catchment area along the flow path, dA_s/ds. It is computed as specific catchment area of flow leaving the cell less the average of the specific catchment areas entering the cell, divided by the flow length across the cell. This attribute has application in some erosion models, where erosion is a function of slope, profile curvature, and dA_s/ds (Moore 1996).

3.1.5.8 Edge Effects and Catchments   Contributing area values can be inaccurate where the edge of the DEM does not coincide with the edge of a catchment unless the flow direction is outward across the border of the DEM. TAPES-G does not identify cells that are contaminated with edge effects. However, it does allow the user to interactively choose drainage basins for analysis instead of analyzing the whole DEM. If requested, TAPES-G will display a flow network highlighting drainage lines with contributing area greater than a user-specified amount, and the user can select (using a point-and-click graphical interface) outlet points that define drainage basins to be analyzed. Drainage boundaries can alternatively be specified using a file containing x and y coordinates of a bounding polygon defining the region to be analyzed.

3.1.6 Flow Width

TAPES-G computes a flow-width attribute, w, that is used to compute specific catchment area, A_s, from contributing area, A:

$$A_s = \frac{A}{w} \tag{3.20}$$

Flow width is reported in terms of cell width and is computed in different ways for different flow algorithms. For the D8 and Rho8 algorithms, flow width is equal to cell width for flow in cardinal directions and cell diagonal for flow in diagonal directions:

$$w = \begin{cases} h & i = 2, 4, 6, 8 \\ \sqrt{2}h & i = 1, 3, 5, 7 \end{cases} \tag{3.21}$$

The flow width for the FD8/FRho8 algorithm is difficult to model well since the flow is dispersed in different fractions to multiple downslope cells. Quinn et al. (1995) used a geometrical construction to determine that flow widths in the cardinal directions should be $0.5h$ and those in diagonal directions $0.354h$. The total flow width for a cell is the sum of the flow widths in the directions of downslope cells, and ranges from $0.354h$ to $3.416h$. TAPES-G originally used the D8 flow width for FD8/FRho8, but as a result of recent work (Gallant and Hutchinson in preparation) this was changed to account for the amount of dispersion of flow. Initially, the effective number of flow directions is calculated from the weights:

$$n_{\text{flow}} = \frac{1}{\sum\limits_{i=1}^{8} F_i^2} \tag{3.22}$$

This n_{flow} varies from a minimum of 1 when all flow is in a single direction to a maximum of 8 when flow is dispersed equally in all eight directions. The flow width is then estimated as

$$w = \begin{cases} h & n_{\text{flow}} < 3 \\ h\left[1 + 3\left(\dfrac{n_{\text{flow}} - 3}{5}\right)\right] & n_{\text{flow}} \geq 3 \end{cases} \tag{3.23}$$

This width is h for flow on convergent and planar areas when n_{flow} is small, but increases to a maximum of $4h$ when n_{flow} is large. This method takes no account of whether the flow is in a cardinal or diagonal direction.

For the DEMON algorithm the flow width is the distance across the cell at right angles to the flow direction, which, in turn, is a function of aspect:

$$w = |\sin \psi| + |\cos \psi| \qquad (3.24)$$

where $|x|$ denotes the absolute value of x.

3.1.7 Maximum Flow-Path Length

The maximum flow-path length is the maximum length of all flow paths from the catchment boundary to a given point in the DEM. It can be computed using the D8 or Rho8 algorithms described above, but rather than accumulate areas the algorithm accumulates flow distances across cells, and only the largest flow-path length of all upslope cells is passed on to the downslope cell, instead of the sum. For flow in a cardinal direction ($i = 2, 4, 6$, or 8) the cell flow distance is h (the grid spacing), whereas for diagonal flow ($i = 1, 3, 5$, or 7) the cell flow distance is $\sqrt{2}\, h$. However, flow paths calculated by these methods tend to have a zigzag appearance because the flow-path directions are restricted to the four cardinal and four diagonal directions. Flow paths calculated using the stream tubes constructed by the DEMON algorithm avoid this problem; however, the calculation of flow-path length using this method has not been implemented in TAPES-G.

3.1.8 Downslope Attributes

In some applications it is the shape of the landscape downslope of each point, rather than upslope, that is of particular interest (Table 1.1). For example, Speight (1980) described how soil-water content was influenced by specific dispersal area in combination with specific catchment area. Specific dispersal area is the area downslope of a unit length of contour to a stream line, and represents the available area into which soil water can drain. Downslope path length is another attribute that can be used where proximity to a stream is considered a useful attribute.

The algorithms for contributing area and flow-path length can be adapted to compute downslope attributes by looking uphill rather than downhill and working from the lowest to highest elevations. However, the existing algorithms can be used without modification simply by inverting the DEM. There are two points to be aware of when using this approach. First, some terrain-analysis programs, including TAPES-G, use negative elevations internally for special purposes, so the elevations of the inverted DEM should be positive. This is easily achieved by subtracting the DEM from a constant value larger than the highest elevation of the DEM. Second, the peaks in the surface become sinks when it is inverted, and in this case these sinks are genuine termination points and must not be filled.

3.1.9 Upslope Averages of Terrain Attributes

For processes connected with the accumulation and dispersal of materials across the surface, such as soil development, the average characteristics of the landscape in a site's contributing area may be more significant than the terrain attributes at the site itself (see Chapters 10–12 for additional applications of this type). The mean slope or curvature upslope of the site can be a useful index of processes linked to the sorting and transport of material to the site.

Program UPSUM-G provided with the TAPES-G package performs this type of analysis. It calculates the area-weighted mean value of several of the TAPES-G primary terrain attributes within the contributing area of each cell. For each cell, the area-weighted mean is the sum of the value at that cell plus the values of the inflowing cells multiplied by the contributing areas, divided by the sum of the weights:

$$\overline{v}_j = \frac{\sum\limits_{i \in U_j} w_i v_i}{\sum\limits_{i \in U_j} w_i} \qquad (3.25)$$

where v is the value being averaged, U_j is the set of adjacent cells flowing into cell j (including cell j), and v_i and w_i are the value and the weight for cell i. The weights for the inflowing cells are equal to the contributing areas of these cells, and the weight for cell j, the cell under consideration, is 1.

3.1.10 Other Terrain Attributes

While the attributes discussed here are the most frequently used, a number of other attributes can be derived from DEMs to describe the spatial patterns of landscape geometry and processes. These include the identification of streams and ridges and calculations of distance to nearest stream or ridge; stream ordering; statistical measures of the distribution of elevation within regions or catchments, such as skew and kurtosis; and hypsometric analysis.

3.1.11 TAPES-G Inputs

TAPES-G can accept DEM files in several different formats. The DEM can be either x, y, z triplets or just z (elevation) values. An x, y, z file can have points in any order, while a z-only file may be in either row or column order with the first point in either the northwest or southwest corner. Elevation values may be integer or floating point; if they are integer values a scaling factor can be applied, which is often used to increase the vertical resolution. Finally, the file may be in either ASCII or binary form.

ASCII files have one record per line. Unformatted binary files (integer or floating point) also are record oriented, with each record preceded and followed by a four-byte integer specifying the number of bytes in the record (standard FORTRAN unformatted records). Files containing x, y, z triplets must have one triplet per record, while files containing only z values may have any number of z values per record, so long as each new column or row starts in a new record. Direct access two-byte integer files containing only z values are also accepted. TAPES-G can read ARC/INFO grids exported using the GRIDFLOAT command, which produces a binary floating point file and a separate header file containing metadata such as the size, resolution, and location of the grid.

An input elevation value of 0 indicates a missing value or point outside the analysis region. TAPES-G will not compute attributes for points with 0 elevation, and all elevations for the region to be analyzed should be positive (negative elevations have a special meaning internally). TAPES-G also assumes that coordinates and elevation values are in meters, but other units may be used provided that elevation is in the same units as the x and y coordinates.

The current implementation of TAPES-G requires that the DEM fits into the size limits set when the program is compiled. The arrays are set to 1000 rows and 1000 columns, although these settings can be changed to suit the available memory and size of the DEM. The numbers of rows and columns do not have to be equal. The memory requirements are approximately 80 bytes per grid cell (thus requiring 80 Mbytes of virtual memory in the standard configuration) or 40 bytes per grid cell if the program is compiled without the DEMON module.

3.1.12 TAPES-G Outputs

TAPES-G produces an output attribute file in either binary or ASCII format. The 14 primary topographic attributes that are written to this file are summarized in Table 3.1. Each DEM grid point with all its attributes is written to a single record as either one line in an ASCII file or one unformatted record in a binary file. Cells with missing data can either be omitted from the output file or written with no-data values for all attributes. The no-data value is also used where an attribute is undefined, such as for aspect when the surface is perfectly flat. Other programs using the output of TAPES-G must be aware of the no-data value and take appropriate action when it is encountered.

In addition to the computed terrain attributes, metadata are written to the beginning of the file, describing the contents of the file, the resolution and coordinates of the data, and the options and parameters used by TAPES-G to produce the file. These metadata are used by other programs in the TAPES package to locate the attributes they require, and by TAPESTOARC, which reformats the output data to the form required by ARC/INFO's ASCIIGRID or FLOATGRID command, permitting visualization and further analysis within the ARC/INFO Geographic Information System (GIS).

Two utility programs are provided in the TAPES-G package for simple viewing and analysis of the results. The program ATTRIBUTES provides color visualization of the attributes contained in the TAPES-G output file, with interactive control of the rendering of the data. The program FREQ computes histograms and performs simple statistical analysis of the results. ATTRIBUTES is written in C and uses the X Window system while FREQ is written in FORTRAN.

TABLE 3.1   Terrain Attributes in TAPES-G Output File

	Attributes	Units	Definition	Section
1, 2	*x, y*	Usually meters	*x* and *y* coordinates as determined from the DEM	
3	Flow direction	None	Computed using D8 or Rho8 algorithm	3.1.3
4	*z*	Usually meters	Elevation as read from the DEM	
5	Contributing area	Square meters or number of cells	Area draining out of each cell	3.1.5
6	Flow width	Multiple of cell width	Width associated with flow leaving the cell	3.1.6
7	Slope	Percent	Slope in the steepest downslope direction	3.1.2
8	Aspect	Degrees clockwise from north	The direction of the steepest downslope slope	3.1.3
9	Profile curvature	Radians per 100 meters	Curvature of the surface in the direction of steepest descent	3.1.4
10	Plan curvature	Radians per 100 meters	Curvature of contour drawn through the grid point	3.1.4
11	Tangent curvature	Radians per 100 meters	Plan curvature multiplied by sine of slope angle	3.1.4
12	Elevation residual	Usually meters	Difference between original DEM and depressionless DEM	3.1.5
13	Flow-path length	Usually meters	The longest flow path from the catchment divide or edge of DEM to the cell	3.1.7
14	$d(A_s)/ds$	None	Rate of change of specific catchment area along the flow path	3.1.5

3.2   ELEVATION RESIDUAL ANALYSIS

The terrain attributes computed by TAPES-G are all controlled by local variations in topography (i.e., the change in elevation from one cell to the next). In some instances it can be useful to consider the nature of each point with reference to a wider context (e.g., Blaszczynski 1997). One application of this type of analysis is where subsurface water flow does not follow surface shape but responds to relative elevations within a region. In this situation, a small rise in the bottom of a valley might be nearly as wet as its surroundings, even though it has no contributing area for surface flow. Another application is identifying ridges and valleys by comparing their elevations with those in a surrounding area comparable in size to the length of the hillslopes.

The program ELEVRESIDGRID computes several terrain attributes by comparing the elevation at each point with the elevations in a circular window of a specified size. The analysis ignores hydrological connectivity of the surface topography, considering instead the relationship of a point to the surrounding landscape. This analysis is similar to the functions provided in some GIS packages, such as the FOCAL operators in the ARC/INFO GRID module.

A key issue with this type of analysis is the choice of a suitable analysis window size. Ideally, the window size should be comparable to the length scale of the process under study. This size is often the hillslope length.

In the following equations, the set of cells included in the circle about the point being considered is denoted by C, and the number of cells in the circle is n_C.

3.2.1 Mean Elevation

Mean elevation in the circular window is primarily used as a reference against which to compare the elevation at the central point of the window. However, the mean elevation can be useful when a smoothed DEM is required:

$$\bar{z} = \frac{1}{n_C} \sum_{i \in C} z_i \tag{3.26}$$

3.2.2 Difference from Mean Elevation

The difference between the elevation at the center of the window and the mean elevation in the window is a measure of the relative topographic position of the central point. Its range depends on the range in elevation within the size of the window:

$$\text{diff} = z_0 - \bar{z} \tag{3.27}$$

This terrain attribute can be useful in connection with processes that are sensitive to local differences from regional elevations, such as groundwater flow (Roberts et al. 1997) and susceptibility to lightning strikes (Cary 1998).

3.2.3 Standard Deviation of Elevation

The standard deviation of elevation is a measure of the variability of elevation within the circular window. It is useful as a measure of local relief at the scale specified by the radius of the window:

$$\text{SD} = \text{sqrt}\left(\frac{1}{n_C - 1} \sum_{i \in C} (z_i - \bar{z})^2\right) \tag{3.28}$$

Standard deviation is a measure of elevation variability within the window. When the window size is similar to the hillslope length, standard deviation provides a measure of local relief. For much larger window sizes, this attribute measures the roughness of the landscape at that coarser scale.

3.2.4   Elevation Range

Elevation range measures the full range of elevations within the circular window:

$$\text{range} = \max_{i \in C} z_i - \min_{i \in C} z_i \qquad (3.29)$$

It is similar to standard deviation in that it measures local relief, but has the disadvantage of sometimes changing abruptly from place to place as individual high and low points enter and leave the window. This can result in visible overlapping circular regions in the results.

3.2.5   Deviation from Mean Elevation

The deviation from the mean is the difference from the mean divided by the standard deviation:

$$\text{dev} = \frac{z_0 - \bar{z}}{\text{SD}} \qquad (3.30)$$

This attribute measures the relative topographic position as a fraction of local relief, and so is normalized to the local surface roughness. The deviation mostly ranges between -1 and $+1$, but values outside this range are possible; large values may indicate irregularities in the DEM, where the elevation is significantly outside the typical range in that region.

3.2.6   Percentile

Percentile is a ranking of the point at the center of the window relative to all the points in the window. It is calculated simply by counting the number of points lower than the central point:

$$\text{pctl} = \frac{100}{n_C} \operatorname*{count}_{i \in C}(z_i < z_0) \qquad (3.31)$$

Percentile ranges from 0 to 100, with a value of 0 indicating that the point is the lowest in the circle, 100 indicating that it is the highest, and 50 indicating that half the points in the circle are lower. Percentile is similar to deviation, but its well-defined range makes its application simple and robust. It has been used as an index of local topographic position, which, as Chapter 16 shows, can be a useful explanatory variable for the spatial patterns of vegetation species.

3.2.7   Other Attributes

A number of other attributes can be computed using this type of analysis for various purposes. An alternative to percentile ranking that is easier to implement in the ARC/INFO GIS is the elevation as a percentage of the elevation range:

$$\text{pctg} = 100 \; \frac{z_i - z_{min}}{z_{max} - z_{min}} \tag{3.32}$$

Like percentile, this variable always ranges from 0 to 100; unlike percentile it does not account for the distribution of values within the circle so it is not as sensitive to the shape of the surface. It also suffers from the same circular artifacts as elevation range due to the use of minimum and maximum elevations.

3.2.8 ELEVRESIDGRID Examples

Figures 3.13–3.15 show the difference from mean elevation, percentile, and standard deviation calculated for the Cottonwood Creek DEM using a window radius of 200 m. The attributes calculated at this scale are much more generalized than the local attributes computed by TAPES-G and show the variations of topography at a broader scale, roughly matching the length of the hillslopes.

Figure 3.13, difference from mean elevation, shows the generalized structure of ridges and valleys within the catchment. The brightness or darkness indicates the intensity of highness or lowness: for example, there are some low hilltops along the western catchment boundary that are barely noticeable in this image indicating that they do not protrude much above the surrounding landscape. Conversely, the valley of the main

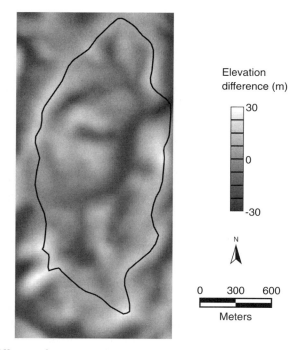

Figure 3.13. Difference from mean elevation for the Cottonwood Creek DEM using a radius of 200 m, with the catchment boundary overlaid.

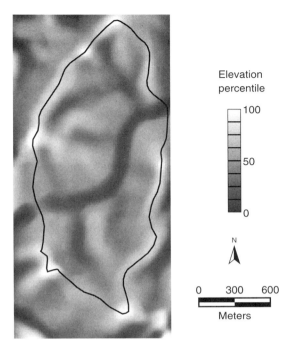

Figure 3.14. Percentile for the Cottonwood Creek DEM using a radius of 200 m, with the catchment boundary overlaid.

drainage line is seen to be deeply incised in its middle portion where its difference from mean elevation becomes more negative than either upstream or downstream.

Figure 3.14, elevation percentile, shows the structure of the catchment in a similar manner to Figure 3.13 but picks out different features of the surface. Being based on a ranking rather than an elevation difference, percentile is insensitive to relief, and so shows all significant hilltops as bright spots regardless of how high they are relative to their surroundings. Similarly, the valleys are more uniformly identified as the lowest part of the landscape without indicating their relative depths.

Figure 3.15, standard deviation, shows the pattern of local relief across the catchment. The valley bottoms and, to a lesser extent, the catchment boundary are areas of lower relief, while the steep hillslopes to the east of the main channel are areas of high relief. The highest standard deviation is found on areas with long, steep slopes, such as on the hill south of the catchment outlet and at the south end of the DEM outside the catchment.

3.3 TAPES-C: TERRAIN ANALYSIS ON CONTOUR DEMS

The contour DEM structure is an alternative representation to the gridded DEM that uses contours and flow lines to delineate elements aligned with the flow direction of

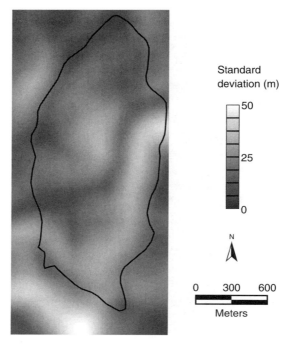

Figure 3.15. Standard deviation of elevation for the Cottonwood Creek DEM using a radius of 200 m, with the catchment boundary overlaid.

the surface, as shown in Figure 3.16. Elements are bounded above and below by contour lines and to the sides by flow lines. Each element is considered to be a unit of landscape, in the same way as the cell of a DEM, but the elements are of variable size and shape. The advantage of this representation is that the flow connectivity is always from one element to its downslope neighbors so the determination of contributing area is trivial. Dynamic hydrologic modeling within this geometric framework is also simpler than grid or triangular geometries because the connectivity of flows is encapsulated in the network geometry; this is the primary use of these terrain representations. The main disadvantage of this approach is that the element network is relatively difficult to construct in the first place.

The method of partitioning catchments into flow elements was pioneered by Onstad and Brakensiek (1968). They referred to it as the "stream-path" or "stream-tube" analogy and implemented the method manually. Speight (1974, 1980) also used manual methods to produce a series of flow lines from which specific catchment and dispersal areas could be estimated.

The automated creation of the network of elements from contour data is not a trivial process and requires well-organized input data and careful handling of critical points in the landscape: peaks, saddles, channel confluences, and ridge junctions.

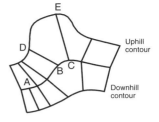

Figure 3.16. A set of elements formed by contours and flow lines. Proceeding uphill, flow lines are terminated (*A*) and added (*B*, *C*) to maintain even spacing. Lines are constructed using either a minimum distance (*BD*) or orthogonal (*CE*) criterion.

Dawes and Short (1994) discuss in detail the use of the topology of the network of lines and critical points to assist the construction of element networks.

There are currently two software implementations of this technique, TAPES-C and TOPOG, which were both developed from the slope line algorithm of O'Loughlin (1986). TAPES-C is described in Moore et al. (1988b) and Moore and Grayson (1991); TOPOG's implementation is described in Vertessy et al. (1993) and Dawes and Short (1994). The five subsections that follow describe the algorithms used by TAPES-C; the topographic attributes TAPES-C calculates; the differences between TAPES-C and TOPOG; and TAPES-C's inputs and outputs.

3.3.1 Element Construction

Creating a network of elements from contour data is fundamentally the construction of regularly spaced flow lines between adjacent contours. These lines are constructed for each pair of adjacent contour lines working from the lowest to highest contour, and the main challenge for the automated algorithms is to select well-chosen lines that provide a uniform and error-free tessellation of the surface.

A user-specified distance controls the spacing of the flow lines, and on the first (lowest) contour, flow lines are placed at that specified distance apart or as close to that distance as the spacing of points along the contour permits. On subsequent higher contours the upper ends of the flow lines from the lower contour are used as the start points of the next set of flow lines provided they are neither too close nor too far apart. This approach results in continuous lines across the surface with each element connecting directly to the element below. The connections from an element to its downslope neighbors are recorded as the elements are constructed.

If a starting point is too close to the previous start point, it is skipped to prevent cluttered lines and narrow elements (although this is optional), as shown at *A* in Figure 3.16. If the points are too far apart, one or more new lines are started between the existing lines to prevent very wide elements as shown at *B* and *C* in Figure 3.16. In TAPES-C a new line is inserted if the distance between the existing start points is more than 1.3 times the user-defined spacing; a line is terminated if the start points are less than half the specified spacing.

Once a start point is chosen for a flow line, its end point on the next highest contour must be found. There are two criteria for choosing this end point:

1. *Minimum distance:* the end point is the point on the next highest contour closest to the start point.
2. *Orthogonal:* the end point is chosen so that the flow line is at right angles to the lower contour.

TAPES-C uses the minimum distance method in divergent (ridge) areas and the orthogonal method in convergent (valley) areas. A user-defined threshold of contour curvature is used to identify convergent areas. This combination of techniques helps overcome the errors introduced by using straight flow lines and always working uphill. The line *CE* in Figure 3.16 is an orthogonal line, while the remaining lines are constructed using the minimum distance criterion.

Peaks and saddles introduce special cases into this construction. Peaks must be explicitly defined within any closed contour that does not have another contour within it, and must be given an elevation higher than the surrounding contour. Flow lines up from the innermost contour all connect to the enclosed peak. Peak points may also be located in flat ridge areas to facilitate the construction of well-shaped elements. Unlike peaks in closed contours, these ridge points connect to a higher contour and multiple ridge points may be strung together, as shown in Figure 3.17.

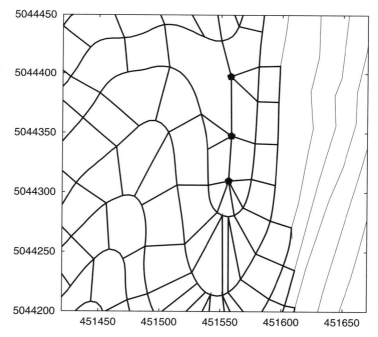

Figure 3.17. Part of the TAPES-C element network constructed from the Cottonwood Creek DEM showing a ridge and valley. Connected high points form a ridge line to facilitate element construction.

Saddle points are required between adjacent contours of the same elevation. These saddles must be given the elevation corresponding to the upslope contour to which they connect. The saddle then forms the connection between the two separate segments of contour at this elevation. TAPES-C constructs lines from the saddle to the closest points on both contour segments before commencing element construction. When one of these points is reached while scanning along the upper contour for a candidate upper end point, the scan switches to the other contour segment.

Figure 3.18 shows the element network constructed for a tributary of the Cottonwood Creek catchment, including the high points, saddle points, and ridge points, and the inferred channel network. Figure 3.19 shows contributing area computed from the element network. The same gray-scale legend has been used here as in the grid-based contributing area images in Figures 3.7–3.11 to allow comparison. One of the deficiencies of the contour-based element network is the large valley bottom elements resulting from the construction of flow lines in an upslope direction. This results in high contributing areas spread across relatively wide valley bottoms, in a similar manner to the FD8 grid algorithm (see Section 3.1.5.4 and Figure 3.10). This

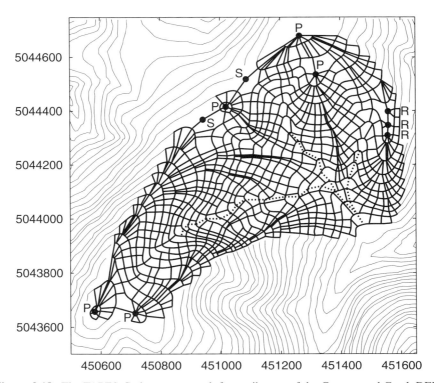

Figure 3.18. The TAPES-C element network for a tributary of the Cottonwood Creek DEM at the northern end of the catchment. The dashed line is the inferred channel based on a contributing area threshold of 5 ha. The labels indicate user-supplied peak (P), saddle (S), and ridge (R) points.

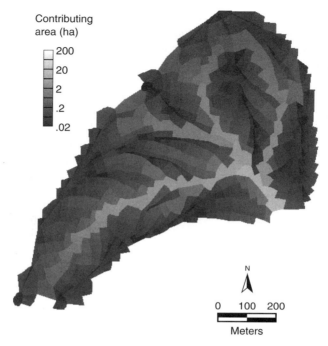

Figure 3.19. Contributing area computed from the TAPES-C element network of Figure 3.18.

can be a particular problem when used for hydrologic modeling, but the inclusion of channels within high contributing area cells allows the model to represent the convergence of flow to a narrow channel in those cells, rather than spread across the whole element.

Note that because TAPES-C and TOPOG are designed primarily for dynamic hydrological modeling they expect to produce element networks for whole catchments. These programs are not designed to construct element networks on arbitrary pieces of topography. The programs can also fail if provided with improper input data such as contours oriented in different directions and mislocated peaks or saddles. Excessive contour spacing also causes difficulties and can be remedied by providing additional contours.

3.3.2 Computed Terrain Attributes

After constructing all the elements, TAPES-C computes the attributes of each element. The area of the element, A_e, and its width at the upper and lower ends, W_u and W_l, are calculated from the coordinates of the element boundary. The effective length of the element, reported as overland flow distance, is the area divided by the average width:

$$l = \frac{A_e}{0.5(W_u + W_l)} \tag{3.33}$$

Two slopes are calculated for each element: the average slope across the element and the slope down the center of the element. Average slope is the elevation change (contour interval) divided by the element length:

$$S = \frac{z_u - z_l}{l} \tag{3.34}$$

Slope down the center of the element is useful as an estimate of the slope of the channel in valley bottom elements. The center line of the element is defined as the line from the point of maximum curvature on the upper contour to the point of maximum curvature on the lower contour, and its length is l_c, the channel flow distance. The channel slope is

$$S_c = \frac{z_u - z_l}{l_c} \tag{3.35}$$

Aspect is computed as the orthogonal direction to a straight line drawn across the lower end of the element (the line BC in Figure 3.16). Plan curvature is not calculated explicitly, but convergence and divergence is represented by the change in width from upper to lower contour. Plan curvature at the center of the element can be approximated by

$$K_c = \frac{2(W_l - W_u)}{l(W_l + W_u)} \tag{3.36}$$

with the sign convention of positive curvatures being convex.

Contributing area is determined by accumulating element areas from highest to lowest elevations using the previously recorded links between elements. Where the contributing area is greater than a user-specified threshold, TAPES-C constructs a channel line down the center of the element joining the points of maximum contour curvature on the upper and lower contour segments.

3.3.3 Differences Between TAPES-C and TOPOG Element Networks

While using similar principles of element construction, TAPES-C and TOPOG differ in their implementation of the methods. TOPOG always uses minimum distance to select the upper end of each flow line. This avoids some problems that occur with the orthogonal lines but sometimes causes problems, such as crossing over valley bottoms where the contours are quite asymmetric. TOPOG constructs drainage lines down from prespecified channel heads before constructing elements, which both increases efficiency and gives better control of elements in valley bottoms, reducing the need for the orthogonal method. As a result of this construction, channels in TOPOG are located between valley bottom cells, rather than in the center of valley bottom cells as in TAPES-C. TOPOG does not handle connected peaks, and it con-

TABLE 3.2   Terrain Attributes in TAPES-C Output File

	Attributes	Units	Definition
1	Contour number	None	Index of contour starting from 1 at lowest contour
2	Element number	None	Index of element on each contour
3	Downslope connection	None	Index of downslope element number to which the element connects
4	Contributing area	Usually square meters	Area draining across lower contour of the element
5–7	Bottom center x, y, and z	Usually meters	Location and elevation of the center of the lower contour
8	Element area	Usually square meters	Area of the element, not including inflowing elements
9	Element slope	Percent	Average slope across the element
10	Downhill width	Usually meters	Length of lower contour segment
11	Uphill width	Usually meters	Length of upper contour segment
12	Overland flow distance	Usually meters	Average length of cell
13	Channel flow distance	Usually meters	Length of channel line
14–16	Centroid x, y, and z	Usually meters	Location and elevation of element centroid
17	Channel slope	Percent	Slope of channel line
18	Aspect	Degrees clockwise from north	Orientation of element as determined from lower contour
19, 20	Max curvature x and y	Usually meters	Location of maximum curvature point on lower contour

structs lines from saddles up and down to adjacent contours before constructing elements. TOPOG is a more robust software product than TAPES-C, having benefited from a more sustained software development effort. TOPOG can be obtained from CSIRO Land and Water through *http://www.clw.csiro.au/topog.* The program is free but registration is required before the software can be downloaded.

3.3.4   Inputs

The input data for TAPES-C consists of contour data supplemented by high points and saddle points. These data must be collected into a particular format readable by TAPES-C, with the contour data ordered from lowest to highest elevations, oriented consistently (with uphill on either the left or right of all contours), and with evenly spaced samples. These resampling, reordering, and reformatting tasks are performed by two support programs, PLOTCON and PREPROC, provided with the TAPES-C package. An interactive contour-editing program, EDITCON, is also provided with the package. TAPES-C also requires some user intervention during running to identify whether particular peak points connect to an uphill contour.

3.3.5 Outputs

TAPES-C produces an output file in the same format as TAPES-G, with a list of attributes for each element (Table 3.2). Metadata have not been implemented in TAPES-C output files at this time. A separate vertex file provides the shapes of each element, in the same order as the attribute file.

3.4 CONCLUSIONS

This chapter described the commonly used primary terrain attributes that are calculated directly from elevation data. The most commonly used are slope, aspect, contributing area, specific catchment area, plan curvature, and profile curvature. These attributes and others are calculated by the TAPES-G program from regular grid DEMs and by the TAPES-C program from contour DEMs. The following chapter describes various secondary terrain attributes, combinations of two or more primary topographic attributes, that are frequently used as surrogates of physical processes.

CHAPTER 4

Secondary Topographic Attributes

John P. Wilson and John C. Gallant

4.1 INTRODUCTION

This chapter describes four grid-based programs for calculating secondary topographic attributes: EROS, SRAD, WET, and DYNWET. These programs were originally developed by Moore (1992) and subsequently modified and added to by various contributors (Wilson and Gallant 1998). The four grid-based programs compute a series of secondary attributes that combine two or more primary attributes and can be used to characterize the spatial variability of specific hydrological, geomorphological, and ecological processes occurring in landscapes. The computed topographic attributes are based on simplified representations of the underlying physics of the processes in question but include the key factors that modulate system behavior (e.g., topography) (Moore and Hutchinson 1991). With this approach we sacrifice some physical sophistication to allow improved estimates of spatial patterns in landscapes (Moore et al. 1993d). The methods are able to handle variations in the availability of possible input data and the spatial resolution of those data. Care must be taken in developing and using these techniques because simplifying assumptions can increase rather than resolve computational complexity. This possibility looms large for systems with many variables and for models with many simplifications where several variables of the original model participate in the simplifying assumptions (Denning 1990, Moore and Hutchinson 1991, Grayson et al. 1993).

The inputs, outputs, and estimation methods employed by EROS, SRAD, WET, and DYNWET are described in the sections that follow together with an illustrative application of each program for the same Cottonwood Creek catchment introduced in Chapters 2 and 3. The programs are written in FORTRAN77 and C for Unix systems, and produce output files in the same format as TAPES-G, including metadata describing the options and parameters specified when running the programs. The file format and tools for displaying, analyzing, and converting the output files are described in Section 3.1.12.

Terrain Analysis: Principles and Applications, Edited by John P. Wilson and John C. Gallant.
ISBN 0-471-32188-5   © 2000 John Wiley & Sons, Inc.

4.2 EROS

This program calculates a pair of simple erosion indices that account for the major hydrological and topographic attributes affecting erosion (Wilson and Gallant 1996). These indices incorporate a dimensionless sediment transport capacity that is a non-linear function of specific discharge and slope (Chapter 1; Moore and Burch 1986a–c). They are derived from the transport capacity limiting sediment flux in the Hairsine–Rose (Hairsine and Rose 1991, 1992a, b), WEPP (Laflen et al. 1991a, b), and catchment evolution (Willgoose et al. 1991) erosion theories (Moore et al. 1992). Both indices can be easily extended to three-dimensional terrain, they can account for different runoff producing mechanisms and soil properties using a spatially variable weighting function, and they can be implemented within a Geographic Information System (GIS) (Moore and Wilson, 1992, 1994).

4.2.1 Estimation Methods

The first erosion index in EROS calculates the spatial distribution of soil loss potential with a simple dimensionless stream power or sediment transport capacity index, T_c, which can be written as

$$T_{cj} = \left(\frac{\sum_{i\epsilon Cj} (\mu_i a_i)/b_j}{22.13} \right)^m \left(\frac{\sin \beta_j}{0.0896} \right)^n \tag{4.1}$$

where μ_i is a weighting coefficient ($0 \leq \mu_i \leq 1$) that is dependent on the runoff generation mechanism and soil properties (i.e., infiltration rates), a_i is the area of the ith cell, b_j is the width of each cell, β_j is the slope in degrees, m and n are constants (0.6 and 1.3, respectively), and C_j is the set of elements that are hydrologically connected to cell j (i.e., the catchment area of the cell, including the current cell j). When $\mu = 0$, no rainfall excess is generated on that cell; when $\mu = 1$, all of the precipitation on the cell becomes rainfall excess.

The second index represents the change in sediment transport capacity across a grid cell, ΔT_c, and provides a possible measure of the erosion or deposition potential in each cell (Moore and Burch 1986a). The change in sediment transport capacity across hydrologically connected grid cells can be written as

$$\Delta T_{cj} = \phi \left[\left(\sum_{i\epsilon C_{j-}} \frac{\mu_i a_i}{b_{j-}} \right)^m (\sin \beta_{j-})^n - \left(\sum_{i\epsilon Cj} \frac{\mu_i a_i}{b_j} \right)^m (\sin \beta_j)^n \right] \tag{4.2}$$

where ϕ is a constant, subscript j signifies the outlet of cell j and $j-$ signifies the inlet to cell j, and C_{j-} is the set of elements that are hydrologically connected to cell j excluding the current cell (Moore et al. 1992). Positive values of ΔT_{cj} indicate net deposition and negative values indicate net erosion.

Three options are provided in EROS for estimating the weighting coefficients in the above pair of equations. The first option assumes that rainfall excess is generated uniformly over the entire catchment (i.e., $\mu_i = \mu = 1$). The second option assumes saturation overland flow in which overland flow occurs only in zones of saturation in landscapes. The final option assumes Hortonian overland flow in which μ_i is spatially variable and a function of the infiltration characteristics of the soil (Wilson and Gallant 1996). The choice of the best option for a particular application (catchment) is difficult because the runoff production system involves the interaction of atmosphere, land geology and geomorphology, vegetation, soils, and people. The final choice of runoff method often relies on intuition and circumstantial evidence (Pilgrim and Cordery 1993).

The uniform rainfall excess method is often used because no additional knowledge of runoff behavior and infiltration rates is required. The erosion indices can be computed directly from the slope and drainage area attributes calculated in TAPES-G with simplified versions of Equations 4.1 and 4.2 in this instance:

$$T_c = \left(\frac{A_s}{22.13}\right)^m \left(\frac{\sin\beta}{0.0896}\right)^n \tag{4.3}$$

$$\Delta T_{cj} = \phi \, [A_{sj-}^m(\sin \beta_{j-})^n - A_{sj}^m(\sin \beta_j)^n] \tag{4.4}$$

where β is the slope (in degrees) and A_s is the specific catchment area or drainage area per unit width orthogonal to a flow line (m^2/m). Equation 4.3 is equivalent to the length–slope factor in the Revised Universal Soil Loss Equation (Renard et al. 1991) for planar slopes with lengths <100 m and gradients < 14°, but it is simpler to use and conceptually easier to understand (Moore and Wilson 1992, 1994). This index can also explicitly account for flow convergence and divergence, and an extension can be used to account for both detachment- and transport-limited soil loss rates (see Chapter 1 and Wilson and Lorang 1999 for additional details).

The saturation overland flow method should be selected if runoff is to be determined by surface saturation controlled by a user-specified critical wetness index. For this case, $\mu_i = 0$ when $\ln(A_s/\tan \beta) < W_c$ and $\mu_i = 1.0$ when $\ln(A_s/\tan \beta) \geq W_c$, where $\ln(A_s/\tan \beta)$ is a topographic wetness index (Chapter 1; Moore et al. 1988a) and W_c is a user-specified critical wetness index that identifies the location of zones of surface saturation in the landscape. Moore et al. (1992) used a critical wetness index of 6.0 and found that $T_c > 2.5$ showed good agreement with areas of degradation observed in a 9.6-ha catchment in Queensland, Australia during 1979–80 when vegetation was sparse. The wetness index values are calculated for each grid point and compared with the user-specified critical wetness index to determine the weights used in Equation 4.1 with this option.

The third and final (Hortonian overland flow) method incorporated in EROS should be selected if runoff is to be determined by infiltration excess. This option requires a file with two or more soil polygons and weights. The program then performs a series of point-in-polygon overlays to assign weights to individual DEM grid points. The drainage areas are recalculated using these weights, the flow directions copied from TAPES-G, and one of the D8, Rho8, or FD8/FRho8 flow-routing methods discussed in Chapter 3. The slope values copied from TAPES-G and the new drainage areas are then used to calculate T_c in Equation 4.1.

4.2.2 Inputs

EROS requires the *x, y, z* (elevation), slope gradient, flow direction, and drainage area attributes from TAPES-G (Chapter 3; Gallant and Wilson 1996). It looks for metadata on the input file to determine the grid cell size, missing value, and field numbers, and assumes standard TAPES-G format if there are no metadata. EROS recognizes an elevation value of 0 as a missing value, and also recognizes TAPES-G's missing value of -9999.0. It uses the same missing value in its output file. EROS assumes that the coordinates and elevation are in meters, but there is no problem using other units as long as the elevation units are the same as the units of the *x* and *y* coordinates. No additional inputs are required so long as the rainfall excess runoff method is chosen. However, the user will be asked to specify the critical topographic wetness index if the second (saturation overland flow) method is chosen, and a soil boundary file containing soil polygons and weights is required if the third (infiltration excess overland flow) option is chosen.

The critical topographic wetness index required for the second option can be estimated from the fraction of precipitation converted to runoff and a cumulative frequency plot of the steady-state topographic wetness index. The fraction of precipitation converted to runoff may be obtained from long-term rainfall–runoff records, published reports (e.g., Parrett and Hull 1985), or the U.S. Soil Conservation Service (SCS) runoff curve number method. The SCS method is sometimes used in conjunction with the WET program and is discussed in more detail in Section 4.4.2. The steady-state topographic wetness index required here can be calculated with either DYNWET or WET and plotted against catchment area (as in Figure 4.4) to identify the critical topographic wetness index for a particular catchment.

The soil boundary file required for the Hortonian overland flow option specifies (1) an integer indicating the number of soils for each model run; (2) the number of vertices, soil number, and weight for each polygon; and (3) the *x, y* coordinates delineating the boundary of each polygon. The second and third items are repeated for each soil polygon, and soil polygons must be labeled with integers starting from 1. The soil polygons may be able to be acquired from one of several digital geographic soil databases produced in the past few years or converted from an existing soil map to a digital file (Wilson 1999a). The weights are more difficult to acquire and will usually have to be derived by the EROS user from two additional types of information as follows.

We need to know something about the infiltration rates of soils when thoroughly wetted and the time pattern of precipitation intensity to estimate these weights. The assignment of soil series to hydrologic soil groups gives a rough guide to infiltration rates in the United States (Pilgrim and Cordery 1993, Rawls et al. 1993). More precise inputs will usually require site-specific sampling, field measurement, and interpolation. Some variant of kriging is often used with field measurements (hard data) and other (soft) data to generate soil infiltration maps for farm fields and catchments in these situations (e.g., Rogowski and Hoover 1996). Rainfall intensity information can be presented in hyetographs and it is usually assumed to be the same over all points in a catchment (Pilgrim and Cordery 1993). The time pattern of rainfall intensity can then be combined with information about the areal extent and hydrologic behavior of different soil series to estimate the proportion of runoff generated from different soil polygons and determine the final weights used with the Hortonian overland flow method in EROS.

4.2.3 Outputs

EROS produces an output file in either binary or ASCII form with metadata, as described in Section 3.1.12. Each DEM grid point with all its attributes is written to a single record as either one line in an ASCII file or one unformatted record in a binary file. The *x, y,* and *z* fields are copied from the TAPES-G input file, and the weights are determined from the soil polygons or EROS program itself. The drainage area field is similar to the drainage area calculated by TAPES-G except where it is modified by soil weights.

4.3 SRAD

This program calculates potential solar radiation as a function of latitude, slope, aspect, topographic shading, and time of year, and then modifies this estimate using information about monthly average cloudiness and sunshine hours. Temperature is extrapolated across the surface using a method based on Running et al. (1987), Hungerford et al. (1989), Running (1991), and Running and Thornton (1996) that corrects for elevation via a lapse rate, slope–aspect effects via a short-wave radiation ratio, and vegetation effects via a leaf area index. Daily outgoing long-wave irradiance is calculated from surface temperature and daily incoming long-wave irradiance is calculated from air temperature and the fraction of sky visible at each grid point. These short- and long-wave radiation fluxes are then used to estimate the surface energy budget at each grid point for a user-specified period ranging from one day to one year in length. Solar radiation is not widely measured, but the fact that the apparent solar path relative to any location and any surface can be simply and easily calculated with great accuracy suggests radiation indices can be used to compare sites (Fleming 1987). These estimates are valuable because the surface energy budget exerts a large impact on the evaporation and photosynthesis processes occurring at the land surface and is highly dependent on topography. Vegetation diversity and biomass production are related to radiation input (e.g., Austin et al. 1984, Tajchman and Lacey 1986, Moore et al. 1993e, Franklin 1995).

4.3.1 Estimation Methods

4.3.1.1 Short-Wave Radiation   The general approach used by SRAD to estimate short-wave radiation at both flat and inclined sites incorporates four steps. First, the potential or extraterrestrial irradiance on a horizontal surface just outside the earth's atmosphere is calculated. Second, a series of instantaneous clear-sky, short-wave radiation fluxes is calculated for each of the DEM grid points at 12-min intervals from sunrise to sunset. Direct beam and diffuse fluxes are calculated for flat sites and direct beam, circumsolar diffuse, isotropic diffuse, and reflected fluxes are calculated for sloping sites at this stage. Third, these instantaneous values are summed to obtain daily totals and these values are adjusted to account for the effects of cloudiness. Fourth, the daily values are summed over the estimation period specified by the user and divided by the number of days to estimate average daily values in each period.

4.3.1.1.1 Extraterrestrial Radiation   SRAD combines a series of fundamental angles, as defined in astronomy and related to one another by means of spherical trigonometry, with the solar constant to estimate daily amounts of direct sunlight incident on a horizontal surface just outside the earth's atmosphere. The amount of sunlight incident at a point just outside the atmosphere, R_{oh}, depends on the time of year, time of day, and latitude as follows:

$$R_{oh} = \frac{I}{r^2} \cos z \qquad (4.5)$$

where I is the solar constant (see Table 4.1 for additional details), r is the ratio of the earth–sun distance to its mean, and z is the zenith angle (Gates 1980, Fleming 1987). The magnitude of r^2 varies continuously throughout the year from 1.0344 on 3 January to 0.9674 on 5 July, but never deviates more than 3.5% from 1.0 (Gates 1980). This ratio is calculated in SRAD as a function of day number.

The zenith angle is the angle between the solar beam and the normal to the surface and can be computed from the following equation:

$$\cos z = \sin \phi \sin \delta + \cos \phi \cos \delta \cos h \qquad (4.6)$$

where ϕ is the latitude of the observer (degrees, negative in the southern hemisphere), δ is the declination of the sun, and h is the hour angle of the sun from solar noon (i.e., the angular distance from the meridian of the observer) (Lee 1978). The solar declination (δ) measures the seasonally varying latitude of the sun's path across the sky, north and south of the equator. It varies from $-23.5°$ at the northern winter solstice (22 December) to $+23.5°$ at the northern summer solstice (22 June). Declination is independent of calendar year and the latitude of the observer and is a function only of time of year (Gates 1980). The hour angle (h) measures the difference in time from solar noon expressed as $15°$ per hour of difference. Some authors substitute the solar elevation or altitude (a) in place of the zenith angle because this variable represents the height of the sun above the horizon for an observer at a specific location and is the complement of the zenith angle (i.e., $\sin a = \cos z$). Both the zenith angle and solar altitude are functions of latitude, time of year, and time of day (Gates 1980).

SRAD calculates instantaneous values of R_{oh} (and the other short-wave radiation fluxes described below) at time steps of 12 min organized symmetrically around noon and sums these to obtain daily totals. Fleming (1987) relied on results from

Monteith (no reference cited) and concluded that this approach yielded sufficiently accurate daily estimates for most of the types of hydrological, geomorphological, and ecological applications considered in this book. The units used for the radiation values in SRAD are chosen at run time together with the value of the solar constant from the list of options summarized in Table 4.1.

4.3.1.1.2 Direct-Beam and Diffuse Radiation at Horizontal Sites Under Clear Skies
The amount of solar energy reaching the ground is reduced because the atmosphere is semitransparent to solar radiation. The molecular constituents of the atmosphere, along with water droplets, dust, and other particulate matter, scatter the direct solar beam and create a hemispherical (diffuse) source of radiant energy (Lee 1978). In addition, the direct and diffuse irradiance are both reduced as a result of direct absorption and reflection to space along the light beam's path through the atmosphere to the ground (Linacre 1992). We therefore need to know the transmission properties of the atmosphere to estimate the amount of extraterrestrial or potential direct solar radiation that traverses the earth's atmosphere and is incident at the ground (Gates 1980). Two approaches are provided in SRAD to estimate these fluxes: One uses a lumped transmittance approach and the other calculates individual transmittance components.

The lumped approach assumes that the attenuation of the direct solar beam in passing through a homogeneous, cloudless atmosphere can be described by the following formula, often named after Beer, though first formulated by Pierre Bouguer in 1760:

$$R_{\text{dirh}} = R_{\text{oh}} \, \tau^m \qquad (4.7)$$

where R_{dirh} is the direct-beam, short-wave irradiance incident on flat surfaces under clear skies, τ is the transmission coefficient or fraction of radiation incident at the top of the atmosphere, which reaches the ground along the vertical (or zenith) path (i.e., the shortest path length between outer space and the ground surface), and m is the ratio of the path length in the direction of the sun at zenith angle z to the path length in the vertical direction (Gates 1980, Linacre 1992).

The local transmission coefficient for each grid cell τ is calculated in SRAD from the elevation, monthly transmission coefficient at sea level, and transmissivity lapse rate as follows:

$$\tau = \tau_{\text{sl}} + t_{\text{lapse}} * \text{elev} \qquad (4.8)$$

TABLE 4.1 SRAD Irradiance Units

Solar Constant	Corresponding Irradiance Units
1.9 cal/cm^2/min	cal/cm^2/day
119.4 langley/h	langley/day = cal/cm^2/day
4.871 MJ/m^2/h	MJ/m^2/day
1354 W/m^2	W/m^2

where τ_{sl} is the transmission coefficient at sea level, t_{lapse} is the transmissivity lapse rate, and elev is the elevation at a specific grid point. This equation mimics the common situation in which transmittance is greater at higher elevations because of the thinner layer of atmosphere that occurs above these locations (Linacre 1992). The quantity m is often called the relative air mass and is given by

$$m = \sec z = \frac{1}{\cos z} \qquad (4.9)$$

where z is the zenith angle that was first used in Equation 4.5. However, this equation works only when the zenith angle is less than about 60°: When the sun is low in the sky, the curvature of the earth reduces the length of the sun's slant rays compared with the depth of the atmosphere in the zenith direction (Robinson 1966, Gates 1980). SRAD relies on the above equation when the zenith angle is less than 60° and reverts to a table obtained from List (1968, 422) that summarizes optical air masses in 1° intervals for sites at higher zenith angles. These values are then corrected to account for the reduction in atmospheric pressure encountered at higher elevations by the factor p/p_o, where p is the atmospheric pressure at the grid cell, and p_o is the standard sea level pressure of 1013.25 mbar (Gates 1980, Fleming 1987).

Equations 4.7 through 4.9 are required to estimate the attenuation of the instantaneous direct-beam, short-wave irradiance incident at the ground surface. Some of the direct-beam radiation is transformed into diffuse radiation and the relationship for instantaneous transmittance to diffuse skylight derived by Liu and Jordan (1960) is used in SRAD to estimate the instantaneous diffuse irradiance R_{difh} as follows:

$$R_{difh} = (0.271 - 0.294\, \tau^m)R_{oh} \qquad (4.10)$$

Equation 4.10 shows that the transmittance to scattered skylight decreases as the direct solar beam transmittance increases (Gates 1980). The distinction between direct-beam and diffuse short-wave irradiance is an important one and we will see shortly how it affects the amount of short-wave irradiance striking a sloping surface (Linacre 1992).

The second approach in SRAD for estimating direct-beam and diffuse short-wave irradiance on horizontal sites under clear skies treats each of the transmittance components separately. The effects of water vapor, dust, and a clear atmosphere can be estimated in SRAD as follows:

$$R_{dirh} = R_{oh}(AW * TW * TD * TDC) \qquad (4.11)$$

where AW accounts for the absorption by water vapor, TW the scattering by water vapor, TD the scattering by dust, and TDC the scattering by air molecules and density fluctuations in a clear-sky atmosphere (Gates 1980). Four equations specified by Fleming (1987) and based on work by Monteith (no reference cited) and Idso (1969) were modified and are used in SRAD to estimate each of these transmittance components as follows:

$$AW = 1 - 0.077 * \left[u * m \left(\frac{p}{p_o} \right) \right]^{0.3} \tag{4.12}$$

$$TW = 0.975^{um(p/po)} \tag{4.13}$$

$$TD = 0.95^{m(p/po)D} \tag{4.14}$$

$$TDC = 0.9^{m(p/po)} + 0.026 * [m \left(\frac{p}{p_o} \right) - 1] \tag{4.15}$$

where m, p, and p_o were defined earlier, u is the water content of a vertical slice of atmosphere in centimeters, and D is an empirically derived dust factor. The original equations specified by Fleming (1987) did not correct the air masses for elevation (like we do in Equations 4.12 through 4.14) because they assumed that local dust and water vapor measurements were available. The dust factor used in Equation 4.14 is especially problematic and can be related to atmospheric turbidity such that $D = 2$ corresponds to 300 ppm dust and $D = 1$ is slightly less than 100 ppm (i.e., the standard conditions at many sites) (Fleming 1987).

The above component atmospheric transmittance model assumes that absorption takes place first followed by scattering, and that the scattering is evenly divided between forward and back scattering. The implementation of this option in SRAD tries to strike a balance between accuracy and input data requirements. As a result, several effects are ignored (i.e., absorption effects of carbon dioxide and ozone) and each of the components that is considered is treated as acting uniformly over the whole solar spectrum when we know that most of these effects are strongly wavelength dependent (Gates 1980). We have also assumed that the scattered radiation gives rise to the diffuse radiation component without further absorption so that

$$R_{difh} = 0.5 * (R_{oh} * AW - R_{dirh}) \tag{4.16}$$

4.3.1.1.3 Circumsolar and Isotropic Diffuse Radiation

Linacre (1992, 152) distinguished "skylight" diffuse radiant energy, which is isotropic (i.e., comes more or less equally from all directions in the sky), and "circumsolar" diffuse radiation, which comes from within approximately 5° of the direct solar beam. The circumsolar diffuse component moves across the sky with the sun and can be separated and added back into the direct-beam component for the calculation of the radiation incident on sloping sites (Fleming 1987). Monthly average values of the fraction of diffuse radiation that is close to the solar disk are obtained from the site-parameter file and used by SRAD to adjust the direct-beam and diffuse radiation fluxes as follows:

$$R_{dirh} = R_{dirh} + R_{difh} * CIRC \tag{4.17}$$

$$R_{difh} = R_{difh} - R_{difh} * CIRC \tag{4.18}$$

where CIRC is the fraction of diffuse radiation derived from within 5° of the direct solar beam (i.e., the circumsolar coefficient). Many other solar radiation models do not separate these components and this omission may generate errors as large as 40% when these models are used to estimate irradiance on sloping sites (Linacre 1992).

4.3.1.1.4 Direct-Beam, Diffuse, and Reflected Radiation at Sloping Sites Under Clear Skies

The flux density of short-wave radiation on a sloping site differs from that on a horizontal surface primarily because the direct-beam radiation is modified. Minor differences also occur because the flux of diffuse sky radiation is affected and there is an added flux of short-wave radiation reflected from adjacent parts of the landscape (Lee 1978, Fleming 1987).

Four site attributes (slope, aspect, horizon, and skyview) are required to estimate the direct-beam and circumsolar irradiance with and without shading, the isotropic diffuse irradiance, and the reflected irradiance on sloping sites. The slope, aspect, and fraction of sky hemisphere that is visible are computed by SRAD for each grid point. The maximum slope and aspect are calculated with the same central finite difference scheme implemented in TAPES-G (see Chapter 3 for details). The one-dimensional horizon algorithm of Dozier et al. (1981) is used to estimate the fraction of the sky hemisphere, v, visible at each grid point. This algorithm constructs profiles across the DEM and determines the horizon angle, H_ϕ, for each grid point in a discrete number of directions, ϕ (usually 16). This algorithm is attractive because its computational efficiency is proportional to the number of grid points in a regular DEM. However, we modified the original algorithm to use bilinear instead of nearest-neighbor interpolation to construct the profiles because the latter method estimates erroneous horizon effects in some steeply sloping areas. The skyview fraction, v, is computed from the horizon angles, H_ϕ, by averaging the cosine of the horizon angles:

$$v = \frac{1}{n}\sum_{\phi=1}^{n} \cos H_\phi \qquad (4.19)$$

The direct beam and circumsolar diffuse radiation on sloping surfaces depend on the solar elevation and the slope's angle to the horizontal (Linacre 1992). Tilting a surface in the meridian plane north or south from the horizontal is the equivalent of going north or south in latitude by the same number of degrees (Gates 1980). In addition, the slope and aspect effects are greatest during winter in middle latitudes and they tend to become negligible toward the equator and the poles (Lee 1978).

The direct irradiance on a sloping surface without shading is calculated in SRAD using the following equations:

$$R_{\text{dirs}} = R_{\text{dirh}} \cos i \qquad (4.20)$$

$$\cos i = A + B \cos h + C \sin h \qquad (4.21)$$

$$A = \sin \delta \sin \phi \cos \beta + \sin \beta \cos \psi \cos \phi \qquad (4.22)$$

$$B = \cos \delta \ (\cos \phi \cos \beta - \sin \phi \cos \psi \sin \beta) \qquad (4.23)$$

$$C = \sin \beta \cos \delta \sin \psi \qquad (4.24)$$

where R_{dirh} in this instance is the direct-beam and circumsolar diffuse radiation on a horizontal surface under clear skies (i.e., as estimated with Equations 4.7 or 4.11 and 4.17), i is the angle between the beam and the normal to the slope, β is the slope angle, and ψ is the aspect (Lee 1978, Linacre 1992). Instantaneous values are calculated in 12-min time steps and these values are multiplied by 12 and summed to estimate daily direct-beam radiation on sloping surfaces without shading (R_{dirsns}). The program also checks whether the sun is obstructed at 12-min intervals and calculates the direct-beam and circumsolar diffuse radiation taking shading into account (R_{dirss}). This approach means of course that $R_{dirss} \leq R_{dirsns}$ at each sloping site.

The "isotropic" diffuse radiation on a sloping site is typically lower than that on a horizontal surface because part of the sky is obscured:

$$R_{difs} = R_{difh}\, \nu \qquad (4.25)$$

where R_{difh} in this instance is the isotropic diffuse irradiance (i.e., as estimated with Equations 4.10 or 4.16 and 4.18), and ν is the sky-view factor (i.e., fraction of sky visible at a specific grid point).

The reductions in direct-beam and diffuse irradiance on sloping sites (compared to horizontal sites) may be partially offset by reflected radiation from other surfaces:

$$R_{ref} = (R_{dirh} + R_{difh})\,(1 - \nu)\,\alpha \qquad (4.26)$$

where R_{dirh} is the direct-beam radiation on a horizontal surface under clear skies (i.e., as estimated with Equation 4.7 or 4.11), R_{difh} is the diffuse radiation on a horizontal surface under clear skies (i.e., as estimated with Equation 4.10 or 4.16), ν is the sky-view factor, and α is the albedo (i.e., fraction of sunlight reflected by the surface). For the sake of simplicity, this equation uses horizontal radiation values rather than attempting to sum the contributions of the spatially varying radiation within the ground-view area. The reflected radiation is calculated at 12-min intervals and summed over the day to estimate reflected radiation from foreground surfaces facing the site. An upward-facing area will see less sky and more ground (from which it receives reflected light) as it is tilted upward from the horizontal (Lee 1978). These inputs are important on many sloping sites and they can vary markedly with the type and status of the surface (Fleming 1987).

4.3.1.1.5 Effect of Overcast Skies   The flux of direct and diffuse short-wave solar radiation incident at the earth's surface is highly variable and difficult to predict when the sky is partially or totally overcast (Linacre 1992). Clouds are highly variable in form, size, density, height, and duration. Very thin transparent cirrus clouds have little influence on global radiation, whereas thick, dark thunderstorm clouds may reduce radiation to $\leq 1\%$ of its clear-sky value (Gates 1980). The approach used by SRAD combines daily short-wave radiation estimates for clear skies with statistical averages of observational data collected over long periods of time.

The accumulated radiation values at the end of each simulated day are combined with the sunshine fraction and cloud transmittance for the month in question to estimate daily incoming short-wave solar radiation on horizontal sites R_{th} as follows:

$$R_{th} = (R_{dirh} + R_{difh}) \left[\frac{n}{N} + \left(1 - \frac{n}{N} \right) \beta \right] \tag{4.27}$$

where n/N is the sunshine fraction (i.e., the observed duration of sunshine out of the maximum possible for that place and date), and β is the cloud transmittance (i.e., the fraction of clear-sky radiation received when the sky is overcast).

For sloping sites, the cloud transmittance in SRAD is adjusted downward to account for the reduction in sky view and upward to account for the enhanced flux of downward and upward diffuse radiation caused by multiple reflection between ground and sky. Gates (1980) reports measurements by Kondratyev (1969) illustrating these effects and they are approximated in SRAD with the following equation:

$$\beta_s = \beta v \left(\frac{R_{tsns}}{R_{tss}} \right) \tag{4.28}$$

where R_{tsns} is the total daily clear-sky, short-wave radiation without shading and R_{tss} is the total daily clear-sky, short-wave radiation with shading on sloping sites. The daily incoming short-wave solar radiation on sloping sites, R_{ts}, is then estimated as follows:

$$R_{ts} = (R_{dirss} + R_{difs}) \left[\frac{n}{N} + \left(1 - \frac{n}{N} \right) \beta_s \right] + R_{ref} \tag{4.29}$$

4.3.1.2 Temperature   SRAD estimates temperatures at each grid point as a function of the mean monthly minimum and maximum air temperatures, surface temperature, minimum, maximum, and average temperature lapse rates, and elevation for a reference station (as specified in the site-parameter file). The radiation effect on temperature is introduced via the short-wave radiation ratio, S, at each grid point:

$$S = \frac{R_{ts}}{R_{th}} \tag{4.30}$$

where R_{th} and R_{ts} are the daily total or global short-wave irradiance on horizontal and sloping sites, respectively. The minimum air, maximum air, and surface temperatures, T, at each grid point are computed using

$$T = T_b - \frac{T_{lapse}(z - z_b)}{1000} + C \left(S - \frac{1}{S} \right) \left(1 - \frac{LAI}{LAI_{max}} \right) \tag{4.31}$$

where z is the elevation of the grid point, z_b is the elevation of the temperature reference station, T_b is the temperature at the reference station (monthly minimum air, maximum air, or surface), T_{lapse} is the monthly temperature lapse rate (°C/1000 m), C is a constant (currently set to 1.0), LAI is the leaf area index at the grid cell, and LAI_{max} is the maximum leaf area index. This equation is a modification of the original pair of equations, since they were discontinuous at $S = 1$. The LAI/radiation ratio corrections are not applied to minimum temperature because these occur during the night (Moore et al. 1993e). Average air temperature is assumed to be the average of the minimum and maximum air temperatures.

4.3.1.3 Long-Wave Radiation

Long-wave radiation estimates are also required to predict the surface energy budget. Long-wave radiation is emitted continuously by the earth's atmosphere and surface. The incoming atmospheric flux, L_{in}, is almost always less than the outgoing surface flux, L_{out}, and this means that the average net daily long-wave irradiance (L_{net}) represents a net loss of energy from the biosphere (Lee 1978).

The daily incoming long-wave irradiance in SRAD is computed from air temperature taking sky view into account:

$$L_{in} = \varepsilon_a \sigma T_a^4 \, v + (1 - v) L_{out} \qquad (4.32)$$

where ε_a is the atmospheric emissivity (a function of air temperature, vapor pressure, and cloudiness), σ is the Stefan–Boltzmann constant, T_a is the mean air temperature, and v is the sky-view factor discussed earlier (Lee 1978). The inclusion of the sky-view factor means that this equation handles both sloping and horizontal sites. Sloping sites see both the sky and adjacent terrain and the sky-view factor, v, accounts for the fraction of the sky that is visible and the $(1 - v)L_{out}$ term shows how a fraction of the outgoing long-wave radiation is added to the incoming component to account for radiation from foreground obstructions.

The daily outgoing long-wave irradiance in SRAD is computed from surface temperature as follows:

$$L_{out} = \varepsilon_s \sigma T_s^4 \qquad (4.33)$$

where ε_s is the surface emissivity coefficient (>0.95 for most natural surfaces) and T_s is the surface temperature (Lee 1978).

4.3.1.4 Net Solar Radiation

The net solar radiation is the quantity of energy available at the ground surface to drive air and soil heating, evaporation, and photosynthesis (Dubayah 1992). The net radiation at each grid cell is estimated in SRAD by adding the incoming and outgoing fluxes for some user-specified period as follows:

$$R_{net} = (1 - \alpha)\, R_{th} + \varepsilon_s L_{in} - L_{out} \qquad \text{(at horizontal sites)} \qquad (4.34)$$

$$R_{net} = (1 - \alpha)\, R_{ts} + \varepsilon_s L_{in} - L_{out} \qquad \text{(at sloping sites)} \qquad (4.35)$$

These equations follow the usual approach in that R_{net} takes a positive sign when energy is transferred to the surface and a negative sign when the direction is reversed

(Lee 1978). This approach means that SRAD will generally predict $R_{net} > 0$ during the summer half-year but R_{net} can be negative in the winter months, particularly at higher latitudes and at locations facing away from the sun that receive no direct beam radiation.

4.3.2 Inputs

Two or three input files are required by SRAD to estimate the radiation fluxes and temperatures described in the previous section: (1) a square-grid DEM; (2) a site parameter file; and optionally (3) a vegetation file specifying the vegetation type present at each DEM grid point.

The DEM may consist of *x, y, z* triplets or just *z* values. The *x, y, z* file can have the points in any order, whereas the *z* file must be in row order (all columns of the first row followed by the first column of the second row, etc.) and may have the first point in the northwest or southwest corner. The *z* values may be either integer or floating point; if they are integer values, a scaling factor may be applied to increase the vertical resolution of the DEM. Finally, the file may be in either ASCII or binary form. The binary forms are either direct-access 2-byte integer files, unformatted integer files, or unformatted floating point files. Direct-access files contain only the data with no record delimiters. Unformatted files contain records with delimiters. ASCII files have one record per line. DEMs containing *x, y, z* data are expected to have one *x, y, z* triplet per record. Files containing only *z* data can have any number of *z* values per record as long as each row starts in a new record. Some additional processing may also be required to create a square-grid DEM prior to running SRAD (as is the case with TAPES-G) and we use ANUDEM (Hutchinson 1988, 1989b) for this purpose (see Chapters 2 and 3 for additional details). The users should also check to see that the DEM they are using fits the size limits set when SRAD is compiled. The arrays are currently set up to handle 1000 rows and 1000 columns, although the users can change these values to suit the memory space of their machine or the size of their study area (DEM). The number of rows and columns do not have to be equal.

The radiation, temperature, and surface condition parameters that must be specified in the site-parameter file vary in terms of availability and difficulty of estimation. These inputs are summarized in Table 4.2 and the discussion here emphasizes data sources and methods of estimation. McKenney et al. (1999) noted that the estimated errors associated with some of these parameters are potentially large or at worst unknown and they quantified the sensitivity of the SRAD outputs to the magnitude of the first four radiation parameters listed below. Their findings are useful because they give some idea as the relative importance of the different inputs and the level of care that is needed when estimating individual inputs.

Five sets of radiation inputs from a nearby station are required: (1) an atmospheric transmission coefficient; (2) circumsolar coefficient; (3) sunshine fraction; (4) cloud transmittance; and (5) the elevation of the reference climate station. Mean monthly or annual values can be used, although mean monthly values are preferred outside the tropics. The user must also specify whether the clear-sky atmospheric transmission coefficients at sea level for each month will be characterized by a single lumped

TABLE 4.2 Site Parameters Required by SRAD

Line	Parameter	Units	Description
1	Latitude minimum and maximum	Decimal degrees	North positive, south negative (the two numbers can be the same)
2	Circumsolar coefficient	None	12 monthly values (see text for details)
3	Albedo	None	Monthly values (see text for details)
4	Cloud transmittance	None	Monthly values (see text for details)
5	Sunshine fraction	None	Ratio of actual sunshine hours to theoretical maximum day length
6	Max air temperature	°C	Monthly average values
7	Min air temperature	°C	Monthly average values
8	Average surface temperature	°C	Monthly average values
9	Avg air temperature lapse rate	°C/1000 m	Monthly average values
10	Min air temperature lapse rate	°C/1000 m	Monthly average values
11	Max air temperature lapse rate	°C/1000 m	Monthly average values
12	NLAI	None	The number of leaf area index annual profiles (see text for details)
13	LAI	None	NLAI lines of monthly leaf area index for each vegetation type in sequence
14	Max LAI	None	Maximum leaf area index, typically 10
	Surface emissivity	None	Typically 0.92 to 0.99
	Transmissivity lapse rate	1/m	Typically 0.00008 (used only with lumped atmospheric transmittance model)
	Elevation of reference station	m	Required for temperature extrapolation
15/16	Atmospheric transmittance	See text	Monthly values (see text for details)

parameter or using water and dust components (see Section 4.3.1.1.2 for additional details).

The lumped transmittance coefficient is a fraction, typically 0.60–0.70, specifying the fraction of solar radiation transmitted by the atmosphere (Gates 1980, Linacre 1992). Mountain locations at high elevations, such as the Sierra Nevada and Rocky Mountains, with clear, dry skies, may have transmittances up to 0.80 (Gates 1980). Several methods have been proposed and used to estimate lumped transmittance coefficients. The most popular method incorporates three steps. First, mean monthly clear-sky irradiance R_{thcs} is estimated from total irradiance (R_{th}) measurements at a nearby climate station:

$$R_{thcs} = \frac{R_{th}}{0.35 + 0.61(n/N)} \tag{4.36}$$

where n/N is the sunshine fraction and 0.35 and 0.61 are constants that vary with latitude (Fritz and MacDonald 1949, List 1968, Fleming 1987). Second, mean monthly clear-sky irradiance is divided by extraterrestrial irradiance to estimate the fraction of extraterrestrial radiation received at this station. Third, this fraction is adjusted with a user-specified transmissivity lapse rate to obtain monthly clear-sky transmittance at sea level. We usually specify 0.00008 per meter for the transmissivity lapse rate. Other methods have used solar radiation data collected at climate stations on clear days that are pooled into monthly groups and divided by extraterrestrial irradiance (e.g., Idso 1969, McKenney et al. 1999) or empirical relationships linking atmospheric transmittance and air temperatures (e.g., Bristow and Campbell 1984).

Alternatively, the user can specify water and dust components and SRAD will calculate the transmittances using the absorption and scattering attenuation coefficients for water and dust summarized in Equations 4.11–4.15. The inputs are specified as the water content of a vertical slice of atmosphere in centimeters, typically 1.5 to 1.7, and a dust factor, where 1 represents a standard value of about 100 ppm and 2 represents a value of about 300 ppm (Fleming 1987). Idso (1969) showed there was an annual variation about the standard value of 1 for the dust component that is correlated with the cube of the monthly windspeed in Phoenix, AZ, and Fleming (1987) recommended using this correction for other semiarid environments as well. Monthly values of precipitible water and aerosol optical depth are reported for many radiation stations.

The circumsolar coefficient, CIRC (as used in Equations 4.17 and 4.18), is the fraction of diffuse radiation originating near the solar disk and is thus subject to topographic effects (slope, aspect, and shadowing). The circumsolar diffuse radiation is typically 5% of direct radiation or about 30% of the isotropic diffuse radiation when the sky is clear, yielding a typical value of 0.25 (i.e., 30/130) for the circumsolar coefficient. However, this coefficient tends to be higher in summer and lower in winter and the following equation can be used with station data to estimate mean monthly values:

$$\mathrm{CIRC} = \frac{R_{\mathrm{dirh}}}{24I} \tag{4.37}$$

where R_{dirh} is the measured mean monthly (or annual) direct irradiance in Wh/m^2, and I is the solar constant (Isard 1986, Linacre 1992).

The sunshine fraction (n/N) used in Equations 4.27 and 4.29 is the ratio of actual sunshine hours to the theoretical maximum at that latitude. These values are reported for many stations. Monthly sunshine hour totals are recorded and can be divided by the maximum number of sunshine hours computed from duration of daylight tables to obtain this fraction in other instances (List 1968, Lee 1978).

The cloud transmittance, β, that appears in the same equations records the ratio of actual radiation to clear-sky radiation during cloudy periods on an average monthly basis. This fraction is seldom reported but can be estimated from total solar irradiance, clear-sky irradiance, and the sunshine fraction as follows:

$$R_{\text{th}} = R_{\text{thcs}} \left[\frac{n}{N} + \beta \left(1 - \frac{n}{N} \right) \right]$$

(4.38)

This approach sets the actual radiation equal to the sum of clear-sky radiation during bright, cloud-free periods (given by n/N) plus some fraction (β) of the clear-sky radiation during cloudy periods ($1 - n/N$). Setting $\beta = 1$ implies that cloudiness does not diminish actual radiation, whereas setting $\beta = 0$ implies that there is no radiation at all during cloudy periods. The actual value lies somewhere between these extremes and is related to the diurnal pattern of cloudiness and average density of cloudiness. Many sites exhibit large monthly variations in terms of cloudiness and the use of site-specific monthly values in place of annual averages will usually produce superior results.

Four sets of temperature inputs from a nearby station are also required for this site-parameter file: (1) minimum and maximum air temperatures; (2) surface temperature; (3) minimum, average, and maximum temperature lapse rates; and (4) the elevation of the reference climate station (Table 4.2). The larger number of climate stations recording temperature information will usually mean that this reference station is different from that used for the radiation inputs. Mean monthly or annual values can be used, although mean monthly values are once again preferred outside the tropics. These inputs are required to predict the spatial variations in both of the long-wave radiation components and net solar radiation.

The mean monthly minimum and maximum air temperatures can be obtained from climate station records. Mean monthly surface temperature is measured at a small number of these stations (Paetzold 1988) and these values will often have to be inferred from either air temperatures and other climatic variables (e.g., Toy et al. 1978) or satellite data (e.g., Brakke and Kanemasu 1981). The mean monthly average daylight, minimum, and maximum temperature lapse rates will have to be estimated from station data as well. These lapse rates vary with season and between different regions (Baker 1944, Glassy and Running 1994). Baker (1944), for example, reported maximum, average, and minimum temperature lapse rates of 6.8, 5.9, and 2.9°C per 1000 m for July, and a single value (2.9°C per 1000 m) for all three lapse rates in January in the mountains of the western United States.

We often use a modified version of a spatial filtering-kernel convolution method (Running and Thornton 1996) with local station data to estimate site-specific mean monthly average daylight, minimum, and maximum temperature lapse rates. This method incorporates a spatial filtering kernel of circular extent and fixed diameter to select a list of qualifying stations and assign weights to each station. There is one point in these regressions for each pair of stations and the weight associated with each regression point is defined as the product of the two individual station weights. The individual station weights, V_{ij}, are defined by

$$V_{ij} = e^{\left(\frac{-D_{ij}^2 \alpha}{R_i^2} \right)} - e^{-\alpha}$$

(4.39)

where R_k is the radius of the circular kernal, D_{ij} is the distance from a qualifying climate station to the center of the kernal grid (i.e., the climate station used for the other temperature inputs), and α is a shape parameter. Running and Thornton (1996) computed the average station density (one station per 1500 km^2) and set $R_k = 201$ km and $\alpha = 4.0$ in Montana. Equation 4.39 defines a Gaussian weighting function within the circular kernel with the greatest weight located at the center of the kernel (0.982) and the weight decreasing radially outward until, at a distance, R_k, from the center of the kernel grid, the weight is zero. The independent variable in the regression is the difference in elevation between the stations in a pair, and the dependent variable is the difference in either mean monthly daylight average, minimum, or maximum temperatures between a pair of stations. Mean monthly maximum and minimum temperatures can be obtained from station records and mean monthly average daylight temperatures can be estimated from these values by assuming that the diurnal daylight trace has a sine form similar to Running et al. (1987). The weights assigned to station pairs must be summed and normalized to that sum prior to building the regression models (Running and Thornton 1996).

Four sets of parameters describing surface conditions must also be specified in the site-parameter input file: (1) the number of leaf area index profiles (NLAI), (2) one or more leaf area index profiles, (3) surface emissivity, and (4) albedo (Table 4.2). The NLAI parameter indicates the number of vegetation types in the optional vegetation type file. This vegetation file (if used) is a grid matching the dimensions of the DEM and should contain, for each grid cell, a type number between 1 and NLAI. This number selects the LAI profile that is to be used by SRAD at each grid cell. The site-parameter file must therefore contain NLAI lines of monthly LAI values. Only one LAI profile is required and used for the whole DEM if no vegetation file is specified (i.e., as in Table 4.7). LAI records the ratio of leaf area to ground cover and is usually estimated from remotely sensed multispectral reflectance data. These values and the maximum leaf area index (we usually specify $LAI_{max} = 10.0$) are used to extrapolate temperature across the DEM because the magnitude of the temperature differences will be modified by the characteristics of the energy exchange surfaces on slopes (McNaughton and Jarvis 1983). Air temperature may be increased by 2°C when a sunward-facing slope of LAI = 1.0 receives twice as much radiation as a flat surface, but the same site may experience no increase when it is assigned a LAI = 5.0 (Running et al. 1987).

The SRAD user must also specify a single surface emissivitity (ε_s) value. Many authors have calculated the mean annual emissivities for common surface types and a single value is used in SRAD because most vegetated surfaces have emissitivities that exceed 0.95 (Lee 1978, Henderson-Sellers and Robinson 1986, Oke 1987). The albedo is the fraction of sunlight reflected from the surface. Lee (1978), List (1968), Iqbal (1983), Henderson-Sellers and Robinson (1986), and Oke (1987) list typical albedo estimates for common surface types. These values range from 0.05 for moist, dark, ploughed earth to 0.90 for fresh, dry snow. Typical values include 0.18–0.25 for most agricultural crops and natural vegetation less than 1 m in height, 0.05–0.15 for coniferous forest, and 0.15–0.20 for deciduous forest (Oke 1987). Houghton (1954) calculated a planetary albedo of 0.34 with a minimum of 0.28 in the subtropics and a maximum of 0.67 at the poles. Mean monthly values should be used in SRAD for midlatitude and high-elevation sites that are covered by snow for part of the year because typical albedo values for snow-covered vegetation may be two to four times higher than summer values (Iqbal 1983). Annual averages (i.e., the same value for each month) may be used in snow-free sites dominated by evergreen vegetation.

When used over large areas, both the estimation of temperature by a lapse rate and the use of uniform cloud conditions are likely to cause substantial errors. To overcome this, the sunshine fraction and temperature parameters can be specified by raster files rather than as single values. The line in the parameter file that would otherwise specify the parameter values is replaced with a line specifying the file name, number of rows, number of columns, cell size, and the x and y coordinates of the center of the lower left (southwestern) cell. The parameter grids may be at different resolutions than the DEM, in which case the value for each DEM grid point is calculated from the parameter grid by bilinear interpolation. Each parameter file must contain one complete grid representing the parameter surface for each calendar month. When a temperature parameter is specified using a grid file, SRAD assumes that the elevation effects are accounted for in that file so the corresponding lapse rate is not required and should be omitted from the parameter file. Temperature corrections due to short-wave radiation ratio are still applied. The sunshine hours parameter surface can be derived from surfaces of monthly average radiation on a horizontal surface: Hutchinson et al. (1984) describes the generation of a radiation surface by interpolation of measured stations, with a transformed rainfall surface providing an index of cloudiness. Gallant (1997) describes an application of SRAD using grids for sunshine hours and temperatures.

McKenney et al. (1999) identified the radiation parameters that had the greatest influence on model output in the Rinker Lake region of Ontario, Canada. They calculated the annual average sunshine fraction from measured sunshine totals, the cloud transmittance from measured radiation data, and the lumped atmospheric transmittance and circumsolar coefficient (using similar methods to those reported here) and used SRAD to predict the global short-wave irradiance. Low and high values were then chosen for each input parameter (one at a time) and eight model runs were performed to quantify the sensitivity of the model output to individual input parameters.

The results were presented as a series of bar graphs that showed overall performance but not grid point by grid point variability (McKenney et al. 1999). Increasing the sunshine fraction and the cloud transmittance from 0.10 to 0.80 (their best site-specific estimates for these two parameters were 0.46 and 0.36, respectively) increased the mean average daily irradiance approximately 250 and 150%, respectively. Increasing the atmospheric transmittance from 0.50 to 0.80 (their best estimate was 0.71) increased the mean average daily irradiance approximately 150%. Increasing the circumsolar coefficient from 0.01 to 0.50 (their best estimate was 0.18) decreased the mean average daily irradiance very slightly. These results seem intuitively correct and match our expectations. Hence, large sunshine fractions imply high frequencies of clear skies, high cloud transmittances imply less dense cloud cover (i.e., less attenuation of direct beam radiation), and the higher atmospheric transmittance indicates less attenuation of direct beam radiation as well. Similarly, the lack of variation in mean average daily irradiance when the circumsolar coefficient was varied is not surprising. This result may be a function of the method of presentation, since this variable would be expected to have markedly different impacts at different sites (depending on slope, aspect, and topographic shading), and its effects may be partially offset by increases in reflected radiation.

Overall, these results indicate that SRAD users should try their best to find or estimate site-specific monthly values for the first three radiation inputs. McKenney et al. (1999) also examined the effect of varying the resolution of the DEM on model output. They used 20- and 100-m DEMs and compared mean daily short-wave irradiance estimates at common x, y grid points. The outputs displayed similar means but the range of estimates was much greater for the 20-m DEM, and in that sense they matched the results obtained in other studies comparing geographic data sets incorporating varying levels of generalization (e.g., Wilson et al. 1996). McKenney et al. (1999) also noted that the computing requirements for the higher resolution DEM and a study area the size of Rinker Lake (900 km^2) were substantial and they concluded that a 100-m DEM would probably be sufficient for many large-area ecological applications.

4.3.3   Outputs

SRAD writes the 14 attributes listed in Table 4.3 to the output file in either ASCII or binary form with metadata as described in Section 3.1.12. Each DEM grid point with all its radiation attributes is written as a single record as either one line in an ASCII file or one unformatted record in a binary file.

4.4   WET

This program predicts soil-water content taking four components of the water balance at the surface into account: precipitation, evaporation, deep drainage, and runoff. Evaporation and deep drainage are both treated as losses, and deep drainage does not contribute to base flow. Runoff comprises both surface and subsurface runoff. The long-term average water balance is estimated using an equilibrium approach and spatially uniform mean precipitation rate. These estimates are important because the soil-water content is one of the limiting factors for plant growth and is also a factor in soil formation and other geomorphological processes (I. D. Moore et al. 1991, 1993b, c). Estimating soil-water content at fine resolution across a large area is a difficult task because of the complex interactions between topography, precipitation, radiation, evaporation, and the movement of water through the soil. In addition, soil-water content and related soil properties may exhibit substantial variability over distances of 1–100 m (e.g., Brutsaert 1986, Sharma et al. 1987). The approach taken in WET is to use simple but physically realistic relationships to capture the dominant topographic influences on long-term average relative soil-water content.

Given this approach, WET can operate at three levels of complexity. At the simplest level (which we shall call level 1), radiation effects are not accounted for and the fraction of precipitation lost through evaporation and deep drainage is assumed to be spatially invariant and is specified by the user. The most complex analysis (level 3)

TABLE 4.3   SRAD Output File Fields

Field	Field Name	Units	Description
1	X	m	x coordinate
2	Y	m	y coordinate
3	Elevation	m	From DEM
4	Short-wave irradiance ratio	None	The ratio of total short-wave irradiance on the sloping surface (R_{tss}) to total short-wave irradiance on a horizontal surface (R_{th})
5	Short-wave irradiance on sloping surface	See Table 4.1	The total short-wave irradiance on the sloping surface corrected for cloud effects and topographic shading (R_{tss})
6	Short-wave irradiance on horizontal surface	See Table 4.1	The total short-wave irradiance on a horizontal surface corrected for cloud effects but not including topographic shading (R_{tsns})
7	Incoming atmospheric long-wave irradiance	See Table 4.1	L_{in}
8	Outgoing surface long-wave irradiance	See Table 4.1	L_{out}
9	Net long-wave irradiance	See Table 4.1	Absorbed incoming long-wave less outgoing long-wave irradiance ($L_{net} = \varepsilon_s L_{in} - L_{out}$)
10	Net irradiance	See Table 4.1	Sum of absorbed short-wave irradiance on sloping surface and net long-wave irradiance (R_{net})
11	Max air temperature	°C	Average maximum air temperature over the analysis period
12	Min air temperature	°C	Average minimum air temperature
13	Avg air temperature	°C	The average of the minimum and maximum air temperatures
14	Surface temperature	°C	Average surface temperature

uses spatially varying net radiation to compute potential evaporation at each grid cell and then determines soil-water content using a set of functional relationships between soil-water content, evapotranspiration, and deep drainage. Both evaporation and deep drainage are dependent on soil-water content using this level 3 analysis. At the intermediate level 2, net radiation is still used to compute potential evaporation but soil-water content is not used to modify the loss rates. Both evaporation and deep drainage proceed at their maximum rate in all grid cells regardless of water content, essentially simulating conditions of maximum soil-water content. The estimation methods, inputs, and outputs at each of these levels are described in more detail below.

4.4.1   Estimation Methods

WET is based on the assumption that the spatial distribution of soil-water content is controlled by the topographic wetness index:

$$\omega = \ln \left(\frac{A_s}{\tan \beta} \right) \qquad (4.40)$$

where A_s is the specific catchment area (catchment area draining across a unit width of contour; m²/m) and b is the slope angle (in degrees). ω is also called the wetness index, topographic index, compound topographic index, and probably other names: We prefer topographic wetness index because the index is intended to represent the topographic control on soil wetness. However, this equation incorporates at least seven key assumptions and limitations (Beven and Kirkby 1979, Moore and Hutchinson 1991, Barling et al. 1994) as follows.

First, this approach assumes that the steady-state downslope subsurface discharge is the product of average recharge and specific catchment area. Second, it assumes that the local hydraulic gradient can be approximated by local slope. Third, it assumes that the saturated hydraulic conductivity of the soil is an exponential function of depth. Fourth, it assumes steady-state conditions. However, the velocity of subsurface flow is sufficiently small that most points in a catchment receive contributions from only a small proportion of their total upslope contributing area, and the subsurface flow regime is in a state of dynamic nonequilibrium (Barling 1992). Barling et al. (1994) proposed a dynamic wetness index, which replaces A_s in Equation 4.40 with an effective specific catchment area based on both topography and drainage time as discussed in Section 4.5 (this variant has not yet been incorporated in the WET program). Fifth, this particular form of the topographic wetness index also assumes spatially uniform soil properties (in particular transmissivity), but this assumption has been justified by Wood et al. (1990), who concluded that the topographic component of the index dominates over the soil transmissivity at the subcatchment scale. Another form of the topographic wetness index that includes transmissivity variations is discussed in Chapter 1. Furthermore, the spatial distribution of topographic attributes may capture the spatial variability of soil properties at the mesoscale because pedogenesis of the soil catena often occurs in response to the way water moves through the landscape in areas with uniform parent material (Moore and Hutchinson 1991). Sixth, this approach implies that the locations in a catchment with the same value of the topographic wetness index will also have the same relationship between the local depth to the water table and the mean depth. Finally, this approach also implies that those points with the same value of the topographic wetness index will respond in a similar way to the same inputs.

The specific catchment area, A_s in Equation 4.40, is calculated using either the D8 or Rho8 algorithm as used by TAPES-G (Chapter 3; Gallant and Wilson 1996). This algorithm can be expressed as

$$A_s = \frac{\displaystyle\sum_{j \in C_j} a_i}{b_j} \qquad (4.41)$$

where a_i is the area of the ith grid cell, b_j is the width of the jth cell, and C_j represents all of the cells upslope of cell j that are hydraulically connected to cell j (the cell's catchment area). The hydraulic connectivity is based on the flow directions determined by either the D8 or Rho8 algorithm in TAPES-G.

Equations 4.40 and 4.41 can be generalized to account for the effects of water loss by modifying the area of each cell by a weighting factor, μ_i, dependent on the water balance at each cell:

$$A_s = \frac{\sum_{i \in C_j} \mu_i a_i}{b_j} \tag{4.42}$$

WET uses a weighting coefficient based on precipitation (P), evapotranspiration (E), and deep drainage (D):

$$\mu = 1 - \frac{E + D}{P} \tag{4.43}$$

in which E, D, and P all have units of millimeters/day. The term $(E + D)/P$ is the fraction of precipitation not converted to runoff.

The different levels of analysis in WET correspond to different methods of computing the E and D terms in Equation 4.43. The level 1 analysis allows the user to specify an $(E + D)/P$ ratio that is applied to the whole area being analyzed. This approach assumes that evapotranspiration and deep drainage do not vary across the landscape and that the only topographic control on soil water is approximated by the wetness index (i.e., Equation 4.40). The $(E + D)/P$ ratio for a catchment can be estimated directly from long-term catchment precipitation and runoff records (as described in Section 4.4.2).

The level 2 analysis incorporates spatially varying net radiation as computed by SRAD to determine the potential evapotranspiration, E_p, at each grid cell. This is used for E and the maximum deep drainage rate specified in the site parameter file is used for D. This pair of assumptions correspond to a permanently saturated surface. Potential evapotranspiration is computed using the function proposed by Priestley and Taylor (1972) for evaporative demand from well-watered vegetation under conditions of minimal advection:

$$E_p = \frac{\alpha_e(R_n - G)}{\lambda(1 + \gamma/\Delta)} \tag{4.44}$$

where R_n is the net radiation, G is the soil heat flux (which can be ignored for periods longer than one day), α_e is an empirical constant ($=1.26$), λ is the latent heat of vaporization of water, Δ is the slope of the saturated vapor pressure curve, and γ is the psychrometric constant (Δ and γ are functions of air temperature and pressure; see Shuttleworth (1993) or another similar work for details). The assumptions used to develop this equation are equivalent to the assumptions used in the Bowen ratio method of measuring evaporation (Shuttleworth 1993), and the corresponding Bowen ratio (the ratio of sensible to latent heat fluxes), $[(R_n - G - \lambda E_p) / \lambda E_p]$ is

$$B_0 = \frac{1 + \gamma/\Delta}{\alpha_e} - 1 \tag{4.45}$$

WET calculates and reports this Bowen ratio and allows the user to alter it if reliable information on the Bowen ratio for saturated surfaces in the study area is available. Note that the Bowen ratio is for saturated surfaces and the evaporation computed from it will be reduced for unsaturated soils when using level 3 analysis. If using level 2 analysis where no soil-water corrections are made to evaporation, a higher Bowen ratio could be supplied if sufficient data are available to determine an average Bowen ratio for the study area.

The level 3 analysis uses the same function to compute potential evapotranspiration but computes actual evapotranspiration, E, as a function of relative soil-water content and potential evapotranspiration. Various functional forms have been proposed to describe this relationship. We use the following parametrically efficient relationship proposed by Kristensen and Jensen (1975), which produces a range of responses under different evaporative demands:

$$E = E_p\,[1 - (1 - \theta)^{C/E_p}] \qquad \text{for } 0 \le \theta \le 1 \tag{4.46}$$

where θ is the relative available soil-water content (ranging from 0.0 to 1.0), E_p is the evaporative demand (mm/day), and C is a constant (about 12 mm/day). θ is determined from the topographic wetness index using one of the following relationships:

$$\theta = \frac{\omega}{\omega_{cr}} \qquad \text{for } \omega < \omega_{cr} \tag{4.47}$$

$$\theta = 1 \qquad \text{for } \omega \ge \omega_{cr} \tag{4.48}$$

The ω_{cr} term used in this last pair of equations is a user-specified critical wetness index corresponding to $\theta = 1.0$ (field capacity). Deep drainage is also made dependent on θ using a standard power-law relationship for hydraulic conductivity (Rawls et al. 1993) with the maximum deep drainage value taking on the role of saturated hydraulic conductivity:

$$D = D_{max}\theta^\beta \tag{4.49}$$

where β is typically between 10 and 15 (Table 4.4).

Because of the implicit relationship $\omega \rightarrow \theta \rightarrow E, D \rightarrow \mu \rightarrow \omega$, an iterative scheme is required to determine ω. WET uses a Newton–Raphson procedure for each cell, starting with cells to which there are no upslope connections (i.e., the tops of hills and ridge lines) and proceeding downslope until all cells are resolved.

4.4.2 Inputs

WET requires output files from two other programs: TAPES-G and SRAD (except for the simplest level 1 analysis where SRAD output is not required). TAPES-G provides slope and flow directions and SRAD provides net radiation estimates for each grid point. WET also requires a small number of site parameters for the study region (Table 4.4). These requirements vary with the level of analysis (1, 2, or 3) that is chosen.

One site parameter, the critical wetness index, is required for all model runs (irrespective of the level of analysis that is chosen). The choice of critical wetness index will affect the magnitude of the relative soil-water content and number of cells that

TABLE 4.4 Site Parameters Required by WET

Parameter	Units	Levels	Description
Critical wetness index, ω_{cr}	None	All	The wetness index corresponding to maximum soil-water content (field capacity), typically 8 to 10
Mean air temperature	°C	2 and 3	Mean air temperature for the period of the analysis, recorded at a location within or near the study area
Mean elevation	m	2 and 3	The elevation of the site from which the mean temperature was recorded
Precipitation	mm/day	2 and 3	Mean precipitation over the study area for the period of analysis
Interception losses	mm/day	2 and 3	Amount of precipitation that does not reach the ground because of interception by vegetation; can be set to zero if not known
Maximum drainage rate, D_{max}	mm/day	2 and 3	The rate at which water is lost by deep drainage when the soil is fully wet (at field capacity)
β	None	2 and 3	The exponent used to relate the soil-water content to drainage rate (Equation 4.49), typically 7 (sandy soils) to 15 (clay soils)
C	mm/day	3	The exponent used to relate the soil-water content to actual evaporation rate (Equation 4.46), typically 10 to 12 mm/day
$(E + D)/P$ ratio	None	1	The fraction of precipitation lost from the soil and not converted to runoff

are saturated. The higher the critical wetness index, the lower the relative soil-water content and the fewer the number of cells classified as saturated.

The only other parameter required for the level 1 analysis is the fraction of precipitation in a catchment that is converted to runoff. This number can be determined from long-term rainfall–runoff records, or the United States Soil Conservation Service (SCS) runoff curve number model (Rawls et al. 1993). When rainfall and runoff records are available, the water balance equation

$$P = R + E + D \tag{4.50}$$

can be used directly, with P being the average annual precipitation and R the average annual runoff. Rearranging Equation 4.50 gives $E + D = P - R$, so

$$\frac{E + D}{P} = \frac{P - R}{P} \tag{4.51}$$

The SCS runoff curve number model can be used to estimate runoff from precipitation and estimates of surface condition. The model estimates runoff as

$$Q = \frac{(P - I_a)^2}{(P - I_a) + S}$$ (4.52)

where Q is the runoff (cm), P is the rainfall (cm), S is the potential maximum retention after runoff begins (cm), and I_a is the initial abstraction (cm) for the period (month, season, year, etc.) in question (Rawls et al. 1993).

The retention term in Equation 4.52 is highly variable from one catchment (storm) to the next and accounts for the water retained in surface depressions, water intercepted by vegetation, evaporation, and infiltration. It is often estimated with an empirical equation that was derived from data for many small agricultural catchments (Rawls et al. 1993):

$$I_a = 0.2S$$ (4.53)

Substituting this equation in Equation 4.52 gives

$$Q = \frac{(P - 0.2S)^2}{P + 0.8S}$$ (4.54)

where the parameter S is related to the soil and cover conditions of the catchment through the curve number CN:

$$S = \frac{1000}{CN} - 10$$ (4.55)

The major factors that determine CN are the hydrologic soil group, cover type, hydrologic condition, treatment, and antecedent runoff condition. In general, the fraction of precipitation converted to runoff will increase as infiltration rate and vegetation cover decrease. Four hydrologic soil groups (labeled A [high-infiltration soils] through D [low-infiltration soils]) and three hydrologic conditions (labeled Poor, Fair, or Good) are usually assigned. These attributes have been determined for most of the soils in the United States and can be obtained from county soil reports for most areas. For other locations, interested readers should consult Rawls et al. (1993) and the U.S. Soil Conservation Service (1985, 1986) for more detailed directions on how to determine the factors affecting CN since these decisions can have a large impact on the final runoff estimates. Composite CN estimates are sometimes computed for catchments containing distinctive soil and land cover map units.

Rawls et al. (1993) have also warned that good judgment and experience based on stream gauge records is often needed to adjust CNs as match local conditions. Pilgrim and Cordery (1993, 9.25–9.26) repeated this advice after summarizing the results of several studies that cast doubt on the accuracy and validity of the SCS method. Mancini and Rosso (1989) showed how the size and spatial arrangement of map units can affect the overall value of CN and the resulting calculated runoff when composite values of CN are estimated because of the nonlinearity of Equations 4.52 and 4.54. The SCS method should be used only when direct runoff dominates the hydrologic response of a catchment because the rainfall–runoff relationship is sensitive to surface soil-water content and the dominant runoff processes operating in the catchment. These relationships can cause problems because of the complex nature of the runoff production system (see Section 4.2.1 for additional details).

For level 2 analysis, six parameters are required in place of the $(E + D)/P$ fraction: mean air temperature from a nearby climate station, elevation of the climate station, mean precipitation, storage deficit, maximum drainage rate, D_{max}, and the exponent β in Equation 4.49 (Table 4.4). The mean air temperature and mean precipitation for the period of interest can be estimated from long-term records for a nearby station. The temperature and reference elevation values are used in the calculation of the Bowen ratio (i.e., Equation 4.45). The storage deficit, maximum drainage rate, and exponent β can be estimated from local knowledge and/or published reports (see Moore et al. (1993e) and Section 4.5 for examples). Selecting larger values for the storage deficit and maximum drainage rate will reduce the relative soil-water content (varying one input at a time). Choosing larger values of β will reduce the deep drainage losses and thereby increase the relative soil-water content.

The level 3 analysis requires one additional exponent (C) that relates the soil-water content to the actual evapotranspiration rate in addition to the one input parameter required at all three levels and the six input parameters required for the level 2 analysis (Table 4.4). We typically set $C = 12$ mm/day. Larger values of C will increase the actual evapotranspiration estimated with Equation 4.46 and thereby reduce the magnitude of the relative soil-water content (assuming everything else is held constant).

4.4.3 Outputs

WET produces an output file in either ASCII or binary form with metadata as described in Section 3.1.12. The computed topographic attributes for each DEM grid point are written as one line in an ASCII file or one unformatted record in a binary file. The attributes written to this file are summarized in Table 4.5. Note that there are six fields in the output file from a level 1 analysis, eight from level 2, and 10 from level 3.

4.5 DYNWET-G

This program calculates a spatially distributed topographic wetness index based on either a steady-state or quasi-dynamic subsurface flow assumption using terrain attributes computed by TAPES-G (Moore 1992). A contour-based version of the program, DYNWET-C, uses the outputs of TAPES-C and is based on the same princi-

TABLE 4.5   WET Output File Fields

Field	Field Name	Units	Levels	Description
1, 2	X, Y	m	All	x and y coordinates as specified in TAPES-G file
3	Elevation	m	All	Elevation from TAPES-G file
4	Topographic wetness index	None	All	The topographic index used to determine soil-water content, modified by losses through evapotranspiration and deep drainage
5	Relative soil-water content	None	All	Soil-water content as a fraction of field capacity; the minimum value of 0 indicates very dry soil and the maximum value of 1 indicates wet soil
6	Effective drainage area	m^2	All	The area upslope of the grid cell, reduced to account for water loss by evapotranspiration and deep drainage
7	Potential evapotranspiration	mm/day	2 and 3	The evapotranspiration under non-limiting soil-water conditions (maximum soil wetness)
8	Runoff	None	2 and 3	The water running off the grid cell to the downslope grid cell, equal to the precipitation rate multiplied by the effective drainage area, less actual evapotranspiration and deep drainage
9	Actual evapotranspiration	mm/day	3	The water lost from the grid cell by evapotranspiration, computed from potential evapotranspiration and the relative soil-water content
10	Deep drainage	mm/day	3	The water lost from the grid cell by deep drainage, computed from the maximum deep drainage rate and the relative soil-water content

ples as DYNWET-G, but will not be described in detail. The role of the wetness index in characterizing the spatial distribution of soil-water content and the location of zones of surface saturation (i.e., zones of partial area runoff) was noted earlier (see Chapter 1 and Section 4.4 of this chapter for additional details). The steady-state assumption incorporated in WET implies that the specific catchment area is an appropriate surrogate for the subsurface flow rate. This will be true only if the recharge to the surface horizons occurs at a constant rate for the length of time required for every point in the catchment to reach subsurface drainage equilibrium (Moore 1992, Moore et al. 1993f, Barling et al. 1994). This state of affairs suggests that we are most likely to find a good relationship between the WET equilibrium approach and soil-water content in humid regions where frequent and substantial rainfalls keep the soil in a wet condition (e.g., Troch et al. 1993). We are likely to find that the lowest points in the catchment are the wettest and that soil-water content consistently decreases as flow lines are retraced upslope toward the catchment divide in these types of environments.

In drier environments, the velocity of subsurface flow is so small that most points will receive contributions from only a small proportion of their upslope contributing area and the subsurface flow regime is characterized by a state of dynamic nonequilibrium (Barling et al. 1994). For example, a 300-m-long slope would need 60–120 days of continuous recharge for every point to reach subsurface drainage equilibrium based on soil saturated hydraulic conductivities of 0.1–0.2 m/h (Kirkby and Chorley 1967). Over shorter time periods, the subsurface flows will only increase linearly downslope in very narrow zones close to the drainage divide and local features will play an important role in determining the distribution of soil-water content and formation of zones of soil saturation in these instances. Barling (1992), studying a 7-ha hillslope near Wagga Wagga in New South Wales, Australia, found that subsurface flow was affected by only a small proportion of the contributing area directly upslope during storm events and that the initial surface saturation did not occur at the points with the largest $\ln(A_s/\tan \beta)$ values.

The quasi-dynamic topographic wetness index calculated by DYNWET is an alternative index that recognizes that the subsurface flow regime in a catchment rarely, if ever, reaches steady state (i.e., that every point is experiencing drainage from its entire upslope contributing area). The index uses an effective specific catchment area A_e that is limited by the velocity of water movement and the time between wetting events. Barling et al. (1994) computed both the steady-state and quasi-dynamic topographic wetness indices for the Wagga Wagga site and compared the results with the observed distribution of soil water and locations of saturated sources areas. The results indicated that there was a low correlation between the two wetness indices ($R^2 = 0.47$) even after reasonably long drainage periods ($t \leq 120$ days) and that the measured patterns of soil-water content closely matched the distribution of the quasi-dynamic ($A_e/\tan \beta$) index. These results show how the derivation of spatially and temporally varying indices, such as the quasi-dynamic topographic wetness index calculated by DYNWET, may be useful in some landscapes, although more work is required to demonstrate their applicability in other types of landscapes.

4.5.1 Estimation Methods

DYNWET calculates both the steady-state topographic wetness index (Equation 4.40) based on A_s and the quasi-dynamic wetness index using the effective specific catchment area, A_e. The derivation of A_e starts with the kinematic wave equation for subsurface flow:

$$q = K_s \tan \beta \tag{4.56}$$

where q is the flux density or flow per unit area of the subsurface flow (m/s), K_s is the soil saturated hydraulic conductivity (m/s), and β is the slope of the impermeable boundary (which is normally assumed to be the same as the slope of the soil surface) (Beven 1981, Sloan and Moore 1984). Barling et al. (1994) showed how this equation can be used to estimate the interstitial velocity of subsurface flow:

$$v = \frac{q}{\eta} = \frac{K_s}{\eta} \tan \beta \tag{4.57}$$

where η is the effective porosity. Iida (1984) had earlier introduced the concept of time–area curves and used this equation to estimate the time required for water to travel from point E to point F along a flow line or stream tube as follows:

$$t_{EF} = \int_F^E \frac{ds}{v} = \int_F^E \frac{\eta}{K_s \tan \beta} ds \tag{4.58}$$

where s is the horizontal distance between E and F. This equation can be used to delineate isolines showing the time required for water to reach a unit width of contour for its entire upslope area. Figure 4.1 shows that for time t, the area bounded by the unit width of contour, the two orthogonal slope lines, and the isoline is defined by $a(t)$. At the maximum time, t_s (i.e., the time required to reach steady-state equilibrium), $a(t_s)$ is equal to the specific upslope area (A_s). Barling et al. (1994) have argued that the time–area concept accounts for the character of the land surface in the upslope contributing area and the time taken for subsurface drainage to redistribute soil water in a simple but physically realistic way.

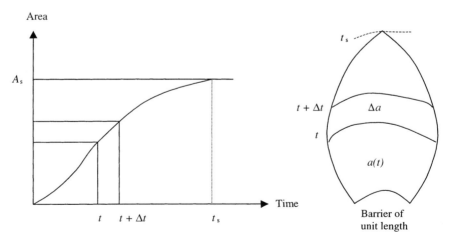

Figure 4.1. Schematic diagrams showing (a) the time–area curve, and (b) the area, $a(t)$, surrounded by the barrier, two orthogonal slope lines, and the isoline of time, t (after Iida 1984 and Barling et al. 1994).

These ideas form the basis of the quasi-dynamic topographic wetness index that is computed with DYNWET-G. The program calculates the time required for subsurface flow to travel the length of each grid cell using Equation 4.58 and a series of topographic (flow direction, slope, area) and site (drainable porosity, saturated hydraulic conductivity) attributes for each cell. The effective contributing area, A_i, for a grid cell is determined by following each drainage path in an upslope direction, accumulating travel time until it reaches the specified drainage time. The effective specific catchment area is then the ratio of A_i to flow width:

$$A_e = \frac{A_i}{b} \qquad (4.59)$$

Due to the need to trace individual contributing flow paths, the more sophisticated FD8 and DEMON flow algorithms have not been implemented in DYNWET-G, so only the D8 or Rho8 flow accumulation algorithms are available for computing effective specific catchment area. The static wetness index computed by DYNWET-G uses contributing area computed by TAPES-G, so the better algorithms are available for that index.

Given this approach, DYNWET essentially extends the time–area concept first proposed and applied to a series of idealized land surfaces by Iida (1984) to grid-based representations of complex natural landscapes and offers an alternative approach for predicting soil-water content to the equilibrium approach incorporated in WET.

4.5.2 Inputs

DYNWET-G requires the output file from TAPES-G and four site parameters. The TAPES output files provide the flow direction, slope, and element area attributes. The user is asked by the program to specify the depth of soil above the impermeable layer, effective or drainable porosity, saturated hydraulic conductivity, and drainage time. The soil variables can be estimated from field measurements (as in Barling et al. 1994) or published values (as discussed with the sample application below). The drainage time is the average number of days between precipitation events and can be estimated from precipitation records at a nearby climate station.

Barling et al. (1994) examined the sensitivity of the quasi-dynamic wetness index to the choice of K_s, η, and drainage time in the Wagga Wagga study. These results showed that the predicted and measured patterns of soil water were highly correlated with the $A_e/\tan \beta$ index over a wide range of user-specified drainage times. This result occurred because the subsurface flow rates are small and the spatial distribution of the $A_e/\tan \beta$ index will not change substantially until several weeks or months have elapsed. Barling et al. (1994) also observed that the velocity of subsurface flow is directly proportional to K_s and inversely proportional to η in Equation 4.58. This structure means that the predicted wetness patterns for a specific combination of soil hydraulic properties and a user-specified drainage time will be identical to wetness patterns predicted by an equivalent set of parameters (e.g., values twice as large). Overall, these findings suggest that different input values will have a substantial impact on the time required to achieve a given wetness pattern without changing the pattern itself and that the mean patterns will not change as quickly as the absolute values (Barling et al. 1994).

4.5.3 Outputs

DYNWET-G produces an output file in either ASCII or binary form with metadata as described in Section 3.1.12. The computed topographic attributes for each DEM grid point are written as one line in an ASCII file or one unformatted record in a binary file. The attributes written to this file are summarized in Table 4.6.

4.6 SAMPLE APPLICATION

EROS, SRAD, WET, and DYNWET-G were applied to the same Cottonwood Creek catchment introduced in Chapters 2 and 3. We chose this catchment for the model runs because it was familiar to the first author and it is typical of many rangeland areas in the United States in terms of the level of environmental characterization. The terrain-analysis programs listed above were used with the 15-m DEM produced in Chapter 2, several of the topographic attributes computed from this DEM by TAPES-G in Chapter 3, and various site parameters that were obtained or estimated from published sources and/or local knowledge.

The Cottonwood Creek catchment covers approximately 197 ha and is part of the Montana State University Red Bluff Research Ranch located near Norris, MT (45°

TABLE 4.6 DYNWET Output File Fields

Field	Field Name	Units	Description
1	X	m	x coordinate
2	Y	m	y coordinate
3	Elevation	m	From DEM
4	Quasi-dynamic topographic wetness index	None	This index incorporates an effective catchment area that is a function of topography and drainage time. This approach means that most cells receive contributions from a small proportion of their total potential upslope contributing areas.
5	Steady-state topographic wetness index	None	This index assumes that the steady-state downslope subsurface discharge is the product of average discharge and specific catchment area.
6	Effective upslope contributing area	m^2	Subsurface upslope contributing area for some user-specified drainage time

33′ N, 111° 38′ W). It contains moderate to steep slopes (Figure 3.2) and the relief
spans approximately 330 m (from 1642 m at the catchment outlet to 1969 m at the
southeast corner of the catchment). The main channel is spring fed and runs year
round. It is flanked by small, intermittent seeps that feed water laterally into this
channel and several ephemeral tributaries (Aspie 1989). The soils are deep, moderately well drained, and formed in colluvium and material derived from gneiss, schist,
and granite. Most of the soils have loamy or sandy loamy surface textures (Boast and
Shelito 1989). The climate is semiarid (25–70 cm annual precipitation) and the vegetation is strongly correlated with landscape position and aspect. Grasses cover about
60% of the catchment and occupy south-facing slopes and ridge tops. Maple, aspen,
willow, and snowberry covers about 10% of the catchment and occupy north-facing
slopes and lower stream bottoms. Sagebrush interspersed with small stands of
conifers dominates the remainder of the study area (Jersey 1993).

Figure 4.2 was produced in EROS using the TAPES-G outputs from Chapter 3
and by assuming uniform rainfall excess runoff. We first created a depressionless
DEM in TAPES-G and then used the finite difference slope method and FD8 flow-
routing method with a maximum cross-grading area of 8000 m² to calculate slope
and specific catchment area, respectively. The uniform rainfall excess option is often
chosen in EROS because no additional knowledge of runoff is required. The smallest
sediment transport capacity index values occur along the catchment boundary and

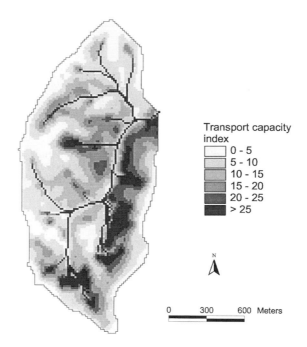

Figure 4.2. Cottonwood Creek map showing the sediment transport capacity index derived
with the uniform rainfall excess runoff method in EROS.

the largest values were estimated in areas with steep slopes and large upslope contributing areas (cf. Figure 4.2 with Figures 3.2 and 3.9). Figure 4.3 shows the change in the sediment transport capacity index across hydrologically connected grid cells. Net deposition areas were predicted along footslopes and in channel areas because negative values are generated when the sediment transport capacity index decreases from one hydrologically connected cell to the next (when moving down the slope). These conditions were predicted in a few other locations along the western boundary and in the southeast corner of the catchment as well (Figure 4.3).

The next pair of diagrams shows why these types of predictions must be used with great care. The patterns shown in Figures 4.2 and 4.3 may not help much with the description of erosion and deposition in this catchment because local observations and monitoring conducted over several years suggest that saturated overland flow is the dominant runoff producing mechanism (e.g., Pogacnik 1985, Aspie 1989). Aspie (1989) measured stream flow at two flumes (labeled B and F in Figure 4.5) for 10 storm events in 1986 and 1987. No saturated zones were observed above flume B. Channel areas constituted 0.8% of this zone. Numerous saturated areas were observed between flumes B and F and these areas combined with the channel constituted 1.6% of this zone. Aspie (1989) explained 80–90% of the storm runoff volumes measured at the two flumes by assuming that all of the precipitation falling on these areas was converted to storm runoff. A critical wetness index of 8.5 was derived from

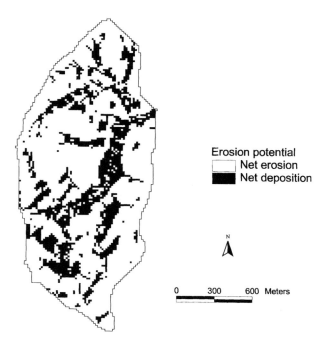

Erosion potential
☐ Net erosion
■ Net deposition

N

0   300   600 Meters

Figure 4.3. Cottonwood Creek map showing the change in sediment transport capacity index derived with the uniform rainfall excess runoff method in EROS.

the cumulative frequency plot of steady-state topographic wetness index reproduced in Figure 4.4 by assuming that approximately 5% of the catchment contributes to storm runoff; the index was used to generate the map reproduced in Figure 4.5. This map produced a very different spatial pattern compared to the first map reproduced in Figure 4.2 and the differences highlight the importance of choosing the appropriate runoff method for the area of interest when using EROS. The shaded areas in Figure 4.5 identify low areas with large upslope contributing areas that are likely to generate saturated overland flow. The cells with zero values show those parts of the catchment dominated by lateral subsurface flow. No overland flow is predicted on the steep slopes with large sediment transport capacity values in Figure 4.2 that are located just east of the main channel in this instance (c.f. Figures 4.2 and 4.5).

The radiation and temperature attributes summarized in Table 4.3 were calculated using SRAD, the 15-m DEM from Chapter 2, and the Madison Range site parameter file (Table 4.7). Thirty-year (1961–1990) monthly records from the Great Falls and Bozeman W6 Experiment Farm climate stations were used to estimate radiation and temperature inputs, respectively. The Great Falls station located 220 km north of the study area is the only primary solar radiation station located in Montana. The U.S. National Solar Radiation Database contains 56 primary stations and 183 secondary stations. Primary stations report measured solar radiation data for at least a portion of

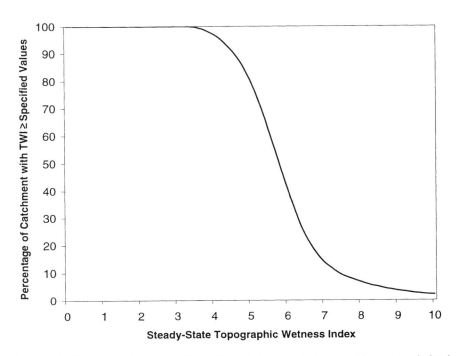

Figure 4.4. Cumulative frequency distribution of steady-state topographic wetness index in the Cottonwood Creek catchment.

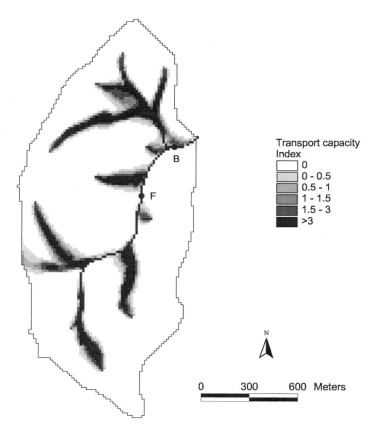

Figure 4.5. Cottonwood Creek map showing the sediment transport capacity index derived with the saturated overland flow runoff method in EROS.

their 30-year record and secondary stations contain modeled solar radiation data (National Renewable Energy Laboratory 1992). The circumsolar, cloud transmittance, and lumped transmittance coefficients were estimated with the equations listed in Section 4.3.2 and the sunshine hours and radiation fluxes reported for this station.

The Bozeman W6 Experiment Farm climate station is located 22 km east of the study area and was chosen because it is the closest station with long-term air and surface (soil) temperature measurements (Munn et al. 1981). Mean monthly minimum and maximum air temperatures were obtained directly from station records and the three sets of lapse rates were estimated with the modified version of the spatial filtering-kernal convolution method described in Section 4.3.2. Mean monthly soil temperatures were reported at depths of 5 or 10 cm for the period 1981–1990 and these data were used to estimate mean monthly temperature gradients and surface temperatures. The recorded temperatures were adjusted for depth on a monthly basis because seasonal variations in soil temperatures are greatest at the surface and

TABLE 4.7   Madison Range SRAD Site Parameter File

45.22	45.22										
0.07	0.10	0.12	0.14	0.16	0.19	0.23	0.20	0.16	0.12	0.08	0.07
0.59	0.57	0.57	0.37	0.19	0.14	0.14	0.14	0.18	0.28	0.49	0.56
0.31	0.30	0.27	0.28	0.28	0.28	0.20	0.22	0.27	0.29	0.32	0.32
0.49	0.56	0.66	0.62	0.62	0.65	0.79	0.76	0.67	0.61	0.46	0.44
0.2	3.1	6.9	12.7	17.9	23.1	27.6	27.1	21.2	14.4	5.6	0.7
−11.6	−8.9	−5.6	−1.4	2.9	6.8	9.3	8.6	4.1	−0.4	−6.2	−11.1
−1.9	−1.8	0.6	5.8	12.2	18.3	24.0	24.3	13.8	7.2	1.5	−1.3
3.9	4.8	6.2	7.1	8.0	6.4	6.2	4.3	6.4	6.5	4.6	4.5
2.5	3.8	5.2	5.2	5.6	4.5	4.4	2.5	3.9	4.1	3.1	3.2
4.8	5.4	6.9	8.3	9.5	7.7	7.4	5.5	8.0	8.0	5.6	5.3
1											
0.2	0.2	0.2	0.5	1.5	2.5	2.5	1.5	1.0	0.5	0.2	0.2
10.0	0.98	0.00008	1455.4								
0.64	0.66	0.63	0.64	0.65	0.68	0.67	0.65	0.65	0.64	0.67	0.65

Note. See Table 4.2 for description of variables recorded in individual rows and columns.

decrease with depth until, at a depth of 10 m or more, they disappear (Smith et al. 1964, Parton and Logan 1981). The land cover information recorded with station temperature measurements may also be important because soil temperatures can be expected to vary with land cover: Munn et al. (1978), for example, found that soil temperatures differed significantly in adjacent high elevation forests and meadows in Montana. The parameter values used to describe the surface conditions in Table 4.7 were derived from published values for midlatitude range sites.

Maps of the net annual solar radiation, net solar radiation in winter and summer, short-wave solar radiation ratio, and mean annual average temperature are reproduced in Figures 4.6–4.8. The net annual solar radiation map shows the major geomorphic features of the catchment. The largest net annual solar radiation values (>10 W/m^2) occur on the south-facing slopes that delineate the northern boundary of the catchment and on the large south-facing slope that dominates the center of the catchment (Figure 4.6). Low values (<10 W/m^2) are predicted on those parts of the catchment with north-facing slopes and/or in areas shaded for part of the day by the surrounding landscape. The net annual solar radiation values reported in Figure 4.6 vary between −19.5 and 15.9 W/m^2, and the mean value of −1.1 W/m^2 indicates that large negative values were slightly more prevalent than positive values. The spatial pattern more or less matches that produced for the short-wave solar radiation index (see Figure 4.8a). This particular map shows the ratio of incident short-wave radiation for each grid point compared to a horizontal point at the same latitude with no topographic shading. Approximately 7.2% of the cells have ratios greater than 1.1 (i.e., they signify south-facing slopes) and 45.4% have ratios less than 0.9 (these cells signify north-facing slopes and/or sites that experience topographic shading). Both of these maps therefore illustrate the impact of surrounding terrain in modifying the radiation budget at specific sites within a catchment.

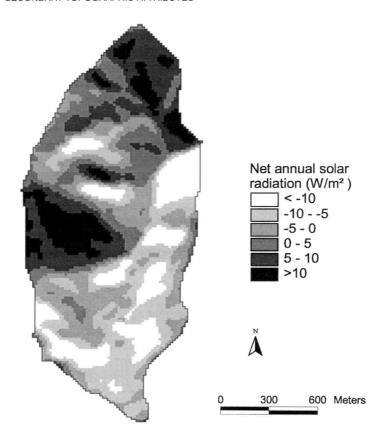

Figure 4.6. Cottonwood Creek net annual solar radiation map.

The same general patterns are repeated in the winter (December–February) and summer (June–August) net solar radiation maps reproduced in Figure 4.7. These maps show net irradiance in six equal area classes. The summer values are much larger than the winter values (144.6 versus −42.3 W/m^2 mean net irradiance) and the spatial patterns are slightly different in areas influenced by topographic shading because of the impact of varying sun angles at these latitudes at different times of the year. Regressing winter net solar radiation against summer net solar radiation produced an R^2 of 0.74 and illustrates how the impact of surrounding terrain in modifying the radiation budget at specific sites is likely to vary seasonally in midlatitude areas such as Cottonwood Creek. The greater variability evident in summer (Figure 4.7b) compared to winter (Figure 4.7.a) is largely a function of magnitude since the coefficient of variation in summer (0.10) is about half as large as that recorded in winter (0.16). The winter values varied from −52.1 to −25.4 W/m^2 (−42.3 W/m^2 mean) and the summer values varied from 102.8 to 167.4 W/m^2 (144.7 W/m^2 mean). These mean catchment values are consistent with the December and June net irradiance estimates reported for southwest Montana by Budyko (1974). Further validation

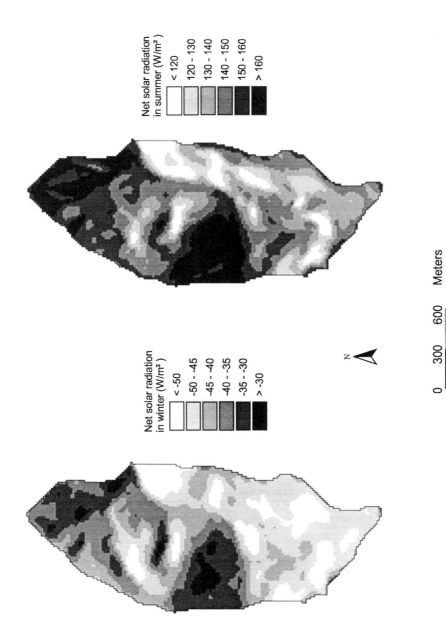

Figure 4.7. Cottonwood Creek maps showing (a) net solar radiation in winter and (b) net solar radiation in summer.

125

was not possible because of the paucity of radiation measurements for horizontal and especially sloping terrain in this region, and the SRAD user can assume that this situation is typical of most other parts of the world as well.

The mean annual average air temperature map reproduced in Figure 4.8b combines the effects of elevation (via the lapse rate) and slope, aspect, and topographic shading (via the short-wave solar radiation index) (Figure 4.8a). Elevation ranged from 1642 to 1969 m and coincided with the locations of relatively high (5.1°C) and low (3.3°C) predicted temperatures, respectively. The highest mean annual average air temperature (5.3°C) was computed at two points with similar elevations (1651 and 1656 m) and short-wave radiation ratios (0.941 and 0.963 respectively). The low temperature (3.0°C) was computed at a point with an elevation of 1821 m and short-wave radiation ratio of 0.983.

The three soil-water content maps reproduced in Figure 4.9 were produced with the three sets of estimation techniques available in WET. The first map utilized a critical steady-state topographic wetness index value of 7.5 and $(E + D)/P$ ratio of 0.90 (i.e., we assumed that 90% of the precipitation was lost from the soil via deep drainage and evapotranspiration and not converted to runoff). The relative soil-water content varied from 0.12 to 1.0 in this instance and the map shows that the higher values generally occurred in cells with large upslope contributing areas (Figure 4.9a). This particular approach (map) assumes that topography controls relative soil-water content and the pattern mimics that shown for the steady-state topographic wetness index in Figure 4.10a.

The relative soil-water content map reproduced in Figure 4.9b was computed with the level 2 estimation techniques and the site parameters listed in Table 4.8. These values ranged from 0.38 to 1.0 and produced subtle variations in spatial patterns compared to the level 1 map. The relative soil water content values predicted with the level 2 estimation methods were approximately 50% larger on average than those predicted with level 1 (i.e., mean relative soil-water content values of 0.45 and 0.69 were predicted in Figure 4.9a and b, respectively). This pair of maps also shows that slightly different patterns were predicted in cells that experienced substantial topographic shading and therefore lower rates of evapotranspiration losses (as illustrated by the steep slopes to the east of the main channel and on north-facing slopes scattered throughout the study area). The final map reproduced in Figure 4.9c was generated with the level 3 estimation techniques and site parameters listed in Table 4.8. This map contains slightly larger values than the second map (0.40, 0.71, and 1.0 minimum, mean and maximum values, respectively) because the rate of loss to deep drainage and evapotranspiration were partially controlled by the relative soil-water content in this instance. The choice of study area and site parameters, however, meant that these differences were very small and the spatial patterns produced with the level 2 and level 3 estimation techniques are very similar.

The last pair of maps reproduced in Figure 4.10 show the steady-state and quasi-dynamic topographic wetness indices calculated with DYNWET-G. The steady-state topographic wetness index (Figure 4.10a) is similar to the WET level 1 soil-water content map (Figure 4.9a) but for the fact that no critical wetness index was specified and used to normalize the computed cell values. These values vary from 4.7 to 14.9

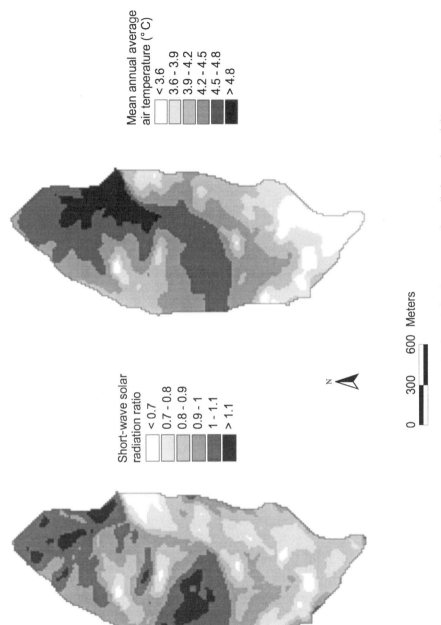

Figure 4.8. Cottonwood Creek maps showing (a) the short-wave solar radiation ratio and (b) mean annual average air temperature.

127

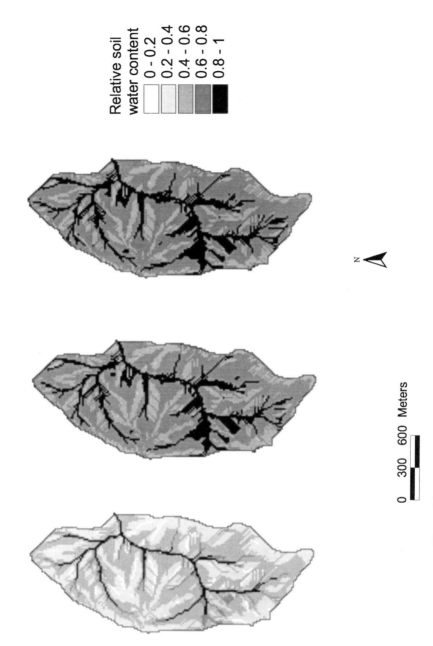

Figure 4.9. Cottonwood Creek maps of relative soil-water content derived with the three sets of estimation techniques available in WET.

TABLE 4.8   Madison Range WET Site Parameter File

Parameter	Level 1	Level 2	Level 3
Critical wetness index, χ_{cr}	7.5	7.5	7.5
Mean air temperature (°C)	—	5.6	5.6
Mean elevation (m)	—	1652	1652
Precipitation (mm/day)	—	1.52	1.52
Interception losses (mm/day)	—	0.18	0.18
Maximum drainage rate D_{max} (mm/day)	—	0.15	0.15
β	—	10	10
C	—	—	12
$(E + D)/P$ ratio	0.90	—	—

and the mean computed for the catchment (6.5) shows that the distribution was positively skewed (as is the case in most catchments) (Figure 4.10a). Figure 4.10b shows the quasi-dynamic topographic wetness index and a subtly different spatial arrangement compared to the steady-state topographic wetness index map in Figure 4.10a. The quasi-dynamic topographic wetness index values vary from 3.4 to 6.7 and the mean is 5.1. The largest quasi-dynamic index values are predicted in topographic hollows in higher elevations (i.e., in local areas with convergent flow lines) and immediately above gently sloping areas near channels (i.e., on footslopes).

The quasi-dynamic topographic wetness index was computed with the same DEM and user-specified soil depth, drainable porosity, saturated hydraulic conductivity, and drainage time values of 1.3 m, 0.4 cm³/cm³, 200 mm/h, and 20 days, respectively. The mean weighted soil depth was estimated from the spatial extent of the different soil mapping units and published soil series descriptions (Boast and Shelito 1989). The mean weighted drainable porosity was estimated from soil texture data reported by Boast and Shelito (1989) (weighted by spatial extent and depth) and some published data specifying typical drainable porosity values by soil texture class (Ratliff et al. 1983). The mean weighted saturated hydraulic conductivity was estimated in a similar fashion with the help of published data reporting typical values of this parameter by soil texture class (Rawls and Brakensiek 1989) and multiplying this estimate by a factor of 3 to account for rapid subsurface pathways. The drainage time represents the spacing between major precipitation and/or snowmelt events and along with saturated hydraulic conductivity and effective porosity controls the size of the effective upslope contributing area that is calculated for each cell in DYNWET-G. The parameters used in this instance predicted maximum travel times (distances) of 24 and 158 m on 10 and 66% slopes, respectively. The use of local site parameters in place of the generalized estimates used here is likely to alter (improve?) the quasi-dynamic topographic wetness index calculated with DYNWET-G. These data are also required to evaluate how well the steady-state and quasi-dynamic topographic

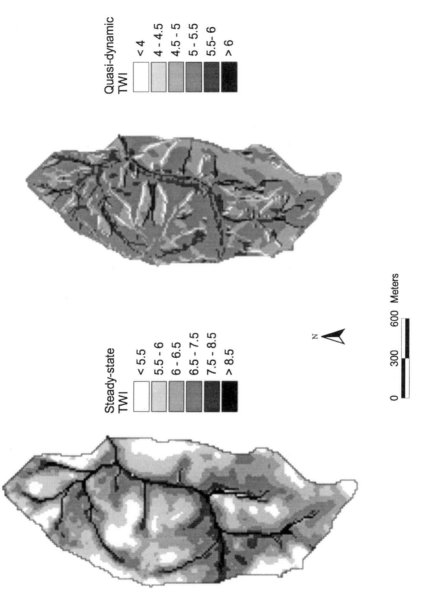

Figure 4.10. Cottonwood Creek maps showing (a) steady-state topographic wetness index and (b) quasi-dynamic topographic wetness index.

Quasi-dynamic TWI

- < 4
- 4 - 4.5
- 4.5 - 5
- 5 - 5.5
- 5.5 - 6
- > 6

Steady-state TWI

- < 5.5
- 5.5 - 6
- 6 - 6.5
- 6.5 - 7.5
- 7.5 - 8.5
- > 8.5

N

0   300   600   Meters

wetness indices are able to represent topographic controls on soil-water distribution in this catchment.

4.7 CONCLUSIONS

Topographic maps are available in most countries and can be used to calculate topographic attributes and plan additional data collection networks for hydrological monitoring, soil survey, and biological survey applications. As additional environmental data become available, they can be used to provide improved estimates of the terrain-based indices. For example, the susceptibility of the landscape to sheet and rill erosion can initially be estimated using only topographic data. As hydrological and soils information becomes available, it can be integrated into the predictions; and finally, as information on vegetation and cover is developed, a physically based erosion model that accounts for detachment and transport processes can be utilized (e.g., Mitasova et al. 1996). The radiation, temperature, evapotranspiration, and soil-water processes can be modeled in similar ways. We therefore visualize different layers of data being developed over time with elevation data and the related topographic attributes constituting a minimum data set. Topographic attributes can be easily estimated from a digital elevation model using any one of a number of terrain analysis methods. Most GIS are based on a pixel or cellular structure so that grid-based methods of terrain analysis can provide the primary geographic data for them and can be easily integrated within their analysis subsystems.

The foregoing discussion of the methods incorporated in EROS, SRAD, WET, and DYNWET-G indicates why care is needed when applying these techniques. The basic approach relies on simplified representations of the underlying physical processes and it is designed to include the key factors, such as topography, that regulate the system's behavior. Factors that are not explicitly included in an index are assumed to have low variance within the landscape. With this approach, we sacrifice some physical sophistication to allow improved estimates of spatial patterns in landscapes (I. D. Moore et al. 1991, 1993f). This approach will be consistent with the level of available data and the precision with which many of the management questions need, and can, be answered in many instances (Moore and Hutchinson 1991, Barling et al. 1994). Most of the chapters that follow explore how one or more of the topographic indices calculated with these tools can be used in hydrological, geomorphological, and biological applications.

CHAPTER 5

Effect of Data Source, Grid Resolution, and Flow-Routing Method on Computed Topographic Attributes

John P. Wilson, Philip L. Repetto, and Robert D. Snyder

5.1 INTRODUCTION

Terrain methods like those incorporated in the TAPES suite of grid-based programs can be used to generate topographic inputs for spatially distributed models of environmental processes and patterns (Moore et al. 1993d, Wilson 1996, Kemp 1998a, b, Burrough and McDonnell 1998). Scientists and managers interested in these types of applications must choose their digital elevation model (DEM) data sources, grid resolutions, and flow-routing methods carefully because past studies have shown that these methods and data can have a large impact on the magnitude and spatial pattern of computed topographic attributes.

Numerous studies have examined the sensitivity of computed terrain attributes to DEM data source and grid resolution, and several have explored what resolution is needed to accurately represent the key hydrologic and geomorphic processes operating in selected landscapes (Quinn et al. 1991, 1995, Wolock and Price 1994, Moore 1996). Zhang and Montgomery (1994), for example, used the D8 single-flow-direction algorithm with 2-, 4-, 10-, 30-, and 90-m DEMs to direct flow across the landscape, TOPOG (O'Loughlin 1986) to examine patterns of surface saturation, and TOPMODEL (Beven and Kirkby 1979) to predict runoff from short-duration storms. DEM grid resolution significantly affected the frequency distributions for the slope, specific catchment area and topographic wetness index attributes, and the hydrographs predicted with TOPMODEL. They recommended using 10-m DEMs for geomorphic and hydrologic applications because the 10-m DEM performed much better than the 30- and 90-m data and only slightly worse than the 2- and 4-m DEMs. Grid sizes of 50 m or more tend to ignore the existence of lower order streams and they

Terrain Analysis: Principles and Applications, Edited by John P. Wilson and John C. Gallant.
ISBN 0-471-32188-5   © 2000 John Wiley & Sons, Inc.

artificially smooth landforms in complex landscapes so that the terrain features that modulate key hydrologic processes are lost (Dikau 1989, Quinn et al. 1991, 1995).

Numerous studies have also examined the effects of flow-routing methods on computed topographic attributes (Quinn et al. 1991, 1995, Costa-Cabral and Burges 1994, Wolock and McCabe 1995, Desmet and Govers 1996a, Moore 1996). In one such study, Desmet and Govers (1996a) compared several single-flow-direction (sfd) and multiple-flow-direction (mfd) algorithms and found that the sfd algorithms generated different spatial and statistical distributions of specific catchment area. The two sfd algorithms produced their own (unique) long, parallel flow lines with sharp boundaries between the major flow lines and surrounding area that matched small (randomly located) planform concavities as well as topographically defined thalwegs. The small local concavities may represent a problem because the original concavities could have been caused by interpolation errors and there is no possibility for the sfd algorithms to spread this flow downslope. Desmet and Govers (1996a) recommended using mfd algorithms in upland areas to minimize these problems and sfd algorithms in valley bottoms where the mfd algorithms tend to spread the contributing area back and forth between the valley bottom and adjoining slopes along the main thalwegs. These recommendations are similar to those of Quinn et al. (1995) and Moore (1996).

The aforementioned studies examined a large number of different DEM data sources, grid resolutions, and flow-routing algorithms. The current study examined two sets of issues:

- The sensitivity of three primary and two secondary topographic attributes to the choice of DEM source data, grid cell resolution, and flow-routing method for a medium-sized forested catchment in southwest Montana. The primary topographic factors were slope gradient, flow-path length, and specific catchment area. The secondary (compound) topographic factors were the Revised Universal Soil Loss Equation length–slope (LS) factor and Moore/Wilson (MW) sediment transport capacity index.
- The topographic controls on soil erosion and the ability of USGS 7.5' 30-m DEMs and the sediment transport capacity index proposed by Moore and Wilson (1992, 1994) to distinguish zones of net erosion and net deposition for a small agricultural catchment in the Palouse region of Idaho.

The focus on topographic attributes that are likely to be used in soil erosion models was guided by three factors. One is the increasing popularity of applications that combine geographic information system (GIS) and soil erosion models to help with the identification and management of erosive lands (Wilson 1996, Wilson and Lorang 1999). Another factor is the inclusion of terrain-analysis tools in GIS and their use in calculating model topographic inputs (Moore and Wilson 1992, 1994, Desmet and Govers 1996b, Wilson and Gallant 1996). The final factor is the possibility that one or more of the topographic attributes described in Chapter 4 may improve model predictions at catchment and larger scales.

5.2   SQUAW CREEK, MONTANA SENSITIVITY ANALYSIS

The 105-km^2 Squaw Creek catchment is located in Gallatin National Forest 45 km south of Bozeman, Montana (lat. 45° 28′ N, long. 111° 15′ W) (Figure 5.1). Elevations vary between 1639 and 3134 m. Most of the 75- to 125-cm/yr annual average precipitation falls as snow during winter and spring, and peak stream discharge and erosion (caused by snowmelt or rain falling on melting snow) usually occurs in May or June. The young soils are highly variable but generally well drained, with zones of clay accumulation and fine to medium textures (Davis and Shovic 1984). Vegetation includes Lodgepole pine (*Pinus contorta*), Douglas fir (*Pseudotsuga menziesii*), Subalpine fir (*Abies lasiocarpa*), Engleman spruce (*Picea engelmanii*), and various bunchgrasses in lower elevation meadows. Approximately 7.5% of the catchment has been clear-cut during the past 35 years.

5.2.1   Methods and Data Sources

5.2.1.1 DEM Data Sources and Preprocessing   The Bozeman, Montana USGS 1 by 2° 3-arc-second DEM and Garnet Mountain, Mount Blackmore, and Fridley Peak, Montana USGS 7.5′ 30-m DEMs were imported, merged, and clipped in ARC/INFO to form a rectangular study area containing the Squaw Creek catchment. Five sets of elevation data were generated from the source DEMs for the study area: (1) The published 30-m DEMs were used as is, and (2) 100- and 200-m grids were created with the TOPOGRID command from the published 30-m and 3-arc-second DEMs. The 100- and 200-m grid resolutions were chosen to match the grid sizes

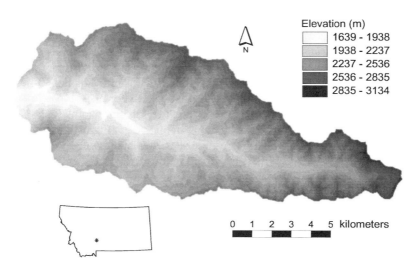

Figure 5.1. Squaw Creek, Montana study area and location maps.

used in previous GIS-based USLE applications (e.g., Hession and Shanholtz 1988, Ventura et al. 1988, James and Hewitt 1995). The TOPOGRID command in ARC/INFO implements an early version of ANUDEM (Hutchinson 1988, 1989b) and offers the advantage of automatic drainage enforcement over other interpolation methods (see Chapter 2 for additional details). The 3-arc-second lattices were converted to point coverages in ARC/INFO and projected to the same Universal Transverse Mercator (UTM) planar coordinate system used for the 30-m DEMs prior to implementing TOPOGRID. This task was added because the conversion of the source data to a planar coordinate system in the GRID module of ARC/INFO would have introduced an additional interpolation step.

5.2.1.2 *Calculation of Primary and Secondary Terrain Attributes*   Three sets of primary and secondary topographic attributes were calculated with TAPES-G, EROS, and the five sets of elevation data described above. Depressionless DEMs were created for each TAPES-G run using the method of Jenson and Domingue (1988). The outputs were transferred to ARC/INFO and clipped to the catchment boundary with the WATERSHED command for further display and analysis.

The first set of terrain analysis runs examined the effect of DEM source and grid spacing on selected topographic attributes. The finite difference, Rho8, and FRho8/Rho8 algorithms discussed in Chapter 3 were used to compute slope, flow direction, and upslope contributing area, respectively. The blue lines from the Garnet Mountain USGS 7.5′ quadrangle were digitized and superimposed on the DEM to estimate the number of cells crossed by streams. The maximum cross-grading area of 20,000 m^2 was then determined by subtracting the area of cells crossed by streams from the total catchment area and dividing this result by the area of the cells crossed by streams. The change in the sediment transport capacity index across a grid cell (ΔT_{cj}) was calculated in EROS (Equation 4.2) and used to delineate net deposition (positive values) and net erosion cells (negative values). The two erosion indices were only compared in net erosion cells consistent with the assumptions used in the USLE and RUSLE models (Wilson and Lorang 1999), and the division of the catchment into net erosion and channel/net deposition cells was based on the 30-m results in every instance. This approach provided a common base for the comparisons and is consistent with the patterns observed in the field (i.e., stream widths seldom if ever exceed 30 m in the Squaw Creek catchment as delineated in Figure 5.1).

The MW sediment transport capacity index was calculated in EROS assuming that rainfall excess overland flow is generated uniformly over the entire catchment. The RUSLE LS factor values at the cell centroids were calculated in ARC/INFO using

$$L = \left(\frac{\lambda}{22.13} \right)^m \tag{5.1}$$

$$S = 10.8 \sin \beta + 0.03 \qquad (\tan \beta < 0.09) \tag{5.2}$$

$$S = 16.8 \sin \beta - 0.5 \qquad (\tan \beta \geq 0.09) \tag{5.3}$$

where L is the slope-length factor, λ is the flow-path length (in meters) ($\lambda = 4$ m when $\lambda \leq 4$ m), $m = F/(1 + F)$, $F = (\sin \beta / 0.0896) / [3(\sin \beta)^{0.8} + 0.56]$ in erosional areas, $F = 0$ in depositional areas, and β is the slope gradient (in degrees) (McCool et al. 1987, 1989). The LS factor becomes unity for the case where the flow-path length is 22.13 m and the slope is 9%. The main difference between the MW sediment transport capacity index and RUSLE slope–length factor is the use of specific catchment area in place of flow-path length (Moore and Wilson 1992). Both the RUSLE slope length and MW sediment transport capacity index values reported in this chapter would have to be multiplied by approximately 1.6 to estimate soil-loss values for individual grid cells (Griffin et al. 1988, Moore and Wilson 1994).

The second set of terrain-analysis runs examined the impact of using the D8, Rho8, FRho8, and DEMON flow-routing algorithms to calculate the specific catchment area and MW sediment transport capacity index attributes. The original USGS 7.5′ 30-m DEMs were used for each of these runs. The D8 run used the D8 method (Section 3.1.5.2) to calculate drainage directions and upslope drainage areas. The other three runs used the Rho8 method (Section 3.1.5.3) to calculate drainage directions and the Rho8, FRho8 (Section 3.1.5.4), and DEMON (Section 3.1.5.5) algorithms to calculate upslope drainage areas. The FRho8 algorithm was used with a maximum cross-grading area of 20,000 m² as above.

The third set of terrain-analysis runs examined the impact of choosing different maximum cross-grading area (i.e., channel initiation) thresholds to disable flow divergence in channel areas when using the FRho8 flow-accumulation algorithm. Three values were arbitrarily chosen—10,000 m², 15,000 m², and 20,000 m²—and used with the USGS 7.5′ 30-m DEMs to examine the sensitivity of the specific catchment area and MW sediment transport capacity index attributes to the magnitude of this threshold.

5.2.1.3 Sensitivity Analysis

The sensitivity of selected topographic attributes to the choice of DEM source and/or grid spacing, method used to calculate upslope contributing areas, and magnitude of the maximum cross-grading area threshold was examined with a series of grids and agreement matrices prepared as follows. First, the values in the 30-m MW sediment transport capacity index grid were divided into five roughly equal area classes using 0–10, 10–20, 20–30, 30–40, and >40 as class limits. Second, the 100- and 200-m grids were resampled using the nearest-neighbor option in ARC/INFO to create 30-m grids. Third, these new RUSLE LS factor and MW sediment transport capacity index grid values were reclassified using the same class limits as were used for the 30-m MW sediment transport capacity index grids. Fourth, the percent agreement between pairs of grids was determined by subtracting the reclassified grids from each other and counting the number of zero cells in the final grids.

5.2.2 Results and Discussion

5.2.2.1 Effect of DEM Source and Grid Resolution on Selected Topographic Attributes

The computed slopes varied with both DEM source and grid

resolution in at least four ways (Table 5.1). First, steep slopes disappeared as the grid spacing was increased from 30 to 200 m for the 30-m DEM source. Second, this same trend occurred when grid spacing was increased from 100 to 200 m for the 3-arc-second DEM source. Third, the quartile values for the 100- and 200-m DEMs derived from the 3-arc-second DEM source were consistently lower than the equivalent DEMs derived from the 30-m DEM source. Fourth, the median slope gradient for the net erosion cells was 11% higher on average compared to channel and net deposition cells because many gently sloping areas at low elevations were classified as channel and net deposition cells.

The computed flow-path lengths also varied with both DEM source and grid resolution. Table 5.2 shows how the second and third quartile values increased with increasing grid spacing in most instances. This result indicates that larger grids produced fewer short flow paths, although the proportion of paths less than two cells increases as cell size is increased and this means there are more short flow paths

TABLE 5.1   Slope Gradients (%) Calculated for Five Different DEM Data Sources and Grid Resolutions

	Minimum	Quartile 1	Quartile 2	Quartile 3	Maximum
Squaw Creek catchment (105.14 km²)					
30-m cells[a]	0	27.1	41.8	56.8	215.7
100-m cells[b]	0.5	23.9	35.8	48.7	115.5
100-m cells[c]	0	19.0	30.1	42.5	121.5
200-m cells[b]	0.6	20.6	30.6	41.8	79.7
200-m cells[c]	0	18.5	27.5	37.6	76.8
Channel and net deposition cells (31.31 km²)					
30-m cells[a]	0	22.5	36.0	50.8	167.1
100-m cells[b]	0.5	22.1	33.3	46.6	111.5
100-m cells[c]	0	18.3	28.7	41.2	105.5
200-m cells[b]	2.3	20.6	29.9	40.5	77.5
200-m cells[c]	0.9	17.8	26.0	37.2	76.8
Net erosion cells (73.83 km²)					
30-m cells[a]	0	29.7	44.3	59.0	215.7
100-m cells[b]	1.1	24.8	36.8	49.5	109.1
100-m cells[c]	0	19.4	30.7	43.0	121.5
200-m cells[b]	0.6	20.6	31.0	42.3	79.7
200-m cells[c]	0	18.9	28.4	37.7	73.3

[a]Original USGS 7.5′ 30-m DEMs.
[b]Derived from published USGS 7.5′ 30-m DEMs.
[c]Derived from published USGS 1 by 2° 3-arc-second DEM.
Note. Depositional and erosional areas as derived with change in Moore/Wilson sediment transport capacity index and 30-m DEM.

TABLE 5.2   Flow-Path Lengths (m) Calculated for Five Different DEM Data Sources and Grid Resolutions

	Minimum	Quartile 1	Quartile 2	Quartile 3	Maximum
		Squaw Creek catchment (105.14 km²)			
30-m cells[a]	15	15	78	196	24,066
100-m cells[b]	50	50	159	396	22,801
100-m cells[c]	50	50	218	542	24,691
200-m cells[b]	100	100	217	672	22,951
200-m cells[c]	100	100	318	769	22,371
		Channel and net deposition cells (31.31 km²)			
30-m cells[a]	15	95	216	490	24,066
100-m cells[b]	50	139	323	726	22,701
100-m cells[c]	50	97	329	717	24,691
200-m cells[b]	100	100	369	960	22,268
200-m cells[c]	100	105	424	1032	22,371
		Net erosion cells (73.83 km²)			
30-m cells[a]	15	15	50	122	1,054
100-m cells[b]	50	50	85	284	22,801
100-m cells[c]	50	50	182	477	24,591
200-m cells[b]	100	100	122	544	22,951
200-m cells[c]	100	100	261	685	22,171

[a]Original USGS 7.5′ 30-m DEMs.
[b]Derived from published USGS 7.5′ 30 m-DEMs.
[c]Derived from published USGS 1 by 2° 3-arc-second DEM.

measured in terms of the number of cells. The minimum values were set to one-half of the grid spacing in this study to match the way in which this variable is calculated in RUSLE. The identical first quartile values reported for the net erosion cells and study area as a whole indicates that flow was initiated in at least 25% of the cells in these instances. The higher values calculated for the first three quartiles in the channel and net deposition cells show how these cells tend to occur further downslope. The median flow-path lengths in channel and net deposition cells were 2.4 times larger on average than the equivalent values in net erosion cells. The maximum value depends on the size of the drainage basin and measures the distance from the farthest point to the watershed outlet (22,150–24,700 m in Squaw Creek). Different grid resolutions will generate the same maximum flow-path length only if the same drainage paths are delineated in the basin as grid spacing is changed. The separation of channel/net deposition and net erosion cells split the cells with short and long flow paths into two groups and removed most of the long flow paths from the latter category for the 30-m DEM (Table 5.2). The failure to achieve this result for the larger grid spacings can be attributed to the use of the 30-m MW sediment transport capacity index to distinguish channel/net deposition and net erosion cells in these instances. Large flow-path-length values were assigned to as many as 44 30-m contiguous cells using this approach; however, this approach also ensured that the same areas were compared with one another in later comparisons.

The computed specific catchment area results summarized in Table 5.3 are more difficult to interpret. The minimum values are a function of grid spacing and represent the area of each cell divided by flow width. The flow width is equal to the cell width or 1.4 times the cell width (depending on the direction of flow). The quartile values reported for the whole catchment show that larger upslope contributing areas were more often than not computed at larger grid spacings since these values must be multiplied by the width of the grid cells at each resolution to calculate upslope contributing areas. The same explanation also means that the maximum values reported for the 100- and 200-m DEMs in the entire catchment represent approximately equal upslope contributing areas (as would be expected) at the outlet cell. The separation of net erosion and channel/net deposition cells split the 30-m cells into two groups with small (erosion) and large (deposition) specific catchment areas (similar to flow-path

TABLE 5.3   Specific Catchment Areas (m^2/m) Calculated for Five Different DEM Data Sources and Grid Resolutions

	Minimum	Quartile 1	Quartile 2	Quartile 3	Maximum
Squaw Creek catchment (105.14 km^2)					
30-m cells[a]	21	62	122	238	3,564,870
100-m cells[b]	71	73	114	413	1,053,300
100-m cells[c]	71	74	154	581	1,054,000
200-m cells[b]	141	155	280	819	527,600
200-m cells[c]	141	163	347	991	527,600
Channel and net deposition cells (31.31 km^2)					
30-m cells[a]	21	150	273	1090	3,564,870
100-m cells[b]	71	99	296	1365	1,052,700
100-m cells[c]	71	85	281	919	1,054,000
200-m cells[b]	141	179	412	1636	512,000
200-m cells[c]	141	204	462	1426	527,600
Net erosion cells (73.83 km^2)					
30-m cells[a]	21	49	89	160	660
100-m cells[b]	71	71	93	216	1,053,300
100-m cells[c]	71	71	125	496	1,053,000
200-m cells[b]	141	149	216	600	527,600
200-m cells[c]	141	152	298	757	525,400

[a]Original USGS 7.5′ 30-m DEMs.
[b]Derived from published USGS 7.5′ 30-m DEMs.
[c]Derived from published USGS 1 by 2° 3-arc-second DEM.

length). The steadily increasing specific catchment area quartile values calculated for the net erosion cells indicates that the sensitivity of the computed values to DEM source and/or grid resolution was 2–3 times higher in these areas.

The last result is important because GIS-based applications of RUSLE should be restricted to these (upland) net erosion areas (Wilson and Lorang 1999). The cumulative frequency distributions for slope gradient, flow-path length, and specific catchment area in these cells reproduced in Figure 5.2 show how coarser source data and larger grid spacings tend to have a "smoothing" effect on computed topographic surfaces. Short, steep slopes and small topographic features tend to disappear with increasing grid size. Panuska et al. (1991), Quinn et al. (1991, 1995), Wolock and Price (1994), Zhang and Montgomery (1994), and Moore (1996) have reported similar results for a variety of landscapes. This result also has important consequences for the RUSLE LS factor and MW stream power index values as discussed below.

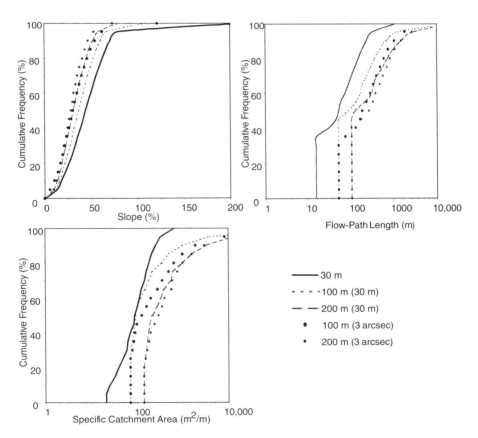

Figure 5.2. Effect of DEM source and grid spacing on (a) slope gradient, (b) flow-path length, and (c) specific catchment area cumulative frequency distributions in net erosion areas.

Coarse grid resolutions tended to produce fewer cells with low RUSLE LS factor and MW sediment transport capacity index values (Table 5.4). The mean values calculated for both variables increased with increasing grid size, and they varied less dramatically with DEM source. The classes reported in Table 5.4 match those used in Figures 5.4–5.8. The mean values in Table 5.4 show that the computed MW sediment transport capacity index values were 30–70% higher than the equivalent RUSLE LS factor values on average and slightly less sensitive to resolution than LS. The paucity of dots on or below the 1:1 line in Figure 5.3 shows that there were few planar hillslopes in this catchment with slope gradients and slope lengths where the two equations produce similar values (Moore and Wilson 1992, 1994).

The spatial agreement between pairs of RUSLE LS factor and MW stream power index grids derived from DEMs with different resolutions was poor when the grids were reclassified into five classes representing low, moderately low, medium, moderately high, and high values (Table 5.5). The RUSLE results indicate that less than 50% of the 82,033 net erosion cells identified with the USGS 7.5′ 30-m DEMs were classified consistently in each of the 10 pairwise comparisons. The best result (39.6% agreement) was obtained with the 200-m grids derived from the USGS 7.5′ 30-m and 1 by 2° 3-arc-second source data. The first column in the top half of Table 5.5 shows that the overall agreement between the 30-m data and the other resolutions was generally low (<40%). These results also show that the level of agreement declined con-

TABLE 5.4   RUSLE LS Factor and Moore/Wilson Sediment Transport Capacity Index Values Calculated for Five Different DEM Data Sources and/or Grid Resolutions in Net Erosion Areas

	Mean	Percentage of 30-m Cells With Values in Ranges Indicated				
		0–10.0	10.1–20.0	20.1–30.0	30.1–40.0	>40.0
RUSLE LS factor						
30-m cells[a]	15.6	47.3	24.4	13.5	7.4	7.4
100-m cells[b]	24.8	31.5	25.2	12.4	11.2	19.7
100-m cells[c]	26.1	33.3	20.9	13.2	10.3	22.3
200-m cells[b]	31.0	22.0	27.0	13.3	8.9	28.7
200-m cells[c]	29.9	22.8	25.8	11.7	11.4	28.3
Moore/Wilson sediment transport capacity index						
30-m cells[a]	26.3	18.8	26.4	22.7	14.6	17.5
100-m cells[b]	38.1	12.2	19.1	20.2	15.6	32.9
100-m cells[c]	35.9	16.7	18.8	17.5	14.2	32.8
200-m cells[b]	41.4	9.7	15.8	18.3	17.1	39.2
200-m cells[c]	38.7	9.9	16.9	18.2	17.5	37.5

[a]Original USGS 7.5′ 30-m DEMs.
[b]Derived from published USGS 7.5′ 30-m DEMs.
[c]Derived from published USGS 1 by 2° 3-arc-second DEM.
Note. Net erosion areas (cells) as delineated on USGS 7.5′ 30-m DEMs.

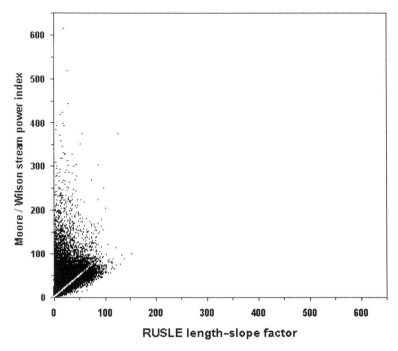

Figure 5.3. Scatterplot of RUSLE LS factor versus Moore/Wilson sediment transport capacity index values at 30-m DEM resolution in net erosion areas.

sistently as the resolution was increased from 30 to 200 m. Similar trends characterize the MW sediment transport capacity index results, although the level of agreement between pairs of maps was about seven percentage points higher on average. Overall, these results show that the different DEM sources and grid resolutions produced RUSLE LS factor and MW stream-power index values with different spatial patterns and statistical properties.

Figure 5.4 shows the spatial pattern of the MW stream-power index calculated from the USGS 7.5′ 30-m DEM using the same classes used in Table 5.4. The broad pattern of high values at higher elevations and low values on footslopes and in channels is obvious but the scale of this map obscures many other important details. The five maps reproduced in Figure 5.5 show the spatial pattern of this index for the square box delineated in Figure 5.4. These maps show four easily identifiable channels flowing from the south into the main channel that traverses the northern section of each of these maps. The three maps on the left show how the relaxation of the resolution from 30 to 100 m and then 200 m assigned larger hillslope areas to the highest MW sediment transport capacity index class. This effect was particularly pronounced for the three westernmost tributaries in the 200-m DEM. These same general patterns are visible in the 100- and 200-m DEMs produced from the 3-arc-second source, although the differences due to the different data sources are small.

TABLE 5.5 Overall Agreement (%) Between RUSLE LS Factor and Moore/Wilson Sediment Transport Capacity Index Grids Based on Five Different DEM Sources and Resolutions in Net Erosion Areas

DEM Data Source and Grid Spacing	30-m Cells[a]	100-m Cells[b]	100-m Cells[c]	200-m Cells[b]	200-m Cells[c]
	RUSLE LS factor				
30-m cells[a]	X				
100-m cells[b]	37.2	X			
100-m cells[c]	31.1	34.9	X		
200-m cells[b]	28.6	36.8	32.9	X	
200-m cells[c]	27.9	35.4	39.2	39.6	X
	Moore/Wilson sediment transport capacity index				
30-m cells[a]	X				
100-m cells[b]	44.2	X			
100-m cells[c]	33.3	43.8	X		
200-m cells[b]	31.5	47.1	41.8	X	
200-m cells[c]	30.6	43.8	48.7	49.1	X

[a]Same as published USGS 7.5′ 30-m DEMs.
[b]Derived from published USGS 7.5′ 30-m DEM.
[c]Derived from published USGS 1 by 2° 3-arc-second DEM.
Note. Net erosion areas (cells) as delineated on USGS 7.5′ 30-m DEMs.

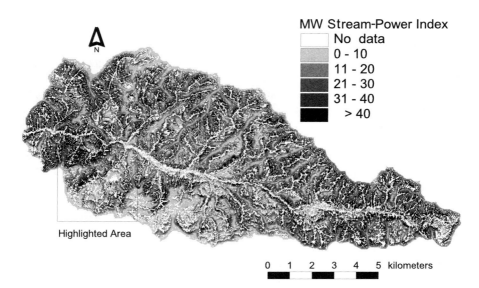

Figure 5.4. Map of Squaw Creek catchment showing Moore/Wilson sediment transport capacity index values after masking out channel and net depositional areas (no data cells).

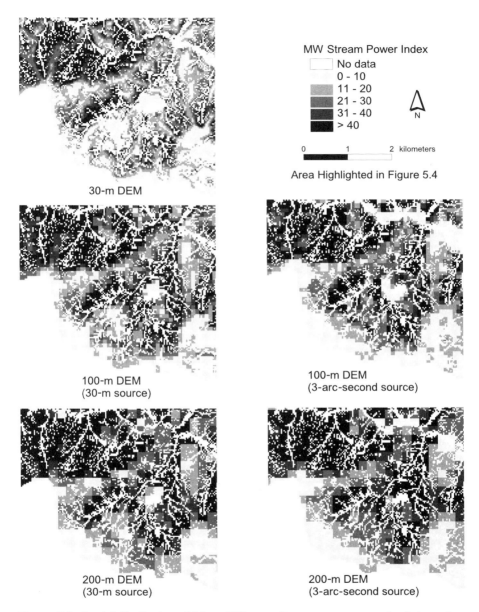

Figure 5.5. Spatial distribution of Moore/Wilson sediment transport capacity index values calculated with five different DEM grid resolutions for catchment area delineated in inset box in Figure 5.4.

Overall, the five maps examined in sequence illustrate the progressive loss of information (topographic detail) as grid size increases similar to Quinn et al. (1991), Zhang and Montgomery (1994), and Moore (1996). These maps also show how depositional areas are predicted immediately below the areas of highest erosion because transport capacity is decreasing in these areas.

Figure 5.6 shows the spatial pattern of differences between the RUSLE LS factor and MW sediment transport capacity index values alongside 100-m contours for the same area as shown in Figure 5.5. The differences between these attributes are expressed as percentages of the predicted RUSLE LS factor values on a cell by cell basis in this map. The largest differences (as expected given the ways in which the two attributes were calculated) occurred in areas experiencing strong convergence (i.e., slopes with concave plan curvatures at lower elevations) and divergence (i.e., slopes with convex plan curvatures at moderate to high elevations). This result indicates how the magnitude of the differences between the two attributes is likely to vary across landscapes.

5.2.2.2 Effect of Flow-Routing Algorithm on Selected Topographic Attributes   Specific catchment area and MW sediment transport capacity index

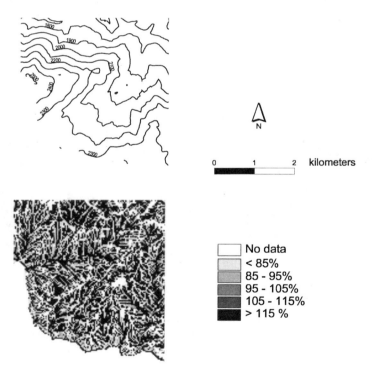

Figure 5.6. Maps showing (a) contour data and (b) spatial distribution of differences between RUSLE LS factor and Moore/Wilson sediment transport capacity for catchment area delineated in inset box in Figure 5.4.

values were also computed for the USGS 7.5′ 30-m DEMs with each of the flow-routing algorithms incorporated in TAPES-G. The top halves of Tables 5.6 and 5.7 show that the two single-flow-direction algorithms (D8 and Rho8), the FRho8 multiple-flow-direction algorithm, and DEMON generated different specific catchment area and MW sediment transport capacity index distributions. DEMON incorporates divergence and convergence by expanding or contracting the width of the stream tubes and therefore behaves like a multiple-flow-direction algorithm. The sfd algorithms initiated flow from 30–40% of the cells and generated much higher proportions of cells with small upslope contributing areas (Table 5.6). This effect translated into larger numbers of cells with low and medium MW sediment transport capacity index values and lower mean values for the sfd algorithms (Table 5.7). These differences had important implications for the spatial patterns (as shown in Figure 5.7).

The lighter gray scales occur in much smaller blocks in the first two maps in Figure 5.7 because the sfd algorithms tend to produce parallel flow paths and therefore take longer to accumulate large upslope contributing areas (i.e., flow). DEMON and the FRho8 mfd algorithm are more sensitive to the subtle convergence and divergence in the higher parts of the landscape. The decrease in banding as mfd algorithms are substituted in place of sfd algorithms noted in other studies (e.g., Desmet and Govers 1996a, Moore 1996) is evident when the four maps in Figure 5.7 are examined in sequence. Table 5.8 summarizes the level of agreement when the MW stream-power index values were reclassified into the five classes previously described, and shows (as would be expected given the previous results) that the two single and two multiple-flow-direction algorithms generated the highest levels of agreement (57–71%).

TABLE 5.6   Specific Catchment Area (m²/m) Attribute Values Calculated With (a) Four Flow-Routing Algorithms and (b) the FRho8 Flow-Routing Algorithm and Three Maximum Cross-Grading Area Thresholds in Net Erosion Areas

		Percentage of 30-m Cells With Values in Ranges Indicated				
	Median	<40	40–70	70–110	110–180	>180
Flow-routing algorithms						
D8	57.2	30.9	27.6	13.3	13.7	14.6
Rho8	41.1	39.3	28.2	11.8	10.6	10.2
FRho8	88.9	13.0	24.9	18.7	23.1	20.3
DEMON	105.4	11.7	22.1	20.1	20.2	25.9
FRho8 and three maximum cross-grading area thresholds						
10,000 m²	85.3	14.9	27.5	21.0	24.5	12.1
15,000 m²	87.4	14.3	25.6	19.2	23.3	17.7
20,000 m²	88.9	13.0	24.9	18.7	23.1	20.3

Note. Net erosion areas (cells) as delineated on USGS 7.5′ 30-m DEMs.

TABLE 5.7   Moore/Wilson Sediment Transport Capacity Index Values Calculated With (a) Four Flow-Routing Algorithms and (b) the FRho8 Flow-Routing Algorithm and Three Maximum Cross-Grading Area Thresholds in Net Erosion Areas

		Percentage of 30-m Cells With Values in Ranges Indicated				
	Mean	0–10.0	10.1–20.0	20.1–30.0	30.1–40.0	>40
Flow-routing algorithms						
D8	19.4	33.7	32.5	17.3	8.0	8.4
Rho8	16.1	43.6	32.4	13.2	5.3	5.6
FRho8	20.3	28.1	31.2	21.6	11.3	7.9
DEMON	21.9	25.9	30.4	21.8	11.6	10.4
FRho8 and three maximum cross-grading area threshold values						
10,000 m^2	19.2	30.0	32.6	21.7	10.0	5.7
15,000 m^2	19.9	28.7	31.4	21.6	10.9	7.3
20,000 m^2	20.3	28.1	31.2	21.6	11.3	7.9

Note. Net erosion areas (cells) as delineated on USGS 7.5′ 30-m DEMs.

5.2.2.3 Effect of Channel Initiation Threshold on Selected Topographic Attributes

Three maximum cross-grading area values were arbitrarily selected and used with the FRho8 flow-routing algorithm and USGS 7.5′ 30-m DEMs in the final set of terrain-analysis runs to evaluate the effect of this threshold on computed MW sediment transport capacity index values. Holmgren (1994) and Quinn et al. (1995) have argued that the choice of channel initiation threshold (maximum cross-grading area in TAPES-G) can have a large impact on computed specific catchment areas for multiple flow-routing algorithms like FRho8.

The bottom halves of Tables 5.6 and 5.7 show how the specific catchment area and MW sediment transport capacity index values varied with the choice of maximum cross-grading area in net erosion areas delineated with EROS and the USGS 7.5′ 30-m DEMs. The mean values increased with increasing values of this threshold. The two smaller thresholds (10,000 and 15,000 m^2) meant that flow dispersion was disabled before any channel cells were encountered (since 20,000 m^2 was used to identify cells crossed by channels). This strategy tended to produce long, parallel flow paths at lower elevations in net erosion areas (Figure 5.8). Table 5.9 summarizes the level of agreement when the MW sediment transport capacity index values were reclassified into the five classes previously described. The results show that the choice of maximum cross-grading area threshold had much less impact on computed topographic attributes than either DEM source and resolution (Table 5.4) or flow-routing method (Table 5.7). These results imply that the maximum cross-grading area threshold used with the FRho8 flow-routing algorithm is less critical than the initial choice of flow-routing algorithm (at least for the range of maximum cross-grading area values tried in this catchment).

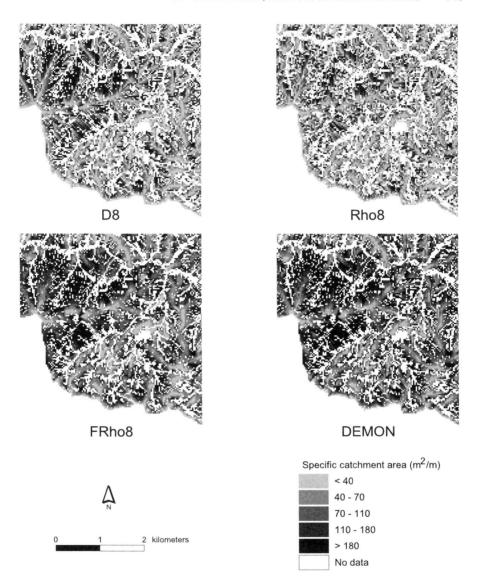

Figure 5.7. Spatial distribution of specific catchment area attribute values calculated using four different flow routing algorithms for catchment area delineated in inset box in Figure 5.4.

TABLE 5.8   Overall Agreement (%) Between Moore/Wilson Sediment Transport Capacity Index Grids Based on Four Different Flow-Routing Algorithms in Net Erosion Areas

Flow-Routing Algorithm	D8	Rho8	FRho8	DEMON
D8	X			
Rho8	56.5	X		
FRho8	55.7	50.9	X	
DEMON	54.0	49.3	70.6	X

Note. Net erosion areas (cells) as delineated on USGS 7.5′ 30-m DEMs.

Overall, these results indicate that the types of topographic attributes utilized in GIS-based soil erosion model applications are sensitive to the choice of DEM source data, grid resolution, and flow-routing method. However, no field data were available for this particular study area to validate model predictions that might incorporate these topographic attributes as inputs. In addition, the GIS-based USLE and RUSLE applications need to be able to distinguish net erosion areas from net deposition areas and cells crossed by channels prior to predicting soil loss (Wilson 1996, Wilson and Lorang 1999). Moore and Wilson (1992, 1994), Desmet and Govers (1996b), and Mitas et al. (1996) have all proposed terrain-based methods for identifying these areas, and the next experiment describes our initial attempt to

- Quantify the relationships between soil redistribution rates and selected topographic attributes.
- Evaluate the ability of one of these methods, the change in the MW sediment transport capacity index from each upslope grid cell to the next downslope cell, to distinguish zones of net erosion and net deposition in the same Idaho farm field used by Busacca et al. (1993).

5.3   IDAHO FARM FIELD MODEL VALIDATION EXPERIMENT

The 52-ha farm field used for this experiment is part of the Missouri Flat Creek catchment in the Palouse River Basin (Figure 5.9). It is located 4 km north of Moscow, Idaho (latitude 46° 46′ N, longitude 116° 59′ W). An embankment carries Idaho State Highway 95 above the field, and has blocked the outlet of this third-order catchment since cesium-137 (Cs-137) was added to the atmosphere via nuclear testing programs in the 1950s. Elevations range from 812 to 852 m and slopes range from 10 to 30%. Mean annual precipitation is 56 cm and occurs mainly as snow and low-intensity rains from November to May. The soils are fine-textured and vary in terms of source material (loess), soil drainage, and wetness (Barker 1981). The field is managed in a wheat–pea–fallow rotation. The largest erosion losses occur when rain falls on partially thawed soils. These conditions have led RUSLE developers to propose several factor values that are unique to this region (Renard et al. 1991).

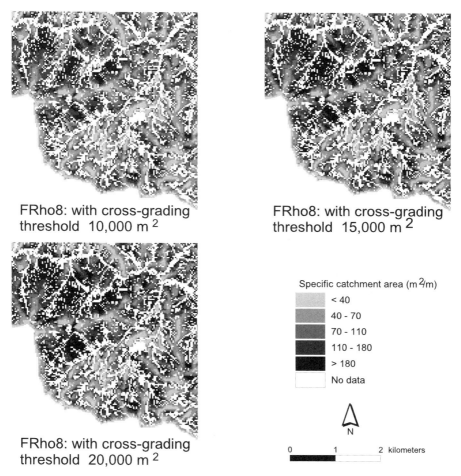

Figure 5.8. FRho8: with cross-grading threshold 10,000 m^2

FRho8: with cross-grading threshold 15,000 m^2

FRho8: with cross-grading threshold 20,000 m^2

Specific catchment area (m^2/m)

- < 40
- 40 - 70
- 70 - 110
- 110 - 180
- > 180
- No data

N

0   1   2 kilometers

Figure 5.8. Spatial distribution of specific catchment area attribute values calculated using FRho8 flow-routing algorithm and three different maximum cross-grading area thresholds for catchment area delineated in inset box in Figure 5.4.

Busacca et al. (1993) compared RUSLE predictions and soil redistribution rates estimated from Cs-137 data. The average erosion rate estimated by RUSLE (31.4 t/ha/yr) was nearly three times the average rate estimated by kriged Cs-137 data (11.6 t/ha/yr) for net erosion areas (60% of catchment; Figure 5.10a). Deposition averaged 18.6 t/ha/yr on net deposition areas (40% of catchment; Figure 5.10b). These results indicate the importance of (1) separating net erosion and net deposition areas prior to the application of RUSLE, and (2) correctly predicting factor values since the authors partially attributed the discrepancies to the simple averaging of K and LS values and their choice of "representative" transects for the LS calculations.

TABLE 5.9   Overall Agreement (%) Between Moore/Wilson Sediment Transport Capacity Index Grids Using FRho8 Flow-Routing Algorithm With Three Maximum Cross-Grading Area Thresholds in Net Erosion Areas

Thresholds	$10,000 \text{ m}^2$	$15,000 \text{ m}^2$	$20,000 \text{ m}^2$
$10,000 \text{ m}^2$	X		
$15,000 \text{ m}^2$	91.8	X	
$20,000 \text{ m}^2$	88.8	95.6	X

Note. Net erosion areas (cells) as delineated on USGS 7.5′ 30-m DEMs.

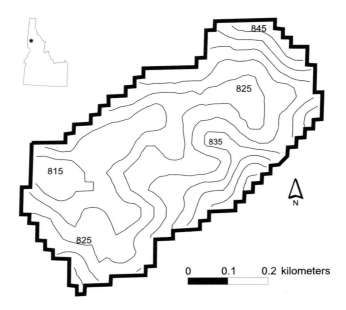

Figure 5.9. Idaho farm field study area and location maps.

5.3.1   Methods and Data Sources

We calculated several primary and secondary topographic attributes with TAPES-G and EROS and compared the results with the kriged soil erosion and deposition estimates of Busacca et al. (1993) (Figure 5.10).

5.3.1.1 Calculation of Primary and Secondary Terrain Attributes   The Robinson Lake, Idaho USGS 7.5′ 30-m DEM was imported into ARC/INFO using the DEMLATTICE command, clipped to form a rectangular study area containing the farm field, converted to an ASCII file, and used as the primary input for TAPES-G runs. A depressionless DEM was computed for each TAPES-G run using the algorithm developed by Jenson and Domingue (1988), and the finite difference, Rho8, and FRho8/Rho8 algorithms discussed in Chapter 3 were used to compute slope gradients, flow directions, and upslope contributing areas, respectively.

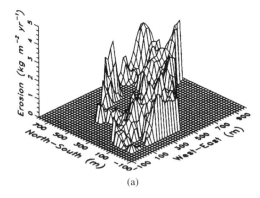

(a)

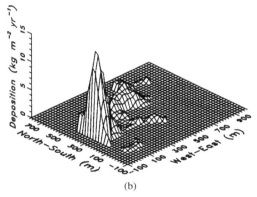

(b)

Figure 5.10. Maps showing (a) soil erosion and (b) soil deposition rates derived from kriged Cs-137 data for Idaho farm field. Reprinted with permission from Busacca, Cook, and Mulla (1993). Comparing landscape-scale estimation of soil erosion in the Palouse using C_s-137 and RUSLE. *Journal of Soil and Water Conservation* 48: 361-7. Copyright © 1993 by Soil and Water Conservation Society.

The maximum cross-grading area (i.e., channel initiation) threshold was determined by trial and error as follows. The 30-m square-grid DEM cells were superimposed on the contour map reproduced in Figure 5.9 and used to manually delineate the third-order channel system in this small, closed catchment prior to the TAPES-G runs (Figure 5.11). This analysis showed that 92 of the DEM cells (16%) contained a channel segment and that each of these cells collected drainage from 5.27 upslope cells (i.e., cells not crossed by channels) on average. A series of TAPES-G runs was performed starting with a maximum cross-grading area of 4745 m² (5.27 × 30 m × 30 m) to find the largest value of this threshold that correctly delineated the sources of each of the eight first-order streams in the ephemeral channel system. A value of 8,000 m² was chosen as the cross-grading area and used in the final TAPES-G run to switch from the FRho8 mfd algorithm to the Rho8 sfd algorithm. A visual comparison of the lines and shaded grid cells in Figure 5.11 shows that this computer-generated stream network contained two additional first-order streams in the southwest corner of the study area that were not delineated in the original hand-drawn network. These streams extended only one cell in length.

The RUSLE LS factor and two sets of MW sediment transport capacity index values were computed with ARC/INFO and EROS, respectively (similar to Squaw Creek). The elevation, slope, flow direction, flow-path length, and upslope drainage area attributes were obtained from the final TAPES-G output file. The two MW sediment transport capacity indices were calculated using the simplest option (i.e., that rainfall excess overland flow was generated uniformly across the entire catchment: μ_i = 1 in Equations 4.1 and 4.2). The elevation, slope, flow-path length, upslope contributing area, plan and profile curvature attributes generated with TAPES-G, and the MW sediment transport capacity index values generated with EROS were converted into ARC/INFO grids for further manipulation and display. Hence, the change in MW stream power index (ΔT_{cj}) grid was reclassified into a binary grid with 0s and 1s representing cells in which net erosion and net deposition were predicted.

5.3.1.2 Cs-137 Data Acquisition and Preprocessing   The Cs-137 soil redistribution data (139 sample locations) collected by Busacca et al. (1993) were obtained from the authors in two x, y, z data files. The x and y fields represented locational coordinates (measured on a north–south, west–east grid relative to an origin chosen by the authors) and the z field represented either elevation (m) or soil redistribution rate (kg/m^2/yr). These files were converted into an ARC/INFO point cover-

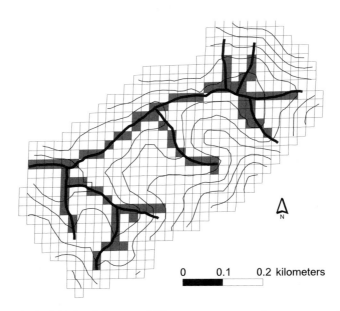

Figure 5.11. Hypothetical (a) hand-drawn and (b) computer-generated stream network maps showing 30-m DEM grid and elevation contours.

age, and the *x* and *y* coordinates were projected into UTM coordinates by calculating the UTM coordinates of the sample point nearest to the catchment outlet with differential GPS and applying the computed XY shift to the original *x, y* data.

5.3.1.3 Statistical Analysis   The POINTINGRID command in ARC/INFO was used to assign primary and secondary topographic attributes to each of the Cs-137 sampling (point) locations. This method meant that several additional fields were added to the point attribute table (PAT) to facilitate comparison of calculated terrain attributes and measured soil redistribution rates.

Three types of statistical tests were performed. First, a correlation matrix was generated to measure the strength of the relationships between pairs of primary (elevation, slope gradient, flow-path length, upslope contributing area, plan, and profile curvature) and secondary (RUSLE LS factor and MW sediment transport capacity index) topographic attributes. Second, the topographic attributes were used as independent variables and the soil redistribution rate was used as the dependent variable in simple (pairwise) regression analyses. The *t*-test statistic was calculated for each correlation coefficient (r) and used to identify terrain/soil redistribution relationships that were significant at the .05 and .10 levels of significance. The final statistical test used the phi correlation coefficient to assess whether the ΔT_{cj} variable calculated in EROS can predict the net erosion/net deposition sites determined with the Cs-137 data. The phi coefficient is used for nonparametric testing of binomial data and produces values between −1 (perfect inverse relationship) and +1 (perfect positive relationship) (Griffith and Amrhein 1991).

5.3.2   Results and Discussion

Terrain attributes were correlated against each other to quantify the relationships between these variables at the 139 Cs-137 sample locations (Table 5.10). The strongest correlation occurred between flow-path length and specific catchment area. This result indicates that the catchment contains mostly planar slopes or alternatively that the shapes (i.e., the convergence and divergence) of the complex slopes that are present in the catchment were not captured with the USGS 7.5′ 30-m DEM, which

TABLE 5.10   Correlation Matrix for Selected Primary and Secondary Topographic Attributes at Cs-137 Sample Points

Topographic Attributes	Elev	Slope	FL	SCA	Planc	Profc	LS	MW
Elevation (m)	X							
Slope gradient (%)	0.36	X						
Flow-path length (m)	−0.44	−0.21	X					
Specific catchment area (m²/m)	−0.38	−0.20	0.95	X				
Plan curvature (1/100 m)	−0.15	0.13	0.35	0.24	X			
Profile curvature (1/100 m)	−0.34	−0.09	0.47	0.43	0.44	X		
RUSLE LS factor	−0.23	0.39	0.68	0.60	0.41	0.37	X	
MW stream-power index	−0.28	0.14	0.80	0.82	0.27	0.32	0.86	X

served as source data. The RUSLE LS factor and MW sediment transport capacity index were also strongly correlated because one used flow-path length and the other incorporated specific catchment area. The stronger correlations between these terms and the flow-path/specific catchment area terms is a little surprising since the RUSLE LS factor and MW sediment transport capacity index values are usually more sensitive to changes in slope gradient (Moore and Wilson 1992, 1994). Overall, these correlation coefficients also demonstrate why it is sometimes difficult to use two or more topographic attributes in multiple regression analyses that explore topographic controls on soil redistribution patterns (Quine and Walling 1993).

Simple bivariate regression was then performed using each of the topographic attributes as the independent variables and soil redistribution as the dependent variable (Table 5.11). The positive correlation coefficients show that long flow-path lengths, large specific catchment areas, and large MW sediment transport capacity index values were correlated with depositional areas (cells) since positive and negative redistribution values represented net deposition and net erosion, respectively. Similarly, the weak negative correlations show that concave profile and plan curvatures were correlated with depositional areas since negative and positive values were used for concave and convex curvatures, respectively. Figure 5.12 shows the predicted upland net erosion, upland net deposition, and channel cells. The general patterns evident in this map are consistent with the bivariate regression results (Table 5.11) in two respects. First, net deposition was predicted in cells immediately upslope from "channel" cells and on footslopes adjacent to "channel" cells in valley bottoms. Second, net erosion (soil loss) was predicted on higher slopes.

The Cs-137 sample sites were assigned to "channel" cells (i.e., cells with upslope contributing areas > 8000 m^2) and either net erosion or net deposition "upland" cells (based on ΔT_{cj} values) for the last part of the analysis. Table 5.12 compares the EROS erosion/deposition predictions and Cs-137 based estimates. The low success rates reported for deposition cells in upland areas and the study area as a whole indicate that the MW sediment transport capacity index had problems identifying the correct result. Overall, the MW index identified net erosion and deposition correctly at only 67 of 120 and 81 of 139 Cs-137 sampling sites in these instances, yielding phi coefficients of 0.01 and 0.11, respectively, and indicating that these results could have

TABLE 5.11   Statistically Significant Regression Coefficients Between Soil Redistribution and Topographic Attributes at the .05 Significance Level

Topographic Attributes	Soil Redistribution
Elevation (m)	
Slope gradient (%)	
Flow-path length (m)	0.187
Specific catchment area (m^2/m)	0.203
Plan curvature (1/100 m)	(−0.151)
Profile curvature (1/100 m)	(−0.166)
M/W stream-power index	0.168

Note. Results in parentheses are significant at the .10 level.

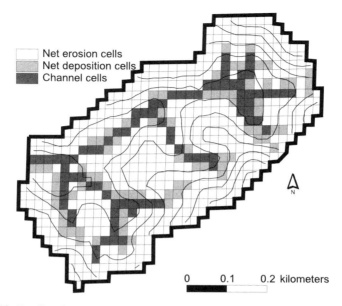

Figure 5.12. Predicted net erosion, net deposition, and channel areas in Idaho farm field.

been easily achieved by chance. The MW sediment transport capacity index predicted net erosion in 77% of the "upland" cells that were identified by the Cs-137
data as depositional. The index performed better in "channel" cells by predicting the
correct status in 14 of 19 cells and yielding a phi coefficient of 0.47. The poor performance of the MW stream-power index may be attributed to several factors. One
likely factor is the coarse resolution of the source DEM. The magnitude of the problem may be increased by the differences between the model validation and model
process representation scales. The failure to incorporate other factors, such as soil
detachment, in the model may have caused additional problems. Finally, the choice
of flow-routing algorithm for calculating the change in the MW stream-power index
across hydrologically connected cells may cause problems in some applications as
well. The likely importance of each of these potential problems is discussed in turn.

The topographic attributes were computed from a standard USGS 7.5′ 30-m DEM
in this experiment. We have already suggested that the shapes of the complex slopes
present in this catchment were probably not captured in this instance. The problem
may be worse than it first appears because the slope gradients and aspects calculated
with these types of terrain analysis techniques are most commonly associated with
cell sizes approximately two times the initial DEM grid resolution (Srinivasan and
Engel 1991, Hodgson 1995, Wilson and Lorang 1999). Hence, Zhang and Montgomery (1994) suggested using 10-m DEMs for this type of application for the reasons noted earlier. Similarly, Quinn et al. (1995) were unable to validate
TOPMODEL predictions against observed point data using a 50-m DEM and sug-

TABLE 5.12   Contingency Table Showing Predicted and Observed Net Erosion/Deposition on Naylor Farm, Idaho

Published Estimates	EROS Model Predictions		Percent Correct	Phi Coefficient
	Net Erosion	Net Deposition		
All Cs-137 sites ($n = 139$)				
Net erosion	62	18	78	
Net deposition	40	19	32	
Overall			58	0.11
Cs-137 sites in upland cells ($n = 120$)				
Net erosion	56	16	78	
Net deposition	37	11	23	
Overall			56	0.01
Cs-137 sites in channel cells ($n = 19$)				
Net erosion	6	2	75	
Net deposition	3	8	73	
Overall			74	0.47

gested that grid sizes of 10 m or less were necessary for this task. Finally, Mitasova et al. (1996) produced similar results to this study when they implemented both versions of the MW sediment transport capacity index in the GRASS GIS and used them with UGSG 7.5′ 30-m DEMs near Yakima, Washington. These results suggest that finer scale DEMs with resolutions of 5 m or less may be required to predict erosion and deposition in agricultural landscapes.

The presence of errors in published DEMs represents another potential problem. Carter (1988) described the causes of errors in DEMs compiled by different methods and numerous researchers have proposed methods to detect, display, and occasionally correct these errors (e.g., Lee et al. 1992, Brown and Bara 1994, Felicisimo 1994, Hunter and Goodchild 1995). Srinivasan and Engel (1991) compared four methods commonly used to calculate slopes from GIS elevation layers with site observations and found that the finite difference method used in this study tended to greatly overestimate slopes in flat and steep study areas. The poor quality of the digital elevation data (derived by the authors from aerial photography) and/or the field and map measurements must have caused this result, since the quad-sheet and field values were calculated with the same rules used in the digital approach. Most of these types of studies have compared computed topographic attributes to reference values. Florinsky (1998) has criticized this approach because the "reference" data may incorporate errors as well, and proposed calculating root mean square errors (RMSEs) for a series of local topographic attributes as a way of characterizing the magnitude and pattern of errors in different parts of the landscape. Burrough et al. (2000a) generated multiple realizations of a DEM and calculated fuzzy memberships to delineate meaningful, spatially coherent landscape classes from gridded DEMs. Either of these methods might have been employed in the current study to avoid some of the pitfalls encountered in using a single DEM for the calculation of topographic attributes.

However, the magnitude of the problems connected with the source DEM may have been compounded by the differences between the DEM (model process representation) and Cs-137 measurement (model validation) scales used in this particular study (Burrough 1996; Wilson 1999a). The soil samples used for the Cs-137 counts were collected from soil pedons that were less than 1 m^2 in spatial extent whereas topographic attributes were computed from a 30-m DEM. These differences almost certainly mean that fine-scale topographic differences expressed with the Cs-137 data were not captured by the DEM. These problems may be avoided in three ways. One solution requires the use of high-resolution DEMs to calculate terrain attributes. Another solution requires the use of multiple DEMs to circumvent the sensitivity of computed topographic attributes to errors in source (DEM) data. The final solution involves the design and implementation of a sampling scheme that will give reliable Cs-137 data for 30-m cells or some other predetermined grid spacing (Burrough 1996, Wilson 1999b).

Another explanation for the failure of EROS to predict net erosion/net deposition areas could be that the model in its present form excludes one or more key variables. The inclusion of soil and land cover variables might improve model performance. Quine and Walling (1993) demonstrated that soil texture was an important factor in soil redistribution and assumptions of spatially constant soil properties should be avoided. Mitas et al. (1996) incorporated the effects of soil and land cover using Manning's n, a detachment rate coefficient (K_d), and a sediment transport coefficient (K_t). They argued that a first-order reaction coefficient, C, which is related to detachment and transport capacity, is the principal parameter that controls where sediments are eroded and deposited. Mitas et al. (1996) performed a series of simulations to illustrate the influence of selected soil and cover parameters. In one such simulation, the transport capacity coefficient, K_t, was held constant and the detachment capacity coefficient, K_d, was increased so that C was increased from 0.0005 to 100. Deposition occurred relatively high up the slope for values of $C > 1$ because the sediment flux reached the sediment transport capacity. This is the transport-limited erosion/deposition case modeled by Moore and Wilson (1992) and Mitasova et al. (1996) and it is typical of sandy soils with relatively high fall velocities. In contrast, nearly all of the detached sediment will be carried to the stream for small values of C, which are likely to occur in clay soils with very low fall velocities and low detachment capacity. The distribution of erosion and sedimentation varied across the landscape in these instances without changing the magnitude of the sediment flux in the stream.

Mitas et al. (1996) also examined changes in soil properties and land cover by varying the transport capacity coefficient, K_t, and holding the detachment coefficient, K_d, constant. The magnitude of sediment delivered to streams and the spatial distribution of erosion/deposition were changed in this simulation. Hence, for $K_t \ll K_d$ and $C \gg 1$ most of the detached sediment was deposited before it reached the streams, and for $K_t \gg K_d$ and $C \ll 1$ there was minor deposition and most of the sediment was delivered to the stream network. The results presented here match the transport-limited erosion case modeled with $C > 1$ that was incorporated in the first simulation. Both the local soil texture and the fact that erosion was often predicted in net deposition areas (Table 5.12) suggest that a value of C just less than one may be more appropriate for this type of landscape. This interpretation also matches that of Foster (1994), who argued that soil loss rates were detachment-limited in most agricultural landscapes in North America. However, site-specific applications of this expanded model will require values for Manning's n, K_d, and K_t that are still under development for the large number and variety of soil and land cover conditions that occur in agricultural landscapes. In addition, this approach almost certainly requires more accurate DEMs than those used for the current study.

Last but not least, the choice of flow direction algorithm may cause problems. Moore (1996) demonstrated that multiple-flow-direction algorithms will generate more net deposition cells than single-flow-direction algorithms because they attempt to model flow dispersion. This is probably not a problem here since the FRho8 mfd algorithm was used and most of the discrepancies involved depositional areas in which net erosion was predicted (Table 5.12). However, the Squaw Creek results show that the choice of flow-routing algorithm can have a substantial effect on the specific catchment area and MW stream-power index estimates. These effects may offset any advantages gained from varying the ratio of the detachment rate and sediment transport capacity coefficients as explained above.

5.4 CONCLUSIONS

The two experiments combined the DEM sources and grid resolutions typically used in GIS-based implementations of the USLE and RUSLE models with the terrain-analysis tools available in the TAPES suite of grid-based programs. The Squaw Creek results confirmed those reported in a series of recent studies: namely, that slope gradient tends to decrease and flow-path length and specific catchment area tend to increase as grid spacing is gradually increased (holding DEM data source constant). These trends were replicated when the 3-arc-second DEMs were used in place of 30-m DEMs as source data (holding DEM grid resolution constant). The delineation of channel cells and net deposition and erosion cells in upland areas divided the study area into two data sets. The net erosion cells occurred at higher elevations and exhibited shorter and steeper slopes on average. In addition, flow was initiated in at least 25% of the cells classified as net erosion cells for all five DEM sources and grid resolutions. Taken as a whole, the results from this and a series of other recent studies suggest that these attributes vary in consistent ways across a large number and variety of landscapes.

The RUSLE LS factor and MW sediment transport capacity index values also increased with increasing grid size. These results extended our knowledge of scale effects in at least two ways. First, they showed what happened to each of these variables when the DEM data source, grid resolution, and flow-routing algorithm were varied. The spatial agreement for RUSLE and MW sediment transport capacity index grids based on different source data was poor (<50%) and different statistical and spatial distributions of each variable were generated when different sources and grid resolutions were used. Second, the two variables produced fundamentally different outputs (cf. Figures 5.3 and 5.6) that are likely to cause additional problems when one variable is substituted for the other in empirical equations (models) like the USLE and RUSLE. Mitasova et al. (1996, 1997), Desmet and Govers (1997), and Wilson and Lorang (1999) offer a more complete discussion of this issue. The single and multiple flow direction algorithms also produced different specific catchment area and MW sediment transport capacity distributions. The mfd algorithms generated smaller numbers of cells with low upslope contributing areas and less banding (Figure 5.7). The choice of the maximum cross-grading area (channel initiation) threshold also affected the performance of the FRho8 mfd algorithm. This threshold must be chosen carefully for this algorithm to avoid the disadvantages of the sfd algorithms.

The Idaho farm field results confirmed the difficulty of using published DEMs for hydrological and geomorphological applications noted by Srinivasan and Engel (1991), Zhang and Montgomery (1994), and Mitasova et al. (1996). The change in the Moore/Wilson sediment transport capacity index across hydrologically connected grid cells could not reliably identify net erosion and deposition cells when used with a USGS 7.5′ 30-m DEM. The coarse resolution of the source DEM, differences between the model process representation (DEM) and model validation (Cs-137 data) scales, and the failure to include other factors, such as soil detachment, were invoked as possible explanations. Fortunately, new opportunities are now available to construct high-resolution DEMs (e.g., Mitasova et al. 1996, Wilson et al. 1998). Additional work is still needed to test and validate the suitability of topographic attributes computed from these sources for representing key environmental processes and patterns (e.g., Mitas et al. 1996, Vieux et al. 1996, Florinsky 1998).

The results of the two experiments reported above contribute to this knowledge in that they highlight the sensitivity of selected primary and secondary topographic attributes to the choice of elevation data source, grid resolution, and flow-routing method. They suggest that these aspects must be handled carefully for digital terrain analysis to succeed with the task at hand. This theme is demonstrated numerous times by the chapters in the remainder of the book that look at hydrological, geomorphological, and ecological applications of terrain analysis in more detail.

CHAPTER 6

Spatial Analysis of Soil-Moisture Deficit and Potential Soil Loss in the Elbe Drainage Basin

Valentina Krysanova, Dirk-Ingmar Müller-Wohlfeil, Wolfgang Cramer, and Alfred Becker

6.1 INTRODUCTION

6.1.1 Background

Human-induced global climate and land-use changes are likely to have direct effects on water and biogeochemical cycles, vegetation structure, and plant productivity. However, the spatial and temporal patterns of direct human impacts may be different in different regions, as are the patterns of temperature and precipitation change predicted by climate models. Several groups have been working on different aspects of regional vulnerability. Some of them are devoted to integrated climate change impact studies in selected regions, like Cohen (1994) in the Mackenzie river basin in Canada and Rosenberg (1993) in Missouri, Iowa, Nebraska, and Kansas—the MINK study— in the United States. Others are focused on certain specific aspects of vulnerability, like Kulshreshtha (1993) on global and regional water resources vulnerability, Tim et al. (1992) on critical non-point-source pollution areas in a small catchment, and Wendland et al. (1993) on groundwater pollution by nitrates in Germany.

The development of tools to predict changes in terrestrial ecosystems is needed to improve the capability of predicting changes for regional situations and in large river basins, because the control measures may be regionally specific. Hydrological assessments in large river basins and development of corresponding tools are also necessary to provide the input and feedback mechanisms to global models of climate and biogeochemistry, where the description of hydrological processes is still weak.

Terrain Analysis: Principles and Applications, Edited by John P. Wilson and John C. Gallant.
ISBN 0-471-32188-5 © 2000 John Wiley & Sons, Inc.

In general, two opposite approaches are used in large-scale models for representation of hydrological processes:

- The bucket-type models, focused on vertical water fluxes (Dickinson 1984, Cramer and Prentice 1988, Henderson-Sellers 1990)
- The river runoff models, focused on the reproduction of runoff hydrographs (see the review in Kuchment 1992)

The first approach is used for representation of the terrestrial hydrological cycle in general circulation models (GCMs) and global biosphere models. Most attention is paid to the description of the vertical exchange of moisture and heat between the land surface and the atmosphere (evapotranspiration), and the lateral flow component is just a residual (precipitation minus evapotranspiration) or not considered at all. A detailed description of land surface processes is included in the so-called "big leaf" models, which use up to about 50 parameters. According to Henderson-Sellers (1990), these models allow the observed fields of evapotranspiration and vertical heat fluxes to be reproduced quite well, but the monthly runoff fields are still very poor. Kuchment (1992) noted that runoff formation evolves in time much more rapidly than evapotranspiration and argued that the absence of a procedure for determining effective precipitation (which generates runoff) may therefore produce errors in the calculation of evapotranspiration in such models.

In the second approach, the main aim is to reproduce river runoff in accordance with observations, and the other components (evapotranspiration and soil moisture) are highly simplified through so-called "effective parameters." The observed runoff can usually be reproduced successfully in such models, although the other fluxes may differ significantly from observed values, by as much as an order of magnitude (Kuchment 1992).

These parametrizations of the hydrological cycle are not adequate to represent the hydrological cycle completely, because they overemphasize some components and practically ignore others. New approaches in this field are urgently needed. In our opinion, the problem of better representation of the hydrological cycle in global climate and biosphere models could be solved if a compromise solution were found in which all the components of water balance are included in a more comprehensive manner. For example, the precipitation–runoff models that deal with spatially distributed input data are principally able to provide output information on water availability at specific locations. One successful attempt to couple water balance and river transport models for the Amazon with a monthly time step was made by Vörösmarty et al. (1989).

None of these approaches normally considers topographic conditions, although they are among the major forcing functions for runoff generation. To include topography as a forcing function for water runoff is attractive, although there are certain difficulties, because the only available topography-based models are largely untested for large basins and/or coarse resolution digital elevation models (DEMs). Our work is a step in this direction: It is an attempt to check whether a simplified topography-based approach can reveal important hydrologic and soil characteristics and their distribution in a river basin. The main focus of the study was on freshwater availability and potential soil erosion in the Elbe drainage basin. Clearly, these two problems are linked to each other. On the one hand, soil erosion influences surface-water quality (because some pollutants are sediment-bound), and, thus, influences the hydrological cycle and water availability in a region. On the other hand, potential erosion is driven by the hydrological cycle (the rainfall factor is one of the most important factors defining potential erosion). Hence, there is a certain interest in studying both processes in the same area and comparing, if possible, the patterns of vulnerability classes.

6.1.2 Study Objectives and Approach

The main objective of the study was to analyze spatial patterns of soil-moisture deficit and potential soil erosion in the Elbe drainage basin and to identify subareas that potentially differ in their sensitivity to global change impacts with respect to freshwater availability and potential erosion. The second objective is to check whether the simplified models, which include only a few key factors determining system behavior, can be efficient preassessment tools for the analysis of the hydrological cycle and water quality in a watershed.

Sensitive areas for freshwater availability can be defined in a number of ways. For example, they can be characterized as areas that have some of the following features in common: low amounts of precipitation, soils with low field capacities and a deep groundwater table, and exposure to high potential evapotranspiration. Another possibility is to quantify river runoff in the lower reach of a drainage basin as a measure of freshwater availability, which would be an appropriate approach in water resource investigations. In general, the dynamics of soil moisture reflect the overall water balance and can be considered as the most important variable defining freshwater availability for vegetation. We accepted the last definition for our study as the most appropriate, though it is clear that the interactions between vegetation and soil moisture can be more complicated (for example, vegetation species in arid and semiarid areas can be well adapted to dry conditions).

Sensitive areas of erosion and non-point-source (NPS) pollution can be defined as those land areas or patches that disproportionally contribute to the erosion and NPS pollution problem. Here, our analysis is restricted to erosion only, despite the existence of simplified methods to estimate potential nutrient pollution (Tim et al. 1992, Krysanova et al. 1995). The reason is that, for this study, the analysis of nitrogen and phosphorus pollution could not be made independently. It should be based on a more detailed estimation of the components of the hydrological cycle in a region (surface and subsurface flows, infiltration). The application of these methods in the Elbe drainage basin, however, requires detailed spatially distributed input data, which are currently not available.

An integrated approach coupling simplified hydrological and erosion models with a Geographic Information System (GIS) was used to delineate vulnerable subregions in terms of water cycling and erosion in the German part of the Elbe drainage basin. The topography-based WET model developed at the Australian National University

(Moore et al. 1993e, Chapter 4) was applied for the Elbe drainage basin (using a DEM with 1-km resolution) to estimate a topographic wetness index for freshwater availability. The simplified USLE approach (Universal Soil Loss Equation) (Wischmeier and Smith 1978) was used for evaluation of potential erosion in the drainage basin.

6.1.3   The Study Region

The Elbe river drainage basin is one of the largest European river basins. In our study, only the German part (two-thirds of the basin, about 96,000 km^2) was considered. Agricultural areas cover about 56% of the total area. A comparatively low amount of precipitation, the widespread occurrence of sandy soils with high infiltration rates, and high water demand (both climatic and anthropogenic) contribute to its high hydrological sensitivity. The Elbe is also one of the most heavily contaminated water courses in Europe, due to ineffective sewage water treatment and lack of non-point-source pollution control. In this study, only the natural (climatic) water demand and water cycling were considered, while the anthropogenic influence on water availability and quality are not yet taken into account.

6.2   FRESHWATER AVAILABILITY

The dynamics of soil moisture represent a component of the overall water balance, and may be the single most important variable defining freshwater availability for vegetation. The goal is to estimate the long-term average monthly soil-moisture dynamics as a component of the water balance, based on long-term average climatic data, topography, land use, and soil data. In other words, using the distributed terrain data and climate parameters as driving forces, we determine the spatial pattern of average monthly soil moisture. After that, averaging the soil-moisture distribution for subareas of interest (larger grid cells, subbasins or administrative units) enables us to delineate vulnerable subregions.

Daily time series would be a more powerful means to evaluate water vulnerability. To our knowledge, however, appropriate tools do not exist for our scale of study. Moreover, most of the dynamic hydrological models with a daily time step work successfully on the patch or small watershed scale, and an attempt to extend their applicability leads to an increase in data requirements. Therefore, taking into account data availability and the scale of study, we restricted the temporal resolution to average monthly conditions.

6.2.1   Method

Topography-based hydrological models like TOPMODEL (Beven and Kirkby 1979) and WET (Moore et al. 1993e) are based on the assumption that local soil-moisture dynamics strongly depend on the size of the upslope area (A_i) drained through an observed point in a catchment, the local surface topographic slope (tan β_i) representing the hydraulic gradient for saturated water flow, and the downslope soil transmissivity (T).

Originally in TOPMODEL, the soil wetness index, χ_i, was calculated using the assumption that the local groundwater table can be related to the catchment mean groundwater table according to the deviation between the local and the mean terrain attributes and soil hydraulic properties. The wetness index, χ_i, was expressed in terms of terrain attributes and soil hydraulic properties as follows:

$$\chi_i = \ln \left(\frac{1}{b \tan \beta} \int dA \right)_i + [\ln (T_e) - \ln (T_i)] \tag{6.1}$$

where A_i is the catchment area (m^2), b_i is the outflow width (m), β_i is the slope angle (degrees), T_i is the local soil transmissivity (m^2/day), and T_e is the average soil transmissivity in the basin (m^2/day). TOPMODEL is best suited for small to medium catchments and allows time series of discharge, soil moisture, and actual evapotranspiration (E) to be simulated.

The WET model is based on a topographic index that is very similar to that of TOPMODEL. A modified version of Equation 6.1 is used to estimate the spatial distribution of the long-term average soil moisture and evaporation (annual, monthly) using an equilibrium approach. For that purpose, an area weighting coefficient, μ_i, representing the proportion of water that is available in the soil, was included (Moore et al. 1993e) in the definition of the wetness index:

$$\chi_i = \ln \left(\frac{1}{b \tan \beta} \int \mu_i dA \right)_i + \ln (P) + [\ln (T_e) - \ln (T_i)] \tag{6.2}$$

The weighting coefficient is dependent on the evaporation, E, precipitation, P, and deep drainage, D, in each elementary unit:

$$\mu_i = 1 - \left(\frac{E + D}{P} \right)_i \tag{6.3}$$

Actual evapotranspiration (E) is dependent on soil moisture

$$E = E_p[1 - (1 - \theta)^{C/E_p}] \tag{6.4}$$

where E_p is potential evaporation, θ is the relative available soil-water content (1 at field capacity, 0 at wilting point), and C is a constant.

Potential evapotranspiration (E_p) is estimated from the Priestley and Taylor (1972) equation for well-watered vegetation under conditions of minimal advection:

$$E_p = \frac{\alpha_e R_n}{\rho \lambda (1 + \gamma/\Delta)} \tag{6.5}$$

where α_e is an empirical constant (=1.26), R_n is the net radiation, ρ is the density of water, λ is the latent heat of vaporization, γ is the psychrometric constant (dependent on air temperature and pressure), and Δ is the slope of the saturation-specific humidity–temperature curve.

The relative available soil-water content is estimated by

$$\theta = \frac{\chi_i}{\chi_{cr}} \qquad (6.6)$$

where χ_{cr} is the critical wetness index at field capacity.

These equations cannot be solved directly, since the topographic wetness index depends on evapotranspiration (Equations 6.2 and 6.3), and evapotranspiration, in turn, depends on the topographic wetness index (Equations 6.4–6.6). Therefore, iterative methods are applied to find the solution. In the WET model, the Newton–Raphson method is used to solve Equations 6.2–6.6 iteratively for every grid cell, beginning with the element of highest elevation and finishing with the element of lowest elevation at the catchment outlet. The three versions of the WET model described in Chapter 4 utilize the outputs of the TAPES-G (accumulation areas) and SRAD (net radiation) programs as inputs. The catchment area A_i in Equation 6.2 was estimated with the quasi-random Rho8 algorithm in TAPES-G in this instance (see Chapter 3 for details).

WET can be used to evaluate the average soil wetness index and evapotranspiration in watersheds. Originally it was applied to a forested region (22 km^2) of the Brindabella Range in southeastern Australia to estimate spatial patterns of average annual soil moisture and evapotranspiration, based on a 25 × 25-m DEM, using the average annual precipitation value for the whole catchment, and without accounting for variation in soil transmissivity. Our application for the Elbe basin differs greatly from this case in spatial and temporal resolution (a 1-km DEM and monthly time-step climatology) and therefore requires modification of the original algorithm.

Our modified method can be represented as a three-step procedure:

Step 1. Calculation of the long-term average monthly soil-moisture index using the modified WET model (Chapter 4; Moore et al. 1993e) with distributed soil-water holding capacity (WHC) and climate (precipitation, *P,* and temperature, *T*) data at 1-km resolution.

Step 2. Delineation of subwatersheds in the drainage basin using the *r.watershed* function in the GRASS GIS (U.S. Army Corps of Engineers 1987)

Step 3. Averaging the soil-moisture index in summer months for subwatersheds and delineation of sensitive subwatersheds regarding water availability assuming a threshold value of the index.

6.2.2 Large-Scale Applications of Topography-Based Models

Studies examining the effects of scale on topography-based models, which may help to identify potential sources and magnitude of errors in low-resolution applications, have mainly been performed for TOPMODEL. In most of these studies the mesh size of the grid varied between 2 and 480 m (Franchini et al. 1996, Bruneau et al. 1995, Quinn et al. 1995, Wolock and McCabe 1995, Wolock and Price 1994, Zhang and Montgomery 1994, Charait and Delleur 1993). There are only a few cases where the mesh exceeded 100 m (Hutchinson and Dowling 1991, Mackey 1994a, Wigmosta et al. 1994). For example, Zheng et al. (1995) investigated the applicability of the topographic index (Equation 6.1) for the state of Montana using a 1-km DEM, and showed that the available soil-water capacity and the topographic index were well correlated.

Errors evolving from the low-resolution application of topography-based models may include the fact that the process dynamics leading to the development of contributing areas may be limited to small areas (Beven 1989, Anderson and Burt 1990). As a result, the use of low-resolution grids for the simulation runs may cause an overestimation of gradients between adjacent grid cells and an underestimation of within-cell dynamics. Further, lateral interactions (routing) between large grid cells may be limited to stream flow and groundwater and not to shallow subsurface flow; both processes are beyond the scope of models like TOPMODEL and WET.

Large-scale DEM smoothing effects on topography may mislead the topography-dependent estimation of dynamic soil-moisture patterns, especially if differences in vegetation and soil physical properties are not considered. This may be particularly important in accounting for intrapatch heterogeneity.

The possible errors in TOPMODEL applications introduced through the use of low spatial resolution data were investigated in a parallel study (Müller-Wohlfeil et al. 1996) for subbasins of the Elbe drainage basin. This study demonstrated that the spatial resolution and area size cause changes in the predicted runoff, and the efficiency (defined according to Nash and Sutcliffe 1979) is reduced significantly. These results confirm conclusions drawn by Wolock and Price (1994) for TOPMODEL applications according to which the ratio of overland flow to total flow increases with decreasing spatial resolution. The main reason for this is due to changes in the mean ln (a/tan β) values that are caused by changes in grid size and area. However, the changes in the topographic indices can to a large extent be compensated by shifting the values for the lateral transmissivity parameter $T0$. Wolock and Price (1994) found that the efficiencies, flow ratios, and long-term average soil-moisture patterns were more or less independent of scale for the catchments and chosen conditions once this adjustment was applied. Figure 6.1 illustrates this relationship and shows the numbers of days with full saturation for different spatial resolutions for the Vils catchment, which is located in the southeastern part of Germany, not far from the Elbe basin. The importance of $T0$ in recalibration can be directly traced to the basic equations used in TOPMODEL for the calculation of base flow.

The use of coarse DEMs for large-scale topography-based approaches in hydrology may be more justified for the WET modeling concept, which was developed for long-term analysis (average annual or monthly) of the topographic effect on soil moisture and evapotranspiration. While TOPMODEL operates on time steps of minutes to days, WET applications use average monthly or annual weather data. While TOPMODEL includes dynamics of different subsurface storages, an a priori assumption has been made for WET about vertical drainage from the unsaturated zone to deep groundwater (through the deep drainage term, D). This means that WET is intended to represent to a larger extent the overall influence of topography *on longer term hydrological behavior of catchments* and different types of lateral flows.

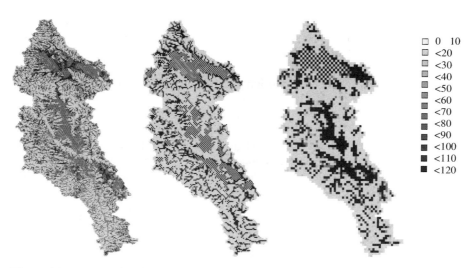

Figure 6.1. Number of days with full soil-moisture saturation in the Vils catchment (southern Germany) for different spatial resolutions (50, 250, and 500 m, from left to right) after a simulation period of 519 days.

6.2.3   Application in the Elbe Basin

The WET model, modified as described below for use in a large drainage basin, was applied to the Elbe drainage basin to calculate the soil wetness indices as monthly average values. The digital elevation model for the Elbe (supplied by the "Institut für Angewandte Geodäsie", Frankfurt, Germany) served as the primary input. Elevation varies from sea level to 1161 m in the mountainous areas of southern Saxony.

First, the topographic index (Equation 6.1) was calculated in GRASS (Figure 6.2). The distribution function of the topographic index (its shape, minimum, and maximum) is comparable with that for other watershed studies (Quinn et al. 1995, Zheng et al. 1995) and the river network is very well represented.

For such a large drainage basin it was necessary to account for spatially distributed precipitation and soil properties. As a basis, we used long-term monthly mean values (years 1950–1980) of temperature and precipitation that were available for 48 meteorological stations inside and near the region. From these averages, climatic surfaces were derived using the partial thin-plate spline algorithm of Hutchinson and Bischof (1993), which gives a good representation of topographic features, provided that elevation is used as a third independent variable along with longitude and latitude. The surfaces were evaluated for the 1-km grid, using the same digital elevation model used with WET. The spatial patterns of mean monthly precipitation are shown in Figure 6.3 (see color insert).

Available soil-water capacity was derived from the GDR soil map (about 80% of the basin belonged to the territory of the former GDR) and the FAO soil map (for the western part of the basin). The program code (WET) was modified to include the dis-

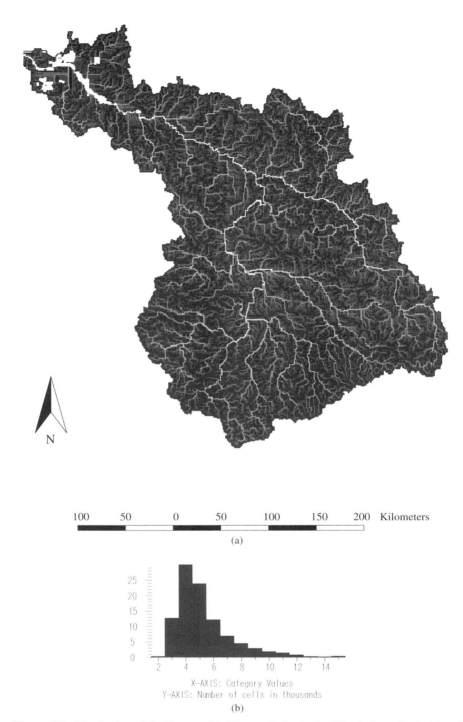

100   50   0   50   100   150   200 Kilometers

(a)

X-AXIS: Category Values
Y-AXIS: Number of cells in thousands

(b)

Figure 6.2. Distribution of the Topographic Wetness Index in the Elbe drainage basin (a) and its histogram (b).

tributed climate and soil parameters. This implementation of WET corresponds to level 3 (as discussed in Chapter 4) since spatially distributed net radiation was used to compute potential evapotranspiration, and soil water was used to modify the loss rates.

The most crucial problem in the WET application was related to calculation of actual evapotranspiration for summer months, when potential evapotranspiration is quite high. The actual evapotranspiration estimated by Equation 6.4, even with smaller C, can be too high in comparison with average monthly precipitation. In such cases the weighting coefficient, μ_i, becomes negative, and in the WET code it is forced to zero:

$$\mu_i = 0 \qquad (6.7)$$

and the actual evapotranspiration must be limited in accordance with Equation 6.3:

$$E = P - D \qquad (6.8)$$

This procedure may be acceptable for calculations on an annual basis or (at least in the temperate zone) for cooler months. But for computations on a monthly basis it should be modified.

There are several possibilities to modify this algorithm. Currently, the evapotranspiration component does not account for vegetation distribution, rooting depth, or root density. Including this additional information would probably improve the evapotranspiration component. On the other hand, it is reasonable to keep the model as simple as possible for larger-scale applications. In our study, we accounted for the soil-moisture distribution of a previous month to get more reasonable results. Instead of forcing the weighting coefficient to zero in such cases, the monthly average actual evapotranspiration was restricted by the average monthly precipitation plus change in the average soil moisture between the current and previous months. By this, we followed the water balance accounting procedure suggested by Thornthwaite and Mather (1957), and introduced a quasi-dynamic feature in the static equilibrium approach. This appeared to be reasonable for temperate zone accounting with the monthly time step.

6.2.4   Results and Comparison With Previous Studies

The results for the long-term monthly average evapotranspiration are shown in Figure 6.4 for three months—April, May, and June. In April, E is still low and almost homogeneous for the entire region. In the summer months, mean monthly E is higher in the south, where precipitation and water-holding capacity of soils are higher. The distribution of the mean monthly soil-wetness index differs mainly between the winter and summer months (Figure 6.5), while the patterns for subsequent summer months are similar. Most of the area is saturated ($\theta = 1$) in winter (Figure 6.5a), while certain parts are under water stress in summer (Figure 6.5b). The saturation in winter months here means that we have soils with moisture content at or above field capac-

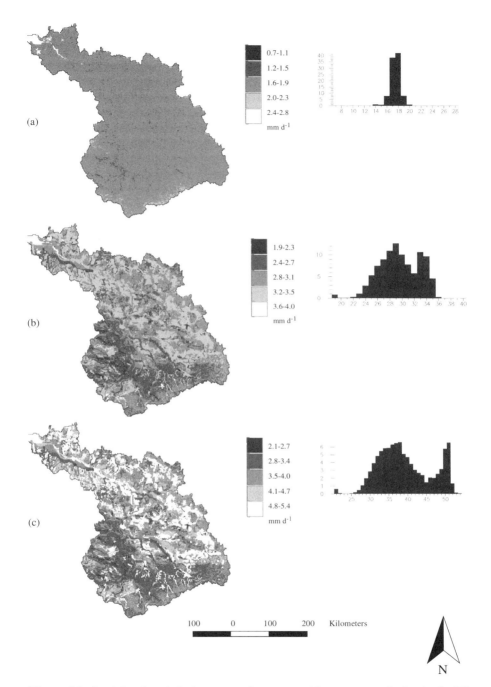

Figure 6.4. Spatial and statistical patterns of mean monthly evapotranspiration in the Elbe drainage basin for (a) April, (b) May, and (c) June. The x axis in the histograms shows category values in 0.1 mm, and the y axis shows the number of cells in thousands. The values on the maps are in mm d^{-1}.

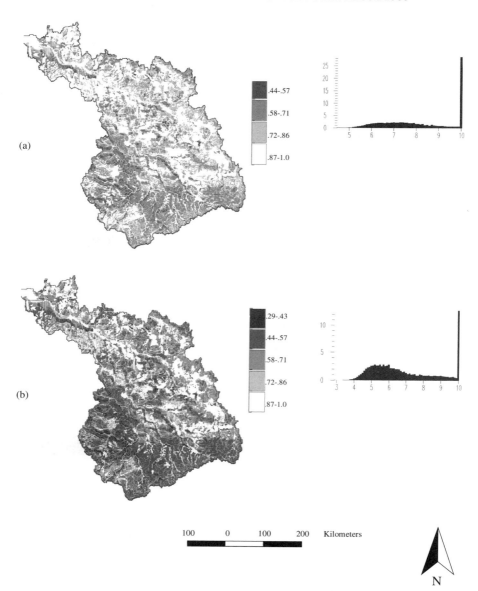

Figure 6.5. Spatial and statistical patterns of mean monthly soil wetness index (relative units, from 0 to 1) in the Elbe drainage basin for (a) January and (b) July. The X-axis in the histograms shows category values in 0.1, and the Y-axis shows the number of cells in thousands.

ity, but the water table in most areas is still below the ground surface. This follows from the model assumptions and derivation of site parameters.

Since we applied the WET model for a much larger basin and with coarser resolution than was done previously, it was necessary to test the overall model behavior against independent observations and calculations in the Elbe basin. The structure of the WET model does not include opportunities to calculate time series of discharge, and so model validation has to be focused on spatial patterns of subsurface moisture (i.e., soil moisture, groundwater). Unfortunately, there are no field observations or remote sensing data available for soil moisture or evapotranspiration in this region. Nevertheless, some indirect methods can be used. We compared our results with an existing map of groundwater depths (WASY GmbH, scale 1:500,000) and with the results of an independent terrestrial modeling project carried out in Germany (Wendland et al. 1993).

The WASY map of groundwater depths is based on a map of groundwater table isolines (scale 1:50,000) which was generalized to a scale 1:500,000 and then gridded (250 × 250-m pixels). The difference between a DEM based on a topographic map at a scale of 1:200,000 with the same resolution and the groundwater grid was calculated to derive a map of groundwater depths. Comparison between the map of mean groundwater depths for eastern Germany and the map of the soil-moisture index generated from our WET application (Figures 6.6 and 6.7, see color insert) shows that in both cases areas of high water content occur mainly along the riparian areas (c.f. Wendland et al. 1993) or in zones of topographic convergence. In contrast, dry areas defined by deep groundwater tables and low wetness indices appear mainly in loess areas in the southern part of the basin.

A similar comparison between a map of mean annual percolation rates (Wendland et al. 1993) based on a 3-km grid and our maps of the wetness index shows that both the percolation values and the wetness indices are low in hilly loess regions (southwestern part of the drainage basin) due to low precipitation rates and relatively high values of evapotranspiration.

As a next step, the subwatersheds in the drainage basin were delineated using GRASS. The procedure *r.watershed* was applied with a 1000-km^2 threshold value (which is defined in GRASS as the minimum size of an exterior watershed basin) to subdivide the drainage basin into 57 subbasins with an average area of about 1700 km^2. The cells with soil wetness index values lower than 50% of field capacity in the summer months (from May to August) were identified, and the mean summer soil moisture deficit was estimated as the difference between 50% of field capacity and predicted soil wetness (Figure 6.8a). Averaging the soil-moisture deficit values for subbasins (area average), we identified several classes and, consequently, the most sensitive subbasins (classes 4 and 5, Figure 6.8b).

These results indicate that a topography-based hydrologic approach, which is focused on long-term dynamics, may be suitable as a tool for estimating temporally averaged subsurface moisture patterns of large regions. This information can help to delineate areas that are potentially vulnerable with respect to water availability, although there might be differences in soil-moisture patterns and groundwater distributions when looking at smaller areas in more detail. The actual amount of soil water

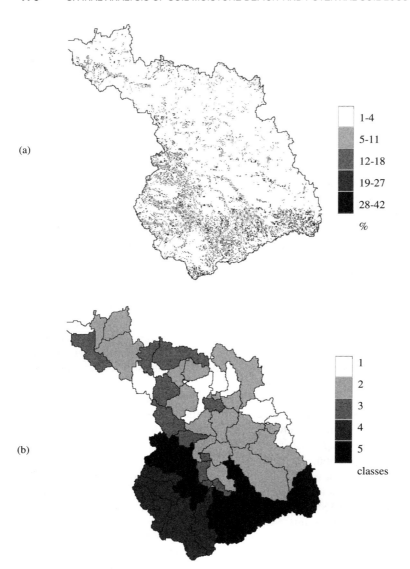

Figure 6.8. Average soil moisture deficit in summer (May–August) measured as the percent differences between 50% field capacity and estimated soil wetness (a) and vulnerability water-availability classes (b) in the Elbe drainage basin.

available to plants should be further related to finer scale patterns in soil, topographic characteristics, and land cover.

Anthropogenic effects on the hydrological cycle (which are significant in this densely populated area) were not considered in this approach. In reality, freshwater availability in such a region is determined not only by natural water supply from the hydrological cycle, but also by human-induced changes. Including these features would influence the results as well.

6.3 EROSION

6.3.1 GIS-Based Approaches for the Analysis of Pollutant Yield in Large Basins

Simplified static pollutant yield models can be integrated with GIS to estimate potential soil loss (water erosion) and phosphorus and nitrogen export from watersheds and to delineate critical subareas of non-point-source pollution. The Universal Soil Loss Equation (USLE) (Wischmeier and Smith 1978) and its modifications (MUSLE, RUSLE) are well-known lumped-parameter methods for estimating net erosion by use of a regression equation involving several major erosion factors: rainfall erosivity, soil erodibility, topography, crop management (or vegetation cover), and erosion control practice. The modifications provide alternative versions and/or methods for calculating the rainfall erosivity factor. The basic method has been used to estimate soil loss in several types of environments in numerous countries (Novotny and Olem 1994).

This method can be modified for use in a watershed (e.g., Tim et al. 1992) and coupled to a raster-based GIS to estimate the long-term average soil loss based on climate data, soil texture, topography, and land use. The combination of several map layers allows the potential soil loss for every grid cell or subarea to be estimated. The actual or realised soil loss is different from the potential, depending on the watershed size, its slope and land cover, because mobilized sediments are routed from cell to cell through the watershed to the river outlet, and retained on the way by vegetation and other barriers.

Phosphorus pollution is closely correlated with soil erosion. The phosphorus export from each unit can be roughly estimated on the basis of the average phosphorus content in the surface soil layer, the sediment yield, and the phosphorus enrichment ratio (which is dependent on soil texture). The nitrogen export from each unit to surface waters can be calculated on the basis of the average nitrogen content in the surface soil layer, and the average annual runoff from the unit area. In the case of nitrogen leakage to groundwater, the average nitrogen content in the soil column and the average annual infiltration from the unit area should be considered. Hence, for nitrogen export the relative importance of the water runoff components (surface, subsurface runoff, infiltration to groundwater) need to be estimated in advance.

On the basis of selected critical (or threshold) values for the soil erosion rate and the N and P export coefficients, the model outputs can be analyzed to identify sub-areas that exhibit low, medium, and high soil non-point-source pollution potential. Due to the lack of spatial data on nutrient content in soils and more detailed data on the components of the soil-water balance, only potential erosion was analyzed in this study.

6.3.2 Methods and Results

A simplified form of the USLE was applied to account for the effects of climate, topography and soil on potential pollution as follows:

$$\mathrm{ER}_i = \sum R_i K_i \mathrm{LS}_i \qquad (6.9)$$

where ER_i is the computed soil loss from a subwatershed (kg/m^2), R_i is the rainfall–runoff (erosivity) factor (N/h), K_i is the soil erodibility factor (kg·h/N/m^2), and LS_i is the slope gradient and slope length factors for land cell i. This approach omitted the land cover (C) and supporting practices (P) factors that vary with human land use.

Since rainfall intensity data are much less available than daily rainfall data, most USLE modifications attempt to find other ways to estimate the rainfall–runoff erosivity factor. The rainfall–runoff erosivity factor (R) was estimated for some parts of the Elbe basin in previous studies by Saupe (1985), Hartmann (1988), Deumlich (1993), and Deumlich and Frielinghaus (1994). In these studies it was shown that R was well correlated with summer precipitation P_s in this region. We therefore used the regression equation from Deumlich and Frielinghaus (1994):

$$R_i = -6.88 + 0.152 P_s \qquad (6.10)$$

to estimate the rainfall–runoff erosivity factor. The spatial patterns (Figure 6.9a) and the range of values correspond well to the results of the above-mentioned investigations.

The combined slope length and slope steepness factor LS was calculated by GRASS using the *r.watershed* program applied to the DEM. The maximum LS values occur in the mountainous southern part of the drainage basin (Figure 6.9b). The soil erodibility factor K was derived from a soil texture map using the soil texture class relationships proposed by Deumlich and Frielinghaus (1994).

The combination of these three factors allowed the potential soil loss to be estimated for the grid cells using GRASS functions. The resulting map of the potential erosion is shown in (Figure 6.10a). The vulnerable subbasins were delineated by averaging the potential erosion for subbasins (Figure 6.10b). The resulting maps show good agreement with previous studies (Saupe 1985, Sauerborn 1993, Deumlich and Frielinghaus 1994). This identification of sensitive subregions for erosion provides a basis for more detailed study of erosion in the most sensitive subbasins in the southern mountainous part of the basin.

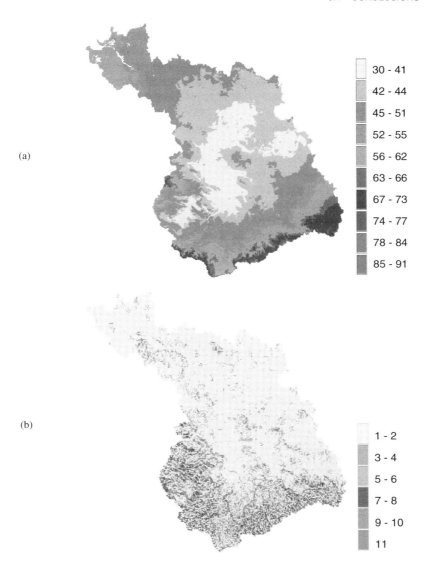

(a)

	30 - 41
	42 - 44
	45 - 51
	52 - 55
	56 - 62
	63 - 66
	67 - 73
	74 - 77
	78 - 84
	85 - 91

(b)

	1 - 2
	3 - 4
	5 - 6
	7 - 8
	9 - 10
	11

Figure 6.9. (a) Rainfall–runoff erosivity factor, *R,* and (b) topographic factor LS in the Elbe drainage basin.

6.4 CONCLUSIONS

The methods described in the paper allow sensitive subareas within a region or relatively large drainage basin to be delineated by patterns of soil moisture and erosion. These can be used for sensitivity assessments in large-scale hydrological and water-quality studies. Clearly, the pattern of vulnerability classes for water availability and erosion was different in the Elbe basin, although the sensitive subbasins belong to the

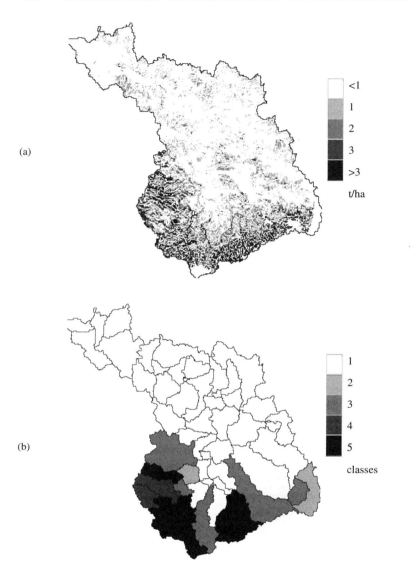

Figure 6.10. Potential erosion based on the combination of rainfall, slope, and soil erodibility factors (a) and vulnerability classes for erosion (b) in the Elbe drainage basin.

same southern part of the region. This confirms our hypothesis about the close inter-linkage of these processes.

Based on the knowledge about vulnerable areas, dynamic process-based models could be applied to better understand the hydrological and biogeochemical processes and to reveal the feedback mechanisms of complex climate–biosphere interactions. Our study demonstrates that even simplified models that include only some key factors to determine system behavior can be effective tools for spatial analysis and sensitivity assessment of the hydrological cycle and water quality in a watershed. Further improvement of the equilibrium water balance modeling with monthly time steps and estimation of potential NPS pollution are possible for regions where distributed data on land cover and soil properties are available.

CHAPTER 7

Mapping Contributing Areas for Stormwater Discharge to Streams Using Terrain Analysis

Jeremy S. Fried, Daniel G. Brown, Mark O. Zweifler, and Michael A. Gold

7.1 INTRODUCTION

Increasing concern about surface water quality has led to the development of best management practices such as grassed waterways, buffer strips, cover crops, and no-till agriculture for mitigating non-point-source pollution from sediment and adsorbed pollutants. Effective implementation of these strategies requires an understanding of the processes by which soils, land use/cover, terrain and land management practices influence the detachment and delivery of sediment to surface waters. Geographic information systems (GIS) provide an environment within which such data can be managed and analyzed, and in which formal models of pollutant delivery processes can be formulated and tested. Distributed watershed models (e.g., ANSWERS, AGNPS) are useful for simulating sediment erosion, deposition, and delivery to surface waters (e.g., Beasley and Huggins 1982, Young et al. 1989), but incur substantial costs for collecting and compiling the spatial and attribute data needed as inputs (Grayson et al. 1992b, João and Walsh 1992). Even when pollution contributing areas can be located, additional data collection (e.g., ownership, management practices, zoning) is usually required to identify feasible mitigation solutions.

Whether areas of potentially high erosion (i.e., sediment production) are identified using a distributed erosion model or some simpler method (e.g., the Revised Universal Soil Loss Equation), collection of spatially distributed erosion potential and management practice data in support of the mitigation effort can be costly. This chapter describes a terrain-analysis approach grounded in hydrological principles that can limit the geographic scope of the data-collection effort. A two-stage process is suggested. In stage 1 (Figure 7.1a), the focus is on collecting and analyzing a lim-

Terrain Analysis: Principles and Applications, Edited by John P. Wilson and John C. Gallant.
ISBN 0-471-32188-5   © 2000 John Wiley & Sons, Inc.

ited set of thematic information about an entire watershed (i.e., terrain, drainage channels, and soils). Terrain analysis is then used to identify potential "hot spots" where sediment from potentially erodible source areas might be delivered to the stream channels. We suggest the term investigative riparian buffer for this zone of potential hot spots. Stage 2 entails a data collection effort spatially focused within this investigative riparian buffer, but more expansive in attributes (e.g., land cover, ownership) to support the ranking of potential remediation sites based on a richer information set (Figure 7.1b). The terrain-analysis models presented here should be useful for making the stage 1 selection.

While a number of authors have suggested that water-quality management activities should focus within a set distance from streams (Duda and Johnson 1985, Maas et al. 1985, Phillips 1989, National Research Council 1993), and that such a distance might well vary within a watershed (Walling 1983), the approach described here extends these ideas to the concept of a variable-width investigative buffer around stream channels (Hunsaker and Levine 1995). The methods we propose can be carried out with readily obtainable data using well-developed hydrological principles embedded in the TAPES-G software to delineate such investigative buffers.

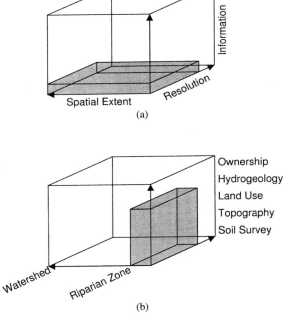

Figure 7.1. (a) For a given budget, investigation of a whole watershed at a resolution suitable for modeling yields little information depth; (b) redeployment of the same investigative effort within the hydrologically active riparian zone enables the collection of a richer attribute data set for support of modeling and planning.

7.1.1 Implications of Sediment for Water Quality

Sediment is the most ubiquitous pollutant of surface waters by both volume and mass (Chesters and Schierow 1985). Eroded soils and other particulates deposited in receiving waters damage aquatic ecosystems, degrade the aesthetic and recreational value of surface waters, increase flooding, and decrease the storage capacity of reservoirs (Novotny and Olem 1994). Nutrients and other chemical pollutants adsorbed to sediments can exacerbate water-quality impacts (National Research Council 1993). In the study site used for this analysis, decomposition of sediments contributed by agricultural runoff and streambank erosion severely limits available dissolved oxygen for aquatic life (Suppnick 1992). Thus, this research addresses a very real problem that is of widespread concern.

7.1.2 Sediment Management with Riparian Buffer Strips

Riparian buffer strips (RBS) have been shown to be effective in reducing the movement of pollutants into streams (Lowrance et al. 1985). RBS consist of a mixture of grasses, forbs, trees, and shrubs adjacent to surface waters that is designed and managed to enhance the capacity of riparian ecosystems to capture and store potential pollutants contained in surface runoff and near-surface throughflow before they can be delivered to surface waters (Dillaha et al. 1989, Osborne and Kovacic 1993, Binford and Buchenau 1993).

Because economic constraints usually preclude installation of RBS along every stream, creek, and drain, a system is needed to prioritize stream reaches by amount of pollutants entering the stream and suitability of RBS as a remediation technique. While considerable effort has been invested in testing alternative RBS designs, the siting issue has received little attention to date. The approach to delineating variable-width investigative riparian buffers presented here is the first step in a strategy to prioritize potential RBS sites. High-priority sites are those stream-side locations where the greatest quantities of sediment-laden runoff are likely to be intercepted at the lowest economic cost and with minimal disruption to riparian landowners' desired uses of their land (Maas et al. 1985, Aull 1980).

7.1.3 Hydrological Principles

The investigative riparian buffers presented here are delineated using models based on a combination of hydrologic principles and ad hoc judgments of the investigators. The principles are, for the most part, well established in the hydrological literature and are briefly summarized here. Where ad hoc judgments were made, these are stated explicitly in the methods or results.

Terrain modeling offers a promising new approach to an old problem: obtaining site-specific estimates of sediment movement processes. Researchers (e.g., Walling 1983, Novotny and Chesters 1989) have long sought to overcome challenges in linking upland erosion rates throughout a watershed to sediment yield in receiving surface waters. They have found that the amount of sediment that shows up in the stream is usually dwarfed by that calculated as eroding upslope. The amount of sediment delivered to a stream can be conceptualized as the product of runoff volume and suspended sediment concentration, but both factors are too heterogenous (both spatially and temporally) to make calculations practical. Disturbance of vegetative cover by human activities (e.g., agriculture, timber harvest, and construction) leaves soils vulnerable to erosion, but topography, soil hydrologic properties, and landscape position largely determine where and how much eroded sediment is actually transported to surface waters via stormwater flows. Eroded sediments typically travel short distances with runoff. Infiltration and reduced velocity, which occur when runoff traverses unsaturated soils, surface depressions, low-gradient slopes or thick vegetation and when rainfall abates, result in deposition of much, if not all, of the sediment load.

Within any given span of time, only a small fraction of sediment eroded within a watershed shows up as sediment in downslope streams. This fraction, formally known as the sediment delivery ratio (Novotny and Chesters 1989), is calculated as the ratio of sediment delivered at the catchment outlet to gross erosion within the watershed. Though the ratio may be a useful conceptual device in discussions of non-point-source pollution, it fails to capture the spatially distributed nature of overland sediment transport capacity, thereby conferring a lopsided importance to areas of high hillslope erosion over sources of high water pollution potential. In fact, highly erodible areas far from the stream may contribute far less pollution than less erodible areas close to the stream (Novotny and Chesters 1989).

The difficulty and expense of directly sampling storm runoff has long hampered watershed-scale sedimentation research (Walling 1983). Indirect sampling based on tracking the cesium-137 that was deposited on all soils by radioactive fallout during the middle of this century has been used effectively in field studies of sediment redistribution (Ritchie and McHenry 1990). Such studies found evidence that the vast majority of eroded particles from upland fields are deposited in upland depressions, grassed field borders, fence lines, hedgerows, and roadside ditches and, for all practical purposes, immobilized indefinitely (Wilkin and Hebel 1982). The research also highlights the importance of riparian condition in determining sediment delivery to streams. For example, replacement of riparian forests with row crop agriculture both places a potential sediment source close to the stream and severely impairs or eliminates the sediment trapping capacity of the riparian zone (Cooper et al. 1987, Lowrance et al. 1988).

An assumption motivating this work is that water-quality modeling and management efforts could be made more effective by refocusing analysis efforts from all source areas of sediment generation in the watershed to those source areas with a high probability of sediment transport to streams via overland flow. The first step in such an approach is delineation of contributing areas for stormwater discharge to streams because tainted water in such areas has the greatest probability of reaching the streams. The partial or variable source area concept of overland flow suggests that most overland flow in humid environments occurs on portions of the landscape where rain falls on saturated soils, generating saturation runoff. The areal zone of soil saturation expands in the presence of precipitation and contracts in its absence (Betson and Marius 1969, Dunne and Black 1970, Dunne 1978).

Partial source areas contributing stormwater discharge to streams are predominantly located in the riparian zone (Maidment 1993). Soil water moves both vertically and horizontally, with lateral movement running parallel to surface gradients (Freeze and Cherry 1979, Dunne 1978). Thus, soil water travels downslope following surface gradients and accumulates in flat areas and depressions where low hydraulic gradients inhibit lateral flow. Relative to the remainder of a watershed, riparian zones have three characteristics that make them likely to receive soil water at rates that exceed their drainage capacity: (1) greater upslope contributing areas, which cause them to receive more lateral subsurface flow than positions closer to the drainage divide; (2) groundwater table levels that are relatively close to the surface; and (3) relatively low gradients. During precipitation events, the areal zone of soil saturation in riparian areas expands upslope in a nonuniform fashion, with more rapid advances upslope in areas of topographic convergence (i.e., water-gathering locations) than in areas consisting of water-dispersing ridge lines or uniform slopes.

By itself, soil saturation is a necessary but not sufficient condition for identifying areas with a high probability of contributing sediment. For example, while low-gradient, riparian wetlands have high soil saturation, they also have high capacity to trap sediment as stormwater runoff slowly filters through them. Delivery of sediments and other non-point-source (NPS) pollutants to surface waters via overland flow involves a two phase process: detachment and transport. The velocity of stormwater flow varies with the terrain as it moves overland toward the stream, and where velocity is low, partial, or complete redeposition of entrained sediment and other pollutants may occur before they can be delivered to receiving waters (Chesters and Schierow 1985). The saturated condition of variable source areas facilitates sediment detachment, but sediments will be moved downslope only if surface runoff has sufficient and sustained stream power to transport the sediment load. Because of their proximity to receiving waters, pollutants detached in variable source areas in and adjacent to riparian zones have a relatively high land to surface water delivery ratio.

Moore et al. (1988a) modeled the distribution of ephemeral gullies on a 7.5-ha, bare-fallow cultivated catchment using these concepts. They found that ephemeral gully locations could be predicted from the magnitudes of the topographic wetness index $\ln (A_s/S)$ and stream-power index $(A_s * S)$, where A_s is the local upslope contributing area per unit width of contour line and S is the local slope at the downslope contributing area contour segment. Topographic wetness and stream power were both greatest at the catchment outlet, and high values of both extended some distance upslope along the "valley floors" of water gathering (i.e., convergent) topography.

The topographic indices $\ln (A_s/S)$ and $A_s * S$ assume "steady-state" conditions, and are referred to here as static topographic wetness and stream power. These indices assume that every point in the catchment has reached subsurface drainage equilibrium, and upslope contributing areas can be quite large. Barling et al. (1994) noted some serious limitations of these static topographic indices for describing dynamic flow processes and predicting specific saturated soil and high stream-power locations. Most catchment locations receive contributions of subsurface flow from only a small proportion of their total upslope area. This proportion depends on antecedent moisture conditions and generally increases during storms. Spatially explicit predictions of the distribution of soil moisture and runoff stream power within a watershed require the ability to describe this "dynamic" upslope contributing area topographic variable.

To this end, Barling developed a quasi-dynamic wetness index that improves on the steady-state wetness index by incorporating both the topography within the upslope contributing area and the time required for subsurface drainage to redistribute soil water. Field observations on the same experimental catchment used in Moore et al. (1988a) verified that the quasi-dynamic topographic wetness index better predicted soil-moisture distributions and location of ephemeral gullies than did static indices (Barling et al. 1994). This study represents a logical extension of Moore's and Barling's work to riparian management.

7.2   DESCRIPTION OF STUDY AREA

This study was conducted at Barnard Drain, a 1700-ha subwatershed of Sycamore Creek near Michigan State University's East Lansing campus. The Sycamore Creek watershed is one of 37 hydrologic units across the United States included in a demonstration project funded by the U.S. Department of Agriculture (USDA) and Environmental Protection Agency (EPA) to provide feedback on non-point-source pollution remediation efforts for small watersheds (USDA-Soil Conservation Service et al. 1990). The most serious pollution problem identified in Sycamore Creek is stream sedimentation with associated sediment oxygen demand and habitat impairment. Streambank and agricultural erosion, as well as urban runoff, have been identified as the primary causes of sedimentation (Suppnick 1992). Topography in the Barnard subwatershed is flat to gently rolling, with slopes ranging from 0 to 28% (mean slope is 1.5%) (Figure 7.2; see color insert). Ninety-five percent of the area has slopes of 5% or less. Land use/cover, as classified from aerial photography, consists of row crop and pasture-based agriculture (67%), forest and wetland (18%), residential/commercial (9%), and miscellaneous other uses (6%).

7.3   METHODS

Analysis involved completion of four principal tasks: acquisition of GIS databases and construction of a digital elevation model of the area, computation of terrain indices via TAPES, formulation and computation of alternative investigative buffer models, and collection and analysis of validation data.

7.3.1   GIS Database and Terrain Model Creation

Our maps of investigative riparian buffers were constructed from three primary GIS data layers obtained for the Barnard Drain subwatershed of Sycamore Creek: streams and drains, topographic contours (10-ft interval), and soils. A GIS coverage (with a base scale of 1:24,000) containing stream and drain center lines for all named streams and drains in Sycamore Creek was obtained from the Michigan Department of Natural Resources (MDNR). The demarcation between streams and drains in this watershed is not distinct. It is not uncommon for streams to be artificially extended upslope as drains, and the biophysical characteristics of some drains are all but indistinguishable from those of streams of comparable order. Contour data (hypsography) were purchased from USGS as digital line graph (DLG) files on 9-track tape. Contour lines for the southern third of the Barnard subwatershed were digitized manually from USGS topographic quadrangle sheets because DLG files for those quads were not yet available. Soil polygons and profile attribute data, originally digitized by USDA–Natural Resource Conservation Service (NRCS), were contributed by NRCS for this analysis. The streams and drains coverage was converted to raster format in preparation for modeling, but hypsography and soils required significant additional processing.

Topographic sinks were coded in the contour coverage by identifying closed contour depressions on the hypsography layer. A digital elevation model (DEM) for the Barnard subwatershed was constructed from the hypsography theme using the TOPOGRID module in ARC/INFO version 7.0.2 (Hutchinson 1989b). TOPOGRID was executed using recommended tolerances with drainage enforcement to the streams and drains layer and the sinks file to create output with a 10-m grid cell size. While the resulting DEM surface appeared to be relatively smooth, histograms of cell frequency by elevation value demonstrated the bias favoring contour elevations that has been widely observed by those who have relied on contour interpolation in regions of gentle topographic relief (e.g., Hutchinson 1989b, Eklundh and Martensson 1995). Contrary to our expectation that this artifact of the interpolation algorithm would lead to locally elevated wetness index values (due to relatively lower slope values in the vicinity of contour lines), no such phenomenon was observed. We therefore accepted and used the results from TOPOGRID as an adequate representation of the topography in the area.

Although TAPES-G does not rely on soil properties to calculate static topographic wetness and stream power, DYNWETG does use saturated hydraulic conductivity (K) and effective porosity (P) information along with slope (S) in calculating dynamic indices via a formula for average lineal velocity (VEL) of soil drainage (i.e., the speed of near-surface throughflow) based on Darcy's law:

$$\text{VEL} = \frac{KS}{P} \tag{7.1}$$

In this calculation, percent slope at the surface, easily derived from a DEM and automatically included in TAPES-G output, serves as a surrogate for hydraulic gradient. Field measurements of K and P, however, are extremely challenging to obtain and these variables are usually quite spatially heterogeneous. DYNWETG accepts K and P as spatially uniform parameters (averages over the study area) or as spatially distributed parameters in the form of raster files. Both uniform and variable soil-parameter assumptions were included in this analysis. Uniform averages and variable raster files of K and P were estimated from surrogate measures in the USDA–NRCS Ingham County Soil Survey digital database associated with the digitized version of the county soil survey map. Permeability, measured in a laboratory environment as the rate of vertical movement of water through a soil column in inches/hour, served as a surrogate for K. The mean of permeability (weighted by horizon thickness) for all surface soil horizons provided the single value per soil type required by DYN-WETG. Because lateral movement of soil water is dominated by near-surface flows, only surface soil horizons were considered. We identified surface horizons as those both above the recorded high water table level and above the first aquiclude (defined here as a fine-textured soil horizon with permeability ≤ 0.005 in/h).

Drainable porosity, estimated from soil texture class (e.g., loamy sand, sandy loam, loam, etc.) and a table relating texture classifications to drainable porosity (Foth 1984; Figures 7.3–7.12), was used as a surrogate for effective porosity in the models described in this chapter. Alternatively, a table that permits a direct lookup of effective porosity based on soil texture is given by Rawls et al. (1982), and those seeking to emulate or extend this terrain-analysis application are advised to consult it.

7.3.2 Generation of Terrain-Analysis Indices

Our analysis begins with the simplest models of topographic wetness and stream power, followed by the more elaborate models. The Barnard DEM was processed twice with TAPES-G: once using the D8 flow-routing algorithm, and once using DEMON. For each TAPES-G run, the outputs slope (S), flow direction, and upslope contributing area (A) were used to calculate static indices and to provide inputs to DYNWETG. Alternatives to the D8 flow-routing algorithm are not currently available in the DYNWETG software. We produced quasi-dynamic indices using both uniform and distributed soil parameter assumptions. A drainage time of 24 h was specified because temporally continuous rainfall precipitation events in this region are rarely of longer duration. The distributed soil parameters were produced as described above. The principal DYNWETG output used in our models was effective upslope contributing area (A_e).

A total of four sets of topographic wetness and stream-power indices, beginning with the simplest, were ultimately generated from upslope contributing area (A), effective upslope contributing area (A_e), and slope (S) to serve as the basis for four models: (1) Static D8, (2) Static DEMON, (3) Dynamic Uniform Soils, and (4) Dynamic Variable Soils. For the static cases, topographic wetness index (TWI) was calculated as TWI = ln (A/S) and stream power (PWR) as PWR = AS, and for the dynamic cases, TWI = ln (A_e/S), and PWR = A_eS.

The contributing area and slope components were transferred to ARC/INFO GRID files using a conversion routine in the TAPES package (TAPESTOARC) and ESRI's ASCIIGRID command. Wetness and stream-power indices were calculated using ARC/INFO GRID from the terrain primitives; however, one could just as easily obtain these from the output files produced by TAPES-G and/or DYNWETG. One caution is that, to produce absolute and comparable results, keeping track of units is essential (e.g., Is slope expressed in degrees, percent, or as rise over run? Was TAPES-G run for areal units or numbers of cells?) because they can have great impact on the magnitude of calculated index values.

7.3.3 Formulation of Investigative Buffer Model

Building on the conclusion of Moore et al. (1988a) that indices of both topographic wetness and stream power provided predictive power in modeling the locations of ephemeral gullies, with relative predictive power of these indices determined by slope position, we sought to build a model that would combine these indices to identify likely stormwater contributing areas. Our approach is similar to Moore's, in that we are combining topographic wetness and stream-power maps to find areas with high levels of both, but differs in that we identify areas of high wetness and high stream power as fuzzy, rather than Boolean, sets (Burrough 1989). Because distributions of stream power were highly skewed, we use a fuzzy logical approach to assign locations to sets called "high stream power" and "high wetness" in a watershed-relative fashion. Cumulative cost distance from the stream is calculated over the cost surface generated by this fuzzy intersection and used to build a buffer defined by a user-selected percentile threshold for cumulative cost distance.

The fuzzy intersection of the topographic wetness and stream-power sets served as input to the calculation of cost distance from the stream channel, which was used to identify variable buffer widths and focus the investigation on near-stream locations. Presumably, areas with high values of both stream power and topographic wetness merit investigation for possible sediment contribution. Because continuity between sediment production and the stream is required for NPS pollution to occur, we used the cumulative cost distance calculation from the stream to construct a buffer that was wider where sites with high stream power and topographic wetness occurred near the stream and narrower where values of stream power or wetness were lower near the stream.

For each model, fuzzy membership surfaces were created for topographic wetness and for stream power using fuzzy sets on 0.01–1.00 intervals specified such that cells with the greatest values of topographic wetness and stream power would be assigned the minimum cost and those with the lowest values (and thus least likely to coincide with detachment and transport) would be assigned the maximum cost. The fuzzy membership functions for each index were defined such that they decline linearly from 1.00 to 0.01 from the 50th to the 95th percentile on the stream-power and topographic wetness distributions, as follows:

$$
\text{Fuzzy Set Value} =
\begin{cases}
1 & \text{if } \text{percentile} < 50 \\[2mm]
1 - 0.99\left(\dfrac{\text{index} - \text{50th percentile}}{\text{95th percentile} - \text{50th percentile}}\right) & \text{if } 50 < \text{percentile} < 95 \\[2mm]
0.01 & \text{if } \text{percentile} \geq 95
\end{cases}
\quad (7.2)
$$

The intersection of two fuzzy sets (i.e., the assignment of a value of membership in the set formed by the intersection of set A and set B) can be either hard, using the MINIMUM operation, or soft, using multiplication (Burrough 1989). We used the soft intersection operation to assign unit cost-distance values. Resultant values were low where topographic wetness and stream power were both high relative to the remainder of the watershed such that when the fuzzy surface was used as a unit cost input to cumulative cost-distance calculation, cumulative cost-distance contours would bulge out from the stream at locations with low fuzzy intersection values.

Cumulative cost-distance values, basically the minimum sum of unit cost distances traversed between a cell and the nearest (cost distance) stream cell, were assigned to each cell by accumulating the cost values as recorded in the fuzzy intersection map starting from the stream. From the cumulative cost-distance map, a map of the investigative riparian buffer was generated by assigning the lowest n percent of values to 1 and all others to 0. The selection of n was dependent on watershed specific criteria and could be sliced using multiple threshold percentiles to obtain bands of progressively diminished concern (e.g., 0–5 percentile, 5–10 percentile, etc.). Note that increases in n do not expand the buffer at a uniform rate or in a self-similar fashion across every segment of stream.

7.3.4 Collection of Validation Data

Two types of validation were performed for these models. Because the traditional approach to investigative riparian buffer siting is based on map interpretation by water-quality experts or field observations of riparian condition, these alternatives make appropriate benchmarks against which model performance can be judged. Although the data sets serve as a point of comparison, they should not be regarded as "ground truth." The interpretations are subject to as many, if not more, deficiencies as the numerical models in representing flow behavior.

7.3.4.1 Expert Interpretation of Topographic Maps
Two experts from the hydrologic unit at the Michigan Department of Environmental Quality (MDEQ) Land and Water Management Division were provided with 1:24,000-scale USGS topographic maps covering Barnard Drain and asked to identify flow paths where they would expect to find surface flow destined for the stream, based only on the clues provided by topography. In essence, they were asked to manually interpret the maps to find coincidence of convergent topography and evidence of sufficient steam power to give surface flow a high probability of reaching the stream. The correspondence between these experts' assessment of the routes that surface flow would follow (i.e., linear features) and the predictions of the wetness indices can be checked to provide a form of validation of either approach.

Delineation of subwatersheds at this scale for engineering purposes and fisheries and water-quality assessment and monitoring is a daily activity for experts at MDEQ's hydrologic unit. Nevertheless, these experts had some difficulty complying with our request because they were accustomed to operating with some design points in mind (e.g., specifying culvert dimensions based on predicted flows) and delineating upslope area on maps to identify subwatersheds. They clearly regarded delineating a subwatershed prior to identification of a specific problem as a novel assignment. It is also worth noting that they repeatedly asked to be allowed to use additional information, such as soil survey data and aerial photo stereo pairs, in making their judgments, probably in search of clues concerning land cover and topography. The flow paths drawn by the MDEQ staff were digitized to create a GIS coverage and compared visually with output from the models.

7.3.4.2 GPS-Referenced Field Observations of Wetness Indicators   To test the correspondence between observed wetness and the terrain-based indices, we surveyed the full length of both sides of Barnard Drain over a period of three days in April 1995, shortly following a series of significant rain events, to identify sites with indications of soil saturation or erosion activity. Using an integrated GPS/data logger, 24 sites (Figure 7.2) showing evidence of wetness were georeferenced as points and attributed as wet spots (soil saturated and/or ponding evident), drain points (evidence of concentrated flow but vegetation still present), and small gullies/bank erosion sites (vegetation and/or topsoil apparently removed by the force of concentrated flow). Relative topographic wetness and stream power for each of these 24 points, which we refer to generically as "wet spots," were calculated to determine whether terrain indices were useful predictors of potential trouble spots identified in the field.

Wet spots were converted to raster grid cells, then expanded from one grid cell (10 m by 10 m) to areas that were three by three grid cells (i.e., 30 m by 30 m) in size using GRID's FOCALMAX operator. This "fuzziness" was added because (1) even with differential correction, GPS-based location estimates can easily be off by one cell in a 10m grid, and (2) the tendency of D8 flow routing to constrain flow paths to even multiples of 45° azimuths could produce artifacts that introduce positional errors of as much as one cell. Because most of the recorded wet spots were within a cell or two of the stream, these three-by-three grid cell areas tended to overlap the stream or, in some cases, portions of the opposite bank. Cells in a rasterized version of the USGS blue-line representation of the stream and cells on the opposite side of the stream were removed from the areal representations of wet spots, leaving a grid of wet patches with a variety of shapes and sizes (between two and nine grid cells).

Average stream-power and topographic wetness values for each wet patch were compared with the distribution of values within a 30-m buffer around the stream. The statistical distributions of stream-power and topographic wetness values for the four models were summarized using percentiles (i.e., percentage of values below a given value) calculated on all cells within the 30-m buffer ($n = 1666$). The percentiles were then calculated for the average of the multiple grid cells in each wet patch. High percentile values would indicate consistency between the model and observed instances of saturated soil and/or evidence of erosion.

7.4   RESULTS

Three types of results were produced for each of the four alternative models: the calculated wetness and stream-power indices, the variable-width investigative buffers, and degree of agreement with the validation datasets. Outputs among the models are contrasted and some potential explanations are explored.

7.4.1 Comparison of Indices

Given the multiple order of magnitude difference between upslope contributing area A (mean = 601, max = 191,859), which represents an accumulation of contributing cells all the way to the top of the watershed, and effective upslope contributing area A_e24 (mean = 5.1, max = 375), which represents only those cells from which water would flow following a 24-h precipitation event, it is not surprising that maps of stream power and topographic wetness calculated from these measures look entirely different (Figure 7.3). Both effective and upslope contributing area are highly positively skewed (Figure 7.4). While topographic wetness (both static and dynamic) is logged and can be approximated by a normal distribution, stream power is not. Dynamic stream power shows more variation outside of the channel network than does static, demonstrates a distinctly checkered pattern with discernable diagonal trends characteristic of D8, and appears to be controlled most strongly by slope in upland areas, and by contributing area near the channels (Figure 7.5).

Differences between topographic wetness maps for the dynamic cases (variable and uniform soils) were great and those between the static cases (DEMON and D8 flow routing) were even more pronounced, but both were dwarfed by the differences between static and dynamic cases (Figures 7.3 and 7.6). The static topographic wet-

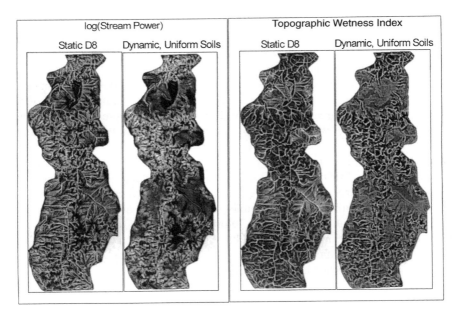

Figure 7.3. Index maps of stream power and topographic wetness based on static and dynamic uniform soil models. Maps were generated using a linear gray stretch such that lighter shades correspond to high index values, with titles and ranges as follows [black, white]: log(Static D8 Stream Power) [≤1.6, ≥6.0]; log(Dynamic Uniform Soil Stream Power) [−6.8, ≥0.4]; Static D8 Topographic Wetness Index [≤7.0, ≥14.6]; Dynamic Uniform Soil Topographic Wetness Index [≤5.9, ≥9.5].

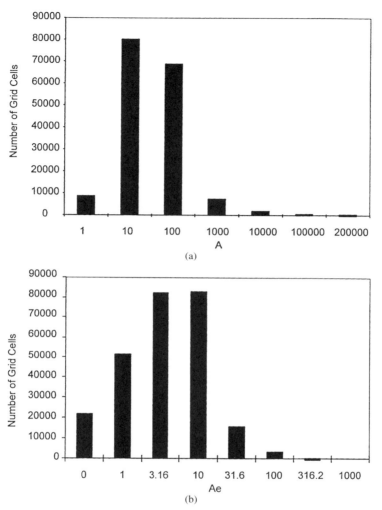

Figure 7.4. Frequency histogram of the number of $10 \times 10\text{-m}^2$ cells by upslope contributing area (measured in cells) for (a) static DEMON and (b) dynamic, uniform soil models for Barnard Drain. Note the log scales on the X axes.

ness model that relied on DEMON flow routing exhibited much smoother spatial transitions between high and low topographic wetness values, which may represent an aesthetic improvement, but there is a substantive difference as well: The drop-off in relative topographic wetness value with distance from the stream is more gradual. While every potential channel is clearly demarcated by high values of static topographic wetness, contiguous areas of high dynamic topographic wetness are less common and somewhat less pronounced.

Slope   Effective Upslope Contributing Area

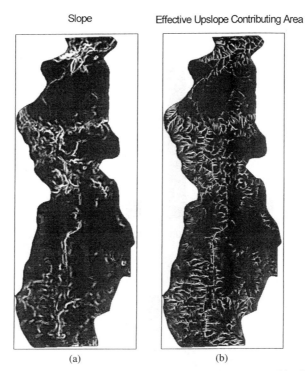

(a)   (b)

Figure 7.5. (a) Percent slope via finite difference method, as computed by TAPESG and (b) A_e (dynamic, uniform soils) for Barnard Drain. Maps were generated using a linear gray stretch such that lighter shades correspond to high index values, with ranges as follows [black, white]: Slope $[0, \geq 9.3]$; A_e $[0, \geq 24.9]$.

7.4.2 Comparison of Variable-Width Investigative Buffers

While it may be possible to identify areas of relatively higher stormwater contribution by visually interpreting the cost-distance map derived from the cost surface created by the fuzzy intersection of topographic wetness and stream power (Figure 7.7), most watershed managers would likely prefer a map product delineating an investigative buffer within which attention should be concentrated. One way of creating such a map is to select a percentile threshold for cost distance, below which is buffer and above which is not. Without further field validation and analysis, it is impossible to arrive at an empirically determined threshold. As a first approximation to facilitate our evaluation of the approach, we chose the fifth percentile as the threshold for the comparison of models (Figure 7.8, see color insert).

All four models generated what could be charitably described as variable-width investigative buffers (Figure 7.8); however, there were striking differences in form among them, and those generated by the static models bore a closer resemblance to a dendritic network than the kind of variable-width investigative buffer envisioned at

Static DEMON   Dynamic, Variable Soils

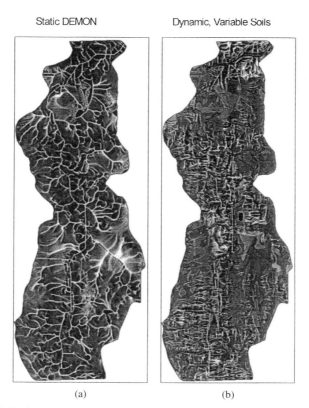

(a)   (b)

Figure 7.6. (a) Static topographic wetness index computed with DEMON flow-routing algorithm, (b) dynamic topographic wetness computed using variable soils and a 24-h drainage time. Maps were generated using a linear gray stretch such that lighter shades correspond to high index values, with ranges as follows [black, white]: Static [≤7.8, ≥15.0]; Dynamic [≤5.9, ≥9.5].

the outset of this investigation. The buffers created using static topographic wetness with both the D8 and DEMON flow-routing algorithms were quite similar to one another (Figures 7.8a and b). The primary observable difference was the presence of numerous strands of buffer oriented in an even multiple of 45°, undoubtedly an artifact of D8.

The dynamic models produced buffers that were much more consolidated than the static models, and that more clearly flagged segments of the stream meriting closer investigation to assess existing sediment filtration capacity and opportunities for remediation. The uniform and variable soils dynamic models produced buffers that were quite different from one another, with the uniform soils buffer tending toward the dendritic structure dominant in the static models (Figure 7.8).

The choice of threshold for cost distance has a significant impact on buffer delineation, and buffer thickness does not expand in a predictable fashion as the percentile

Dynamic, Variable Soils -- Cost Distance

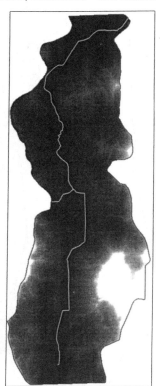

Figure 7.7. Cost distance values generated using dynamic topographic wetness and variable soil parameters for a portion of Barnard Drain displayed using a linear gray stretch with the stream overlaid. Map was generated using a linear gray stretch over ~99% of cell values such that lighter shades correspond to high cost distance while darker shades correspond to the lowest cost-distance values (i.e., cells that are close to the stream and have relatively high topographic wetness and stream power). Black 0, white ≥447.1.

threshold increases. A 15-percentile threshold generates a buffer on far more segments of stream than does a 5-percentile buffer (Figure 7.9). We chose the 15-percentile buffer for the final figure as the most inclusive buffer to illustrate the combination with areal data (e.g., land use).

7.4.3 Concordance With Validation Dataset

7.4.3.1 Flow Lines   Examination of flow paths overlain on static topographic wetness index and topographic contours suggests that some flow paths are misspecified, and that DEMON routing produces strands of high topographic wetness that are far more plausible than those produced via D8 (Figure 7.10, see color insert). The degree of agreement between flow paths and strands of high dynamic topographic wetness is comparatively difficult to assess, in part because dynamic strands tend to be less well defined. Buffer strands coincided with all or portions of the flow paths for all 16 of the flow paths for both static models, for 6 of the flow paths in the variable soil dynamic model, and for 12 of the flow paths in the uniform soil dynamic

Figure 7.9. Cost-distance buffers for 5 (black), 10 (black + dark gray), and 15 (black + dark gray + light gray) percentile thresholds generated from the dynamic, variable soil model.

model. However, for the static models, about one-half of the buffer strands did not correspond with the flow paths delineated by MDNR staff (Figure 7.8). In short, the static models tended to produce many buffer strands, increasing the likelihood that an interpreted flow path would correspond with a buffer strand.

The dynamic models did not generate wider buffers or buffer strands for some of the flow paths for a number of possible reasons. In several cases (4 of 16), it appeared that modeled topographic wetness was low because what might appear to be a flow path encountered a flat or water-dispersing area before reaching the stream (Figure 7.11a, see color insert). Because neither dynamic topographic wetness nor stream power maintained high values all the way to the stream, and because the buffers are constructed as a cost distance from the stream, it is not surprising that these models did not generate wide buffers in such instances. In other cases (3 of 16), lack of coincidence could be attributed to differences between the human interpreters and TOPOGRID in locating drainage routes that sometimes led to differences in interpretation of as much as 50 m (Figure 7.11b). In all cases, the drainage time (24 h) specified for DYNWET may have limited the amount of predicted saturation potential. In contrast to the static models, which assume a drainage event so long that all upslope areas contribute flow to a location, the dynamic models produced relatively few wide spots and strands.

Some of the flow paths that the MDNR interpreters had the most difficulty deciding about involved large subcatchments with low gradients near the bottom of the catchment. The interpreters were uncertain as to whether they should include such flow paths because they were not confident that such flow would ultimately extend to the stream. These were the same flow paths that were not matched by wide buffers in the dynamic models. This result reiterates the importance of viewing the flow lines as a different interpretation of the potential for sites to contribute flow to the stream. It would be a mistake to think of the flow lines as "ground truth."

7.4.3.2 Wet Spots Results of the comparisons with field observations suggest more agreement than differences among the models tested (Table 7.1). Because this point data set represents observed wet spots and field evidence of high flow, one would expect all models to generate high values of topographic wetness and/or stream power for these locations. For every model, expected values of both topographic wetness and stream power for the 24 observed wet spots are in the top half of the distribution of such values within 30 m of the stream. Furthermore, these percentile distributions are all negatively skewed: Dropping even the two or three lowest values produces a considerable boost in average index values.

Although the D8 Static model generated the largest mean topographic wetness percentile (72), the Variable Soil Dynamic model was a close second at 70, and is preferred over the others for reasons related to model assumptions and investigative buffer form outlined above. One surprising result was the substantial difference between the static models with D8 and DEMON flow routing, suggesting that the D8 algorithm does a better job of predicting wet spots than DEMON. This result might suggest that the D8 algorithm tends to produce more realistic portrayals of topographic wetness and stream-power values at individual locations. The primary criticism of the D8 algorithm is its lack of flow dispersal and its poor reproduction of the overall pattern of flow (i.e., its generation of flow lines parallel or diagonal to grid lines). Perhaps the ability of D8 to model flow at individual locations is better than its ability to represent overall patterns. The data presented here are insufficient to draw definitive conclusions about this point. Across all models, an improved correspondence might be established if observed wet areas are digitized as polygon rather than point data.

7.5 DISCUSSION

The considerable differences in form and pattern among buffers produced using different modeling assumptions demonstrate that buffer delineation is highly sensitive to the choice of topographic wetness index and soil parameters (in the case of dynamic topographic wetness), and less sensitive to the choice of flow-routing algorithm (in the case of static topographic wetness). This last finding is encouraging, because the software does not yet exist to implement the DEMON routing algorithm in dynamic topographic wetness index calculations. As noted in Chapters 3 and 5 of this book and elsewhere (e.g., Costa-Cabral and Burgess 1994), the DEMON algorithm provides a more realistic representation of two-dimensional flow and produces

TABLE 7.1 **Relative Index Values for Field-Observed Wet Spots, by Model, as Percentile of All Cells Within a 30-m Buffer of Stream**

Point Number	Point Type	Cells	D8 Pwr	D8 Wet	DEMON Pwr	DEMON Wet	Uniform Pwr	Uniform Wet	Variable Pwr	Variable Wet
408	BANKEROS	3	88	88	64	66	95	96	100	100
107	WETSPOT	3	90	92	55	62	92	98	92	99
124	DRAINPOINT	3	98	90	87	23	99	94	100	99
403	DRAINPOINT	3	89	88	95	86	88	86	99	99
112	WETSPOT	3	90	94	35	30	90	97	88	98
237	DRAINPOINT	3	91	87	94	88	93	87	97	97
505	DRAINPOINT	3	54	53	56	70	56	56	94	97
410	DRAINPOINT	6	88	85	28	82	78	86	87	95
218	DRAINPOINT	6	90	85	89	55	98	77	100	93
108	WETSPOT	3	89	96	88	91	28	88	18	91
233	DRAINPOINT	4	88	88	32	81	44	89	35	91
216	DRAINPOINT	6	93	82	82	16	99	65	100	78
220	DRAINPOINT	6	90	83	23	11	98	60	100	71
411	DRAINPOINT	3	86	87	93	90	55	93	6	69
405	BANKEROS	2	83	83	90	94	33	84	13	67
214	DRAINPOINT	5	92	74	43	4	99	43	100	62
222	DRAINPOINT	9	40	67	43	72	35	61	36	59
117	WETSPOT	9	72	42	69	35	84	47	92	50
122	DRAINPOINT	4	79	28	86	34	87	28	96	48
207	BANKEROS	4	80	20	65	7	82	14	93	35
118	DRAINPOINT	9	27	24	70	59	39	27	56	33
212	DRAINPOINT	5	96	84	76	6	98	43	98	21
228	DRAINPOINT	3	67	55	60	61	52	47	21	17
512	BANKEROS	2	74	60	81	70	80	62	26	11
Mean			81	72	67	54	75	68	73	70

Note. Rows are sorted in descending order by dynamic topographic wetness index for the variable soil case.

more accurate estimates of specific contributing areas. Yet, if the results of the static model comparison are at all valid, a DEMON-based dynamic wetness-based buffer could be expected to be somewhat smoother and less fragmented than those produced by D8, but not have an appreciably different pattern of buffer thickness along the stream.

Of the models tested, the variable-soil dynamic topographic wetness model produced buffers that most resembled the idealized variable-width investigative riparian buffer envisioned at the outset, and that stand the best chance of being interpretable by watershed managers seeking to prioritize sites for riparian buffer strip installation. This model also ranked second in correspondence between field-identified wet spots and high topographic wetness index values.

Transport appeared to be far better represented by the dynamic model, in that buffer widths reflected terrain induced reductions in stream power and topographic wetness present along potential flow paths. Because static models presuppose flow from the entire upslope contributing area (extending to the "top" of the watershed), nearly every potential flow path is represented as a contiguous chain of cells with high topographic wetness and stream power. This surely represents a departure from all but the most infrequent (e.g., 100-year storm) precipitation events.

Although the number of delineated flow paths that corresponded with buffer strands was highest for the static models, there were also many buffer strands resulting from the static models that did not correspond with delineated flow paths (i.e., "false positives"). This tendency of the static models to include far larger portions of the watershed within the investigative buffer stems from their simplistic assumption that each cell receives flow from all cells upslope, regardless of the duration of a precipitation event. Because the dynamic model limits the contributing area, the resulting buffers tend to be concentrated near the main stream and to consist of tighter aggregations of cells which more closely resemble lobes than strands. Under the dynamic model, fewer of the flow paths corresponded with buffer lobes or strands, but, in general, fewer lobes or strands were identified. Although correspondence between dynamic model buffers and interpreted flow lines was poor, the dynamic model more realistically portrays the system and produces buffers with a form closer to our expectations. The general agreement between modeled topographic wetness (and buffers) and flow paths interpreted from topographic maps should instill confidence in the approach in the minds of watershed managers.

For these reasons, the variable-soil, dynamic buffer appears to be the best choice for riparian investigative buffer delineation. However, further work is needed to ascertain the best drainage-time settings and cost-distance thresholds. Additional modeling must be supplemented with more extensive and rigorous field validation of soil moisture that may well involve dynamic monitoring at a fine spatial resolution to relate saturated conditions to storm events of various durations. Ideally, validation would also entail measurement of site-specific water and sediment flow during storm events and spring melt.

Logical extensions of this work include several alternatives for integrating land-use/cover information (Phillips 1989, Xiang 1993). An easy extension to implement would be to prioritize existing land uses within the investigative buffer for riparian buffer-strip installation. A prerequisite to this extension is evaluation of the natural buffering capacity (e.g., sediment trapping capacity) of different land-cover types via field measurements during and following precipitation events. These land types would have to be mapped within the investigative riparian buffer (Figure 7.12, see color insert) so that those land uses with little inherent buffering capacity and those deemed most likely to be net contributors of potential pollutants could be efficiently targeted. However, current land-use/cover information can be difficult to obtain, and in areas of rapid land-use turnover, may need frequent updating (e.g., from recent aerial photographs). A fundamental advantage of the investigative buffer approach is that such information would be needed for an area far smaller than the whole watershed (e.g., the 5–15% of the watershed bounded by the investigative buffer).

A somewhat more difficult extension would address the nature of land-cover (and pollution-generating capacity) *upslope* from each cell. The models presented here were designed to identify areas of surface flow extending to the stream; but no accounting for upslope land cover is considered. Thus, a wide buffer with polluted water contributed from an upslope feedlot or agricultural field is indistinguishable from a wide buffer with clean water contributed entirely from an upslope forested area. If weightings or pollutant loadings can be calculated for different land uses, these could be accumulated in the calculation of dynamic indices with some modification of the DYNWET software. However, any gains in explanatory power resulting from the endogenous inclusion of land use in such a model must be weighed against the costs of the additional data needed to implement it.

Finally, to be effectively implemented, such technical modeling efforts must ultimately be integrated with institutional realities and the preferences of riparian landowners. A combination of terrain modeling and land-use/cover mapping may be useful for selecting riparian segments meriting a high priority for remediation effort, but some kind of regulatory or incentive structure will likely be needed to translate such analyses into remediation on the ground. Ownership patterns (e.g., parcel size distributions) may also be important in the siting decision if costs of education and technical assistance are high on a per landowner basis. Because the topographic predisposition of riparian segments toward being variable source zones of potential pollution is relatively static (at least on human time scales), counties, townships, and drain commissions may find it useful to denote terrain predisposition on property records to facilitate targeting of landowner contacts and assistance efforts.

7.6 CONCLUSIONS

Terrain modeling of contributing areas for stormwater discharge appears to be an application of terrain analysis with great promise for guiding site-specific water pollution remediation efforts. Analysis for large watersheds should not be appreciably more difficult or time-consuming than for small ones, so the potential for scale economies exists. While there is general agreement between maps produced by terrain modeling and those produced by human interpretation of topographic maps, the areas of disagreement are even more interesting. The areas of greatest disagreement tend to be relatively flat, lending credence to the conventional wisdom that flow-path determination by any method can be especially challenging in areas of low relief. Preliminary validation of model outputs against field observations of wet spots provided encouraging evidence of model accuracy.

CHAPTER 8

Soil-Moisture Modeling in Humid Mountainous Landscapes

J. Alan Yeakley, George M. Hornberger, Wayne T. Swank, Paul V. Bolstad, and James M. Vose

8.1 INTRODUCTION

Distribution of soil moisture on hillslopes in upland forested watersheds plays an important role in determining a variety of ecological and physical processes, ranging from vegetation distribution and soil respiration to slope stability and stream-flow dynamics (Hack and Goodlett 1960, Hewlett and Hibbert 1967, Edwards 1975, Olson and Hupp 1986). Hydrologists have studied patterns of hillslope soil moisture in the southern Appalachian mountains for decades. For these humid, deep-soiled watersheds, several general conclusions have emerged: (1) Cove and ridge sites have comparable annual soil-moisture content, and both tend to be wetter than midslope sites; (2) variation in soil moisture is less in cove sites than upper slope sites; and (3) variation in soil moisture decreases with depth (Helvey et al. 1972, Helvey and Patric 1988).

Two general physiographic factors control soil-moisture distribution on hillslopes: topographic features and soil properties. Topographic factors controlling soil-moisture distribution include local slope and upslope contributing area (Beven and Kirkby 1979, Burt and Butcher 1985) as well as plan curvature (Moore et al. 1988b). Soil properties that control soil-moisture distribution include soil textural class and horizon depths (Boyer et al. 1990, Afyuni et al. 1993). Recent work at the Coweeta Hydrologic Laboratory in the southern Appalachians has observed that physiographic controls on soil-moisture content along hillslopes may vary between topography and soil properties, with variation depending on rainfall patterns and depth in the soil profile (Yeakley et al. 1998).

The purpose of this chapter is to present development, calibration, and testing of a terrain-based hillslope hydrology modeling framework to model soil-moisture distribution in forested watersheds in the southern Appalachian mountains. The resolu-

Terrain Analysis: Principles and Applications, Edited by John P. Wilson and John C. Gallant.
ISBN 0-471-32188-5   © 2000 John Wiley & Sons, Inc.

tion of the modeling framework can be set to time scales as small as an hour and to spatial scales as small as $10 \times 10 \text{ m}^2$.

8.2 STUDY SITE

The experimental site was Watershed 2 (WS2) at the Coweeta Hydrologic Laboratory, located in the Coweeta syncline in the eastern part of the southern Appalachian Blue Ridge (Figure 8.1). WS2 is a 12.3-ha first-order drainage, with an elevation range spanning 709 to 1009 m and a mean slope of 30%. Vegetation in WS2 was predominantly an aggrading mixed oak–hickory forest. The forest in WS2 has not been disturbed since the 1920s; before that time, it is likely that the interior areas of the watershed were used for small-scale agriculture.

The soils of Coweeta are predominantly Ultisols and Inceptisols underlain by a deep saprolite layer, derived from mica gneiss parent material. Overall weathering profile thickness (depth to bedrock) averages 6 m (Swank and Douglass 1975). Three soil series have been identified in WS2: Cullasaja–Tuckasegee (fine-loamy, oxidic,

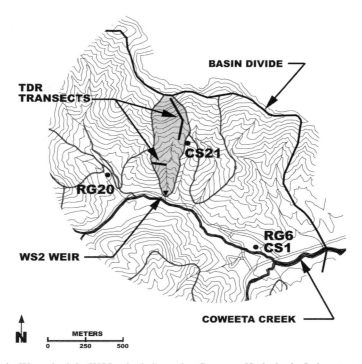

Figure 8.1. Watershed 2 (WS2, shaded) at the Coweeta Hydrologic Laboratory. Contour interval is 50 ft. Time domain reflectometry (TDR) transects included 188 individual soil-moisture measurement sites. Climate stations shown include recording rain gages (RG6, RG20) and meteorological stations (CS1, CS21).

mesic Typic Haplumbrept) at lowest elevations near the stream; Fannin (fine-loamy, micaceous, mesic Typic Hapludult) throughout most of the lower to midelevation areas, including east and west divides; and Chandler (coarse-loamy, micaceous, mesic Typic Dystrochrepts) at the highest elevations near the Coweeta Basin divide (Thomas 1996). Soils in WS2 are characterized by sandy loam to sandy clay loam texture, with relatively high organic matter content in the O and A horizons (Yeakley et al. 1998).

Coweeta's climate has been classified as marine, humid temperature (Cfb) because of high moisture and mild temperatures (Swift et al. 1988). The long-term mean precipitation at RG6 (685 m MSL) is 182 cm. Precipitation increases with elevation along the east–west axis of the Coweeta Basin, about 5% per 100 m. Precipitation at Coweeta peaks in spring—March has a long-term mean of 20.3 cm—while the driest month occurs in autumn, with a long-term mean of 11.2 cm in October. Air temperature at Climate Station 01 (CS1, Figure 8.1) varies from a monthly mean of 3.3°C in January to a monthly mean of 21.6°C in July (Swift et al. 1988).

8.3 MODELING APPROACH

A terrain-based hillslope hydrology modeling framework was developed from existing models. The modeling framework consisted of three modules of code: An aboveground module for climate processing and interception, which was coded based on the primary literature (Rutter et al. 1971, 1975, Swift 1976); terrain-analysis programs (PREPROC and TAPES-C; Moore and Grayson 1991; Chapter 3, present text); and a watershed hydrology module (IHDM4; Beven et al. 1987, Calver 1988, Calver and Cammeraat 1993).

The overall approach used the terrain-analysis programs PREPROC and TAPES-C to determine the boundaries of the hillslope planes in the watershed hydrology model (IHDM4). The basic structure of the IHDM4 is a network of hillslope and channel components. Each hillslope plane contributes headwater or lateral discharge into a channel component at its base. No-flow boundaries are assumed to exist between adjacent hillslope planes. Each hillslope plane is represented by a two-dimensional, vertical cross-section of element nodes. Subsurface flow is simulated using a finite element solution of a 2D Richards equation, which allows the planes to vary in width to account for hillslope flow convergence. The planes may also vary in depth and spatially with respect to soil hydraulic properties. The canopy module either delivers positive fluxes (precipitation) or abstracts negative fluxes (evapotranspiration) at the surface. These fluxes are modified by soil-moisture availability and the fractional root distribution in a given layer. Overland and channel flows are modeled with a kinematic wave solution.

8.3.1 Terrain Analysis

As Moore and Grayson (1991) outlined, there are three primary ways of structuring a network of topographic data: (1) triangulated irregular networks (TINs); (2)

raster or grid networks; and (3) vector- or contour-line-based networks. Of the three, contour-based networks provide more physical realism than grid-based networks that restrict water flow from a given node to only one of eight possible directions. TINs provide physical realism, but require interpretive alignment of the elements, many times based on vector digital elevation maps (DEMs). Moore and Grayson (1991) provided an automated contour-based method (TAPES-C) for partitioning watersheds into natural units bounded by irregularly shaped polygons. These polygons are bounded by equipotential (or contour) lines on two sides and by stream lines, orthogonal to the contours, on the other two sides. Stream lines are assumed to be no-flow boundaries; thus, subsurface flow is constrained to move through a series of elements positioned along a natural gradient. Such a series of cells was termed a "stream tube." By orienting the flow equations of a distributed parameter model along stream tubes, spatial complexity in the equations may be reduced from three dimensions to two, while accomplishing a terrain-based model structure.

Contour-based terrain analysis as developed by Moore et al. (1988b) and Moore and Grayson (1991) required three general steps. First, a contour map of the watershed was digitized, creating a vector DEM. Here the ARC/INFO GIS was used to accomplish this task for WS2. Then a preprocessing program (PREPROC) was used to transform the vector DEM into a north–south-oriented coordinate system. Finally, the program TAPES-C partitioned the watershed into stream-tubes or hillslope units using a constant offset between trajectory (i.e., stream-tube boundary) starting points. Results from a TAPES-C computation for WS2 using a 50-m offset are shown in Figure 8.2.

Further processing was then required to transform the stream-tube output of TAPES-C into a structure suitable for IHDM4. Each hillslope plane in IHDM4 was represented by a two-dimensional vertical cross section of finite-element nodes running longitudinally from watershed divide (or interior high point) to stream. At each cell boundary within a stream tube (e.g., in plane #1, Figure 8.2, there are five cells and six cell boundaries along the stream tube), a vertical set of nodes represents soil layers. At each cell boundary, a constant width was assumed for all soil layers. To fit the TAPES-C output to IHDM4, the no-flow boundaries shown in Figure 8.2 (left) were extended to permanent stream locations using ARC/INFO to derive 15 hillslope planes (Figure 8.2, right). Arcs bounding stream-tube cells were selected to maximize the criteria that the arcs be positioned parallel to the contours and the streamside arc be parallel to the stream. For example, in plane #1 (Figure 8.2), upslope and downslope arcs bracketing a given cell are observed to run along contour lines in the upper slopes, but then gradually shift toward a position parallel to the stream at the streamside arc.

From map view, a hillslope plane in IHDM4 was constrained to a series of adjacent trapezoids beginning at the stream and continuing to the divide. Area and slope for the surface of each cell (i.e., each four-sided polygon) were calculated using ARC/INFO. Subsequently, for each stream tube, widths of the cells were allowed to vary in order to transform the cells into trapezoids while maintaining area and slope for each cell. The transformation proceeds iteratively from the streamside arc up the stream tube using the relation

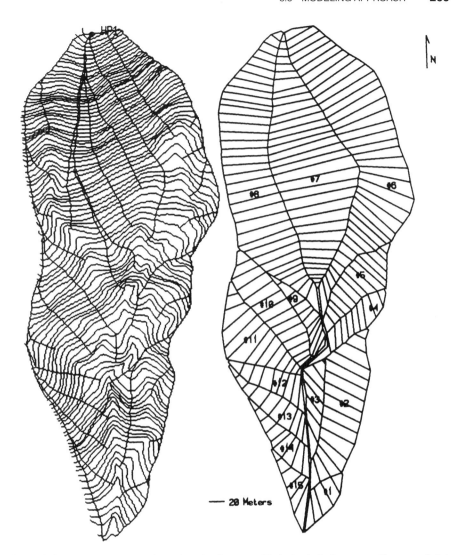

Figure 8.2. Terrain analysis of WS2 at the Coweeta Hydrologic Laboratory. Shown at left is the stream-tube delineation using TAPES-C programs; contour interval is 10 ft, with elevation range 2340 to 3295 ft. Shown at right are the resulting hillslope planes fitted (but untransformed) for IHDM4. Hillslopes #6 and #12 were instrumented with a TDR network for soil-moisture sampling.

$$Y_{N+1} = \frac{\text{Area}_{N,N+1}}{X_{N,N+1}} \tag{8.1}$$

where Y_N is the effective width of arc N, $\text{Area}_{N,N+1}$ is the original area of the polygon bounded by stream-tube orthogonals and arcs N and $N+1$, and $X_{N,N+1}$ is the average distance between arcs N and $N+1$.

8.3.2 Canopy Interception Modeling

The above-ground module used was a modified Rutter interception model (Rutter et al. 1971, 1975, Halladin et al. 1979, Calder 1990). The modified Rutter model follows a dynamic canopy storage (C, in cm) with input of a constant fraction of rainfall determined by leaf area index and vegetation type, and output as evaporation and drainage. The equations of the canopy model are

$$\frac{dC}{dt} = Q - K_c[\exp(b_c C) - 1] \tag{8.2}$$

$$Q = (1 - p)R - E_p * f(C) \tag{8.3}$$

where K_c and b_c are drainage parameters (cm/h), R is the total rainfall (cm/h), E_p is potential evaporation (cm/h, determined by a Penman–Monteith equation with stomatal resistance set to zero), and p is the canopy throughfall fraction. If $C > S$, then $f(C) = 1$, or else if $C < S$, then $f(C) = C/S$, where S is interception storage and corresponds to a completely wet canopy. The model allows for simultaneous evaporation and transpiration from a partially wet canopy ($C < S$).

Rutter et al. (1975) specified the rate of drainage, D, from the canopy as $\ln D = a + b_c C$, where the drainage parameter $K_c = \exp a$. Rutter et al. (1975) found $D = 0.012$ cm/h for a Corsican pine (*Pinus nigra*) stand when $C = S$. For that stand, $a = \ln(0.012) - b_c S$. The drainage parameter b_c was found to equal 22.2 cm/h (Rutter et al. 1975). The numerical value 0.012 cm/h for D was an empirically determined constant for the Corsican pine canopy. For other canopies with different leaf area indices (L), this constant was expected to be $0.012 * (L/L_c)$, where L_c is the leaf area index for the Corsican pine canopy. L/L_c was assumed proportional to S/Sc where $S_c = 0.105$ and was the value of S for Corsican pine. The drainage constant D was then modified to be $0.012*(S/0.105)$. Rutter et al. (1975) provided values for S for oak stands; here for WS2, S was used as a calibration parameter.

The above-ground model is regulated by a water balance given as

$$R = T + E + \Delta C \tag{8.4}$$

where T is throughfall and E is evaporative loss. Transpiration demand is calculated as E_p for that fraction of the canopy which is dry. An effective precipitation is then calculated that is the throughfall amount (which includes direct throughfall as well as drainage) minus the transpiration demand. In the absence of throughfall, effective precipitation at the soil surface is negative, which is input to the hillslope hydrology model (IHDM4) as a sink term at the surface. The sink is regulated by soil-moisture availability times the fractional root distribution in a given layer in the hydrology model as given by Feddes et al. (1976). If positive, that is, if rainfall is occurring, then input to the surface becomes a source term.

8.3.3 Hillslope Hydrology Model

Within a stream tube, a cell is bounded by two vertically layered sets of finite element nodes. The top surface of the soil (i.e., highest set of nodes) is treated as a flux boundary with fluxes controlled by the applied input rates of effective precipitation unless the surface becomes saturated and overland flow develops. The surface boundary then changes to a fixed-head boundary while saturation persists, with the potentials fixed at atmospheric pressure. The change of boundary conditions at the soil surface can occur locally on the slope to enable simulation of a dynamic variable source area (Beven et al. 1987).

For a given stream tube, subsurface flow is given by the Richards equation:

$$BC(\psi)\frac{\partial \phi}{\partial t} - \frac{\partial}{\partial x}\left[BK_x(\psi)\frac{\partial \phi}{\partial x}\right] - \frac{\partial}{\partial z}\left[BK_z(\psi)\frac{\partial \phi}{\partial z}\right] = Q_s \qquad (8.5)$$

where B is a gradually varying stream-tube width, ψ is capillary potential, x is horizontal distance downslope, z is gravity potential (measured vertically from some arbitrary datum), ϕ ($= \psi + z$) is total hydraulic potential, $C(\psi)$ is the specific moisture capacity of the soil (slope of relation between θ and ψ), θ is soil-moisture content by volume, K_x, K_z are saturated hydraulic conductivities in the x, z directions (functions of ψ), Q_s is a source/sink term, and t is time. Implementation requires several assumptions: (1) Water is of constant viscosity and unit density; (2) flow occurs in an isothermal medium; (3) Darcy's law applies with time-invariant parameters; (4) only single-phase water flow in response to hydraulic gradients is considered; (5) the relationship between ψ and θ is locally differentiable (Beven et al. 1987).

If either the infiltration capacity of the soil surface is exceeded by input rates or the soil becomes fully saturated resulting in return flow, then overland flow occurs and is given by

$$B\frac{\partial Q}{\partial t} + c\frac{\partial}{\partial x}[BQ] - Bic = 0 \qquad (8.6)$$

where Q is discharge, i is net lateral inflow rate per unit downslope length, x is distance downslope, and c is kinematic wave velocity defined by dQ/dA, where A is the cross-sectional area of flow. Solution requires a specification between discharge and cross-sectional area, which in IHDM is given as $Q = fs^{0.5} A^b$, where s is local slope angle, f is an effective roughness parameter, and b is a calibration parameter (Beven et al. 1987).

Soil-moisture characteristics are determined using modified Campbell (1974) relationships with parameters based on soil textural differences (Clapp and Hornberger 1978). Actual evapotranspiration (E_a) is given as a function of E_p and soil moisture based on Feddes et al. (1976):

$$E_a = W_r \alpha(\psi) E_p \qquad (8.7)$$

where W_r is a weighting of proportion of root mass for a depth and

$$\alpha(\psi) = \begin{cases} 0 & \text{if } \psi < \psi_w \\ \dfrac{\psi - \psi_w}{\psi_i - \psi_w} & \text{if } \psi_w < \psi < \psi_i \\ 1 & \text{if } \psi_i < \psi < \psi_s \\ 0 & \text{if } w_s < \psi \end{cases} \qquad (8.8)$$

Note here that ψ_s is anaerobiosis point (−0.05 bars), ψ_i is vegetation stress initiation point (−0.3 bars) (Hewlett 1962), and ψ_w is wilting point (−15.0 bars).

At the end of each subsurface time step, inputs from each hillslope section to both overland flow and the channel are calculated. To compute channel flow, IHDM4 uses the kinematic wave equation and power law flow relationship, specified above, for the overland flow solution on stream tubes, except that each channel is assumed to be of uniform width. Four levels of time step occur in IHDM4. The highest level is the input climate data timestep, which here is at one-hour intervals. The next level involves flux exchange between hillslope and channel at a time step equal to or smaller than the climate step. Subsurface and channel flow is calculated at a finer time resolution; following previous work (Calver and Wood 1989), a one-half-hour step was used. Finally, overland flow if it occurs is calculated a fixed number times in each subsurface flow time step (Beven et al. 1987).

8.4 PARAMETERIZATION AND CALIBRATION

8.4.1 Structural Parameters

Plan view structural parameters were determined using the TAPES-C terrain-analysis programs. Vertical structural parameters, that is, soil layer depths, were based on published literature (Gaskin et al. 1989, Vose and Swank 1991), U.S. Natural Resource Conservation Service (NRCS) surveys, and auger measurements. Soils were partitioned into eight layers. The four upper layers were fixed at 25 cm thick each and the bottom four layers were partitioned into four equal thicknesses that varied to bedrock depending on hillslope position.

Depth to the no-flow boundary at the bottom of each stream tube was determined from a reanalysis of drilling data from N. H. Miner (1968, unpublished report, Coweeta Hydrologic Laboratory). Miner did not find a statistically significant relationship between depth to bedrock and percent distance ($\%D_r$) between streams for the entire dataset ($n = 139$ drill sties throughout the basin). Those data were reevaluated here. Data were restricted to sites that had similar aspect (90° to 270°) and elevation range (2400 to 3500 ft MSL) as WS2 ($n = 45$ drill sites). This analysis produced a regression equation with an adjusted $r^2 = .34$ ($p < .001$):

$$D_b = 0.36 * (\%D_r) + 7.9 \tag{8.9}$$

where D_b is slope-corrected depth to bedrock (=ft) and $\%D_r$ is percentage distance (plan view) from stream to ridge. This equation underpredicted depth at the stream (mean of reduced set ($n = 3$) was 10 ft) and overpredicted depth at the divide (mean of reduced set of ridge sites ($n = 8$) was 42 ft), but only by 2 and 1 ft, respectively.

Accordingly, depth at the stream-tube seepage face was set at 3.0 m. For stream tubes in WS2 that did not reach the watershed boundary (#3, #9, see Figure 8.2), this depth was held constant. For these hillslopes, the depth of the bottom four soil layers is a constant 0.5 m. For lower elevation stream tubes (#1, #2, #4, #5, #10–#15), the depth was set to vary from 3.0 m at the seepage face to 6.0 m at the divide. Bottom soil layers thus varied from 0.5 to 1.25 m thick. For the upper elevation streamtubes (#6–#8), thickness varied from 3.0 m at the seepage face to 12.0 m at the divide; thus, bottom layers varied from 0.5 to 2.75 m thick.

8.4.2 Above-Ground Parameters

The distributed model (IHDM4) receives inputs from the above-ground model that accounts for canopy and litter fluxes. The timestep of the aboveground model was set at 1 h for several reasons: (1) variation in intensity and duration of source water due to seasonal variability in storm type (March et al. 1979); (2) canopy evaporation and drainage, which are significantly affected by diurnal fluctuation in solar radiation and temperature; (3) stormflow response time.

Data for calibration were taken from two climate stations and one rain gage for the study period of the simulations (Figure 8.1). Rainfall measurements were conducted on at least an hourly basis (15-min intervals during storms) at RG20 (located at an elevation more representative for WS2 than RG6). Temperature and relative humidity were available on a daily basis from CS21. Windspeed and incident solar radiation were measured for the period 1974–1985 on a daily basis from CS1. From 1986 to 1991, hourly measured solar and wind values were used. CS1 values for solar radiation and windspeed were used because the CS21 data indicated a negative correlation with canopy presence. With the exception of rainfall, climate variables were primarily measured at a daily time step that necessitated several daily to hourly data transformations. Translation of direct solar radiation measured by the horizontal solarimeter at CS1 to the slope and aspect of WS2 was accomplished using the algorithms of Swift (1976).

The Coweeta Basin has been the focus of much research quantifying interception loss for various cover types (Waring et al. 1980). Helvey and Patric (1965) conducted a meta-analysis on results from the eastern hardwood region, deriving equations to predict annual interception loss from total precipitation amount and number of storms. Solving the equations for Coweeta precipitation and number of storms, annual interception loss was found to be 13% of total precipitation. Recent measurement during the Integrated Forest Study for 3 years in WS2 found interception loss to equal 12% of total precipitation.

Above-ground parameters were chosen based primarily on Swift et al. (1975) and Rutter et al. (1971, 1975). Parameters were chosen for three scenarios for the mixed oak–hickory canopy of WS2: summer, winter, and transition (April, October). For the transition period, intermediate values between summer and winter were chosen. Trial and adjustment calibration was performed for annual interception loss; the parameter adjusted during the calibration was interception storage, S. It was specified that S during leafout periods would approximately double S during winter periods after leaf abscission. Values for S were obtained for all seasons that gave a satisfactory prediction for interception loss over an annual period. For a simulation period from 1975 to 1989, interception loss averaged 13.4% of annual precipitation, with a range from 10.7 to 18.7%. Final above-ground parameters are shown in Table 8.1.

TABLE 8.1 Model Parameters

Canopy Characteristics

Variable	Summer	Transition	Winter
Roughness length (z_0)	0.02	0.035	0.05
Zero plane displacement (d)	0.83	0.665	0.50
Albedo	0.22	0.19	0.16
Throughfall fraction (p)	0.47	0.645	0.82
Canopy height (h) [m]	19.	19.	19.
Interception storage (S) [cm]	0.2	0.15	0.1
Drainage parameter (K_c) [cm/h]	3.3e4	7.4e4	1.4e3
Drainage parameter (b_c) [cm/h]	22.2	22.2	22.2

Conductivity by Soil Layer

Layer	Depth [cm]	Sat. K_x [cm/h]	Sat. K_z [cm/h]
1	0–25	0.20	4.0
2	25–50	0.05	1.0
3	50–75	0.015	0.3
4	75–100	0.015	0.3
5	100–150[a]	0.003	0.06
6	150–200[a]	0.003	0.06
7	200–250[a]	0.003	0.06
8	250–300[a]	0.003	0.06

General Soil Characteristics

Texture	Sandy loam
Porosity (Θ_s)	0.45
Wilting point potential (ψ_w)	−15 bars
Stress initiation point potential (ψ_i)	−0.3 bars

Root Zone Characteristics

Depth [cm]	% Active Roots (Summer/Winter)
0–10	47/80
10–20	23/20
20–30	12/0
30–40	11/0
40–80	7/0

Surface Flow Parameters

Regime	Roughness [m$^{0.5}$/h]	Exponent [b]
Overland	2.5e3	1.5
Channel	1.0e5	1.5

[a]Variable bottom depth.

8.4.3 Below-Ground Parameters

Calibration of IHDM4 for soil moisture involved a two-step approach (Hornberger and Cosby 1985): (1) fitting watershed-scale stream-flow prediction to observed across a suite of storm types, and (2) fitting hillslope-scale soil moisture prediction to measured soil moisture data in upper soil layers. In prior applications, IHDM has usually been applied to wet antecedent conditions to predict stormflow in catchments with relatively shallow soils for single storm events (Rogers et al. 1985, Calver 1988, Calver and Wood 1989). As such, both calibration steps here involved an extension of the model to a deeper-soiled catchment over seasonal timescales that included periods of drought, that is, to conditions for which it typically has not been applied.

8.4.3.1 IHDM Calibration Background Hydraulic characteristic parameters (primarily vertical and lateral saturated conductivity, and porosity) are most sensitive for the IHDM (Rogers et al. 1985, Beven et al. 1987, Calver 1988). Another parameter that has previously been sensitive was the overland flow power coefficient, b (Beven et al. 1987). Wood and Calver (1992) discussed how initial moisture conditions affect modeled stream-flow behavior. They drew a distinction between an initial uniform moisture potential and an initial "saturated wedge" positioned at the base of a hillslope plane where $\psi = f(z)$. Simulations using a uniform initial potential over a sequence of storms tended to produce successively higher base flows. Saturated wedge initial conditions produced more stable, or repeatable, base flows. Calver and Wood (1989) also showed that the model was sensitive to the ratio of lateral (downslope) to vertical element spacing. They recommended that the ratio be held to less than 20 overall and that near the seepage face the ratio should approach 1.

8.4.3.2 Soil-Moisture Response The primary concern of this work was to simulate realistic soil-moisture responses. Overland flow has been an important mechanism in previous IHDM runs. For example, Calver (1988) showed that such an overland flow condition persisted for at least 35 h after the onset of the storm, and that overland flow accounted for as much as 89% of the storm flow. For the deep-soiled forested slopes at Coweeta, however, overland flow is rarely seen even during the most intense storms (Hewlett and Hibbert 1963). Here, an approach was taken that would deemphasize the importance of the overland flow component of the model, by adjusting subsurface model parameters.

Soil-moisture characteristic parameters used in the IHDM were those found for sandy loam by Clapp and Hornberger (1978). Hydraulic conductivity was specified to decrease realistically with depth. Primarily, although not entirely, hydraulic conductivity was used to calibrate for stream flow and for the timing of soil-moisture response; porosity was used to calibrate for the level of the soil-moisture response. A severe problem that existed for all early simulation attempts was a "flash" effect for antecedent dry conditions. From examination of nodal values during these storms, it was apparent that a saturated condition formed in the top set of nodes, which persisted for several hours or even days. A solution to this problem was found by dropping the isotropic assumption and assuming rather that vertical saturated conductivity exceeded horizontal saturated conductivity. This assumption reduced the occurrence of overland flow by increasing infiltration rates. Gaskin et al. (1989) in an adjacent watershed at Coweeta showed that vertical solute flux exceeded lateral flux by as much as an order of magnitude. Prior work at Coweeta has shown a dominance of subsurface flow, with virtually no overland flow due to high infiltration capacities in forest soils in the southern Appalachians (Hewlett and Hibbert 1963). Wetting front dynamics on modeled hillslopes during storms matched observations much more closely as a result of this modification.

8.4.3.3 Additional Model Specifications Following Calver and Wood (1989), lateral spacing of the vertical nodes was specified as 0.5 m in the first 10 m from the seepage face, 1.0 m in the next 10 m, 2.5 m in the next 10 m and no more than 20 m for the remainder of the stream tube. This spacing provided a lat/vert ratio = 2 and 1 in the top and bottom layers, respectively, near the stream. The lower 1.0 m of the seepage face of all stream tubes was specified as a fixed-head boundary with the potential set to atmospheric pressure. The upper 2.0 m of the seepage face was set to remain unsaturated and so was a no-flow boundary.

After numerous comparisons, imposing a saturated wedge to begin a simulation (Wood and Calver 1992) provided a more representative response across storm types. A particular problem using uniform initial potential was an increasing stream flow after the recession curve had returned to base flow. While a sufficiently long, continuous simulation (e.g., 4 months) gradually resulted in a wedge forming near the base of the hillslope, this unrealistic post-storm response was yet evident. A saturated wedge was thus imposed to start any simulation by calculating soil-moisture potential as a decreasing function of elevation.

Root structure in the model followed McGinty (1976). Based on that information for WS 2, root mass was specified in the 0- to 10-cm depth as 47%, at 10–20 cm as 23%, at 20–30 cm as 12%, at 30–40 cm as 10%, and at 40–80 cm as 8%. This vertical distribution was held constant for the entire watershed. For the winter months (October–April) subsurface evapotranspiration was restricted to 80% in the top 10 cm and 20% from 10 to 20 cm. This distribution reflected the lack of evapotranspiration originating from deeper layers during periods after leaffall and before spring canopy regrowth.

Several vertical layer spacings (soil depths) were attempted but eventually an eight-layer spacing was adopted. The top layer corresponded to a composite A–BA horizon, while the 2nd and 3rd layers corresponded to the B horizon, the 4th layer to the BC horizon and the bottom four layers to the C, or saprolite, horizon. The soil was prescribed as a sandy loam based on soil textural analysis, Gaskin et al. (1989), and the local NRCS survey (Thomas 1996).

8.4.4   Storm-Scale Calibration

Stream-flow calibration attempted to minimize an objective function of the sum of squared errors (SSE) between observed and simulated stream flow. Calibration also attempted to minimize water balance error as a percentage of precipitation. Superceding these considerations, however, was the need to achieve reasonable soil-moisture infiltration dynamics. Therefore, no single objective function was minimized. The calibration relied more on "eyeball fitting" that minimized error at both the hillslope and watershed scales across a suite of possible storm types.

Three calibration storms were selected from 1975 to largely span the range of possible antecedent values and precipitation intensities (cf. flow frequency distribution curve of Swift et al. 1988): (1) the Jan 75 storm with a base flow of 10 m³/h and a peak flow of nearly 60 m³/h; (2) the Mar 75 storm represented a prior-wet heavy spring storm, with a base flow of 25 m³/h and a peak of 140 m³/h; and (3) the Sep 75 storm represented a series of dry-season, high-intensity storms, with a base flow of 2 m³/h and a peak flow of 120 m³/h. It was possible to obtain a different close-fitting parameter set for each storm type, but it was much more difficult to obtain a single parameter set for all storms. A single invariant parameter set was specified after numerous simulations of the model across all three storm types (Table 8.1).

8.5 VALIDATION

8.5.1 Storm Scale

8.5.1.1 Stream-flow Response   A validation run was conducted for the period that coincided with measurement of hillslope soil moisture, from 1 Nov 91 to 12 Dec 91. This period began at the end of the second most extreme 2-month drought on the 64-year Coweeta record. Two storms occurred, depositing a total of 24.3 cm of rainfall over the validation period. Such a period constituted an extreme test of the model, spanning climatic extremes over a relatively short timeframe.

Figure 8.3 shows model results against the observed hydrograph for WS2 for the validation period. Model efficiency (Hornberger et al. 1992) was found to be less than zero at −0.43 for the period from 10.4 to 42 days. Although timing of model response was reasonably close, the model was incapable of reproducing observed storm-flow magnitudes. Storm-flow peak was initially overpredicted, but subsequent predictions were low. This behavior also occurred in the calibration storms: Very dry antecedent conditions caused a flash effect in stormflow, while wetter antecedent conditions produced an underprediction of storm-flow volume. Total storm-flow volume over the period from 10.4 to 42 days was underpredicted by the model, differing from the observed by −1.3 m³/h over the validation period.

Validation was also conducted on an annual basis to determine how well the model predicted hydrologic components. Two hillslopes (#11, #12) were simulated for the 1975 water year (May 75–Apr 76) and assumed to reflect the water balance at the watershed scale. Precipitation delivered to WS2 for that year was only slightly higher than the long-term mean. This average annual climatic condition allowed comparison with long-term values of evapotranspiration losses, estimated by precipitation minus runoff (P-RO) (Swift et al. 1988). For WS2, Swift et al. (1988) showed P-RO to be 51.8% of mean annual precipitation (177.2 cm). For the water year 1975, predicted evapotranspiration losses (P-RO) equaled 62% of annual precipitation

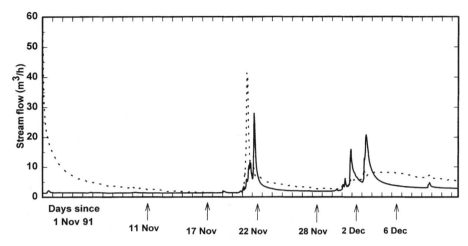

Figure 8.3. Validation run for stream flow over the period 1 Nov–12 Dec 91 at WS2. Solid line, observed data; dashed line, simulated. Arrows indicate soil-moisture validation points (see Figure 8.4).

(186.4 cm in 1975). Thus, for a near-average precipitation year, evapotranspiration was overpredicted and runoff was underpredicted at the watershed scale.

8.5.1.2 Soil-Moisture Response
Validation for hillslope moisture patterns was conducted using the same period following extreme seasonal drought: 1 Nov 91–12 Dec 91. Based on the terrain analysis, a time-domain reflectometry (TDR) transect was established along the center of hillslope plane #12 in WS2 (Figure 8.2). On plane #12, TDR sites were established with three replicates at each of two depths (0–30 cm, 0–90 cm) along an 85-m transect. Fourteen sites were established giving a total of 84 TDR measurement points. A sample point consisted of two 3-mm dia stainless steel welding rods (i.e., TDR rods) set 5 cm apart and inserted vertically into the soil surface. Methods of TDR measurement followed those specified by Topp et al. (1985) and have been described in detail by Yeakley et al. (1998).

Six moisture measurement periods were chosen to span various stages of base flow, peak flow, and recession over the validation period (Figure 8.3). Modeled moisture content for the 0- to 90-cm depth was interpolated from nodal values. Plan distance to stream was used for both model and measured values. Modeled moisture values were six hour averages taken from the period 0600–1200 on the day of measurement. Figure 8.4 shows modeled and measured soil moisture at the 14 sites along the transect for the six measurement periods.

Least-squares regression was used to determine soil-moisture slope. The independent variable was normalized slope distance; the dependent variable was moisture content as a fraction. Two-tailed t tests were used to test whether significant differences existed between both the means and regression coefficients (i.e., slopes) ($p < .05$; Zar 1984). Significant differences between modeled and measured soil-moisture means were found for 3 of the 6 dates (11 Nov, 22 Nov, 28 Nov). Modeled and measured slope estimates, however, were not significantly different.

One source of error in the storm-scale validation was due to thicker BA and B horizons in the footslope soils of the lower transect (about 60 m from ridge, Figure 8.4), which have been shown to have higher clay content and therefore higher water-holding capacity (Yeakley et al. 1998). While the modeling approach used here varied soil hydraulic parameters vertically, variations along the slope were not taken into account.

8.5.2 Seasonal Scale

The accuracy of this modeling approach was tested for a multiannual period. An extensive TDR transect was installed on one of the upper hillslope planes (#6) previously determined by terrain analysis (Figure 8.2). TDR sites were installed every 25 m (plan distance) on a transect 335 m in length that began near the high point (HP1) of the watershed and ended near the emergent point of the permanent stream on plane 6. Each TDR site consisted of measurement points with 4 replicates at each of 2

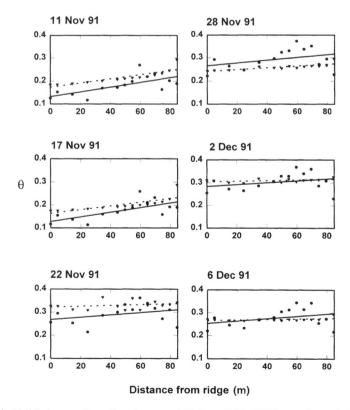

Distance from ridge (m)

Figure 8.4. Validation run for soil moisture on hillslope #12 in WS2 over the period 1 Nov–12 Dec 91. Abscissa is meters from ridge (plan view); ordinate is fractional soil-moisture content. Circles are measured values (mean of 3 replicates); inverted triangles are simulated results. Solid line, regression on measured data; dashed line, regression on simulated results.

depths (0–30 cm, 0–60 cm). Measurements were taken at or near the 20th of each month, beginning Oct 94 and continuing through Oct 96.

Climate data used in the model were taken from the CS1 climate station and the RG6 rain gage (Figure 8.1). Temperature, soil radiation, windspeed, and relative humidity were available on an hourly time step for the period beginning Jan 94 through Oct 96. Precipitation for this entire period was available on a daily time increment. Each daily precipitation value was partitioned into 24 hourly segments for these model simulations; the daily hyetograph for this period is shown in Figure 8.5 (top). The model was simulated for the period Jan 94 through Oct 96 for hillslope plane #6, using model parameters reported in Table 8.1.

8.5.3 Model Comparisons

8.5.3.1 *Mean Hillslope Soil-Moisture Dynamics*   Measured soil moisture for the TDR transect on hillslope plane #6 was averaged for each month for the 0- to 60-cm depth ($n = 56$ total sites) to obtain a mean soil-moisture response for this upper

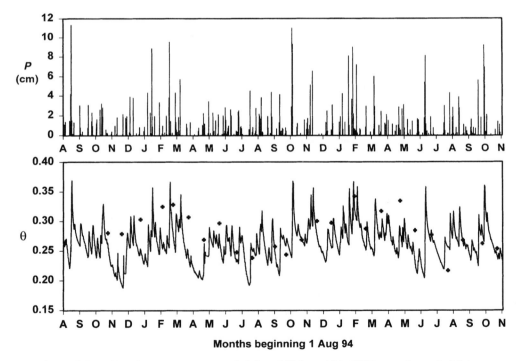

Figure 8.5. Soil-moisture response to rainfall on hillslope #6 in WS2 over the period 1 Aug 94–31 Oct 96. Daily hyetograph for RG 6 is shown in the upper graph. Mean soil moisture over a depth of 0–60 cm for upper 335 m of hillslope #6 on WS2 is shown in the lower graph. Crosses show measured soil moisture (mean of 56 replicates spanning 335 m of hillslope #6) at approximately monthly time intervals; solid line shows simulated results.

elevation hillslope plane over the 25-month period (Figure 8.5, bottom). The hill-slope model for plane #6 consisted of 62 lateral nodes by 9 vertical nodes. Soil-moisture values were obtained for 0–60 cm in depth by taking a depth-weighted average of the top 4 vertical nodes (0, 25, 50, 75 cm) at each of the 62 lateral positions in the model profile. Modeled soil moisture for each TDR measurement point was then determined from the closest two lateral nodes in the model using linear interpolation.

Modeled mean hillslope soil moisture was plotted in comparison with measured mean hillslope soil moisture (Figure 8.5, bottom). The largest discrepancy occurred in Nov 94, with modeled soil moisture 45% lower than was measured. Errors were greatest during the following period through Mar 95. Modeled soil moisture matched measured values better from Apr 95, onward through Oct 96, with monthly errors ranging from 1% to 23%, and averaging 5%, during that remaining period. Comparing the hyetograph with the modeled soil-moisture response in Figure 8.5, it is apparent that the model represented soil-moisture responses to rainfall dynamics on the hillslope reasonably well. Soil-moisture peaks follow shortly after rainfall peaks. In periods of drier weather, mean soil moisture on the hillslope appears to follow a negative exponential, with drainage slowing down with prolonged low rainfall (e.g., during Apr 95).

8.5.3.2 Seasonal Moisture Responses Along the Hillslope   During the final year of the simulation, quarterly comparisons are shown between measured and modeled soil moisture along plane #6 (Figure 8.6). These sample times spanned both a very wet period and a very dry period for this watershed. The measurement point in Jan 96, was during an extended high rainfall event. Modeled soil moisture throughout the hillslope matched measured values reasonably well, although some discrepancies occurred at 110 and 235–310 m from the ridge. The Apr 96, prediction was the poorest, with the model generally predicting values about 20% lower for most of the transect. The modeled points near the ridge and nearer the stream, however, were close to the measured values. During wet periods, the model predicted soil moisture reasonably close at the ridge and near the stream, but generally was lower than measured values on the midslope.

Drier periods were also simulated, and are shown for Jul and Oct 96 (Figure 8.6). For each of these drier months, modeled values were a closer fit than during wetter months overall. Modeled soil moisture for ridge soils for the driest month, July 1996, was significantly higher than actual values. The model predicted a persistent near-saturated condition on the lower end of the hillslope, even during drier months. Generally, during drier months, the model predicted midslope soil moisture reasonably well, but did worse near the ridge and nearer the stream.

8.6   DISCUSSION

The modeling framework is physically based, incorporating realistic structure both at the watershed scale in terms of terrain variation and at the plot scale in terms of variation in hydraulic parameters vertically in the soil profile. In terms of vegetation

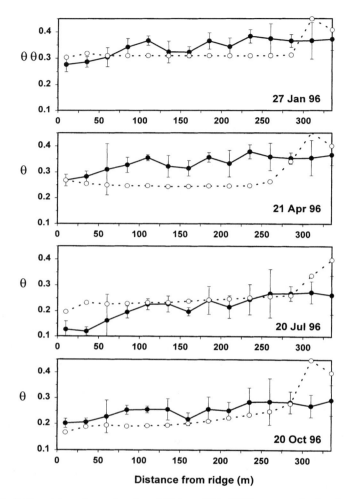

Figure 8.6. Soil-moisture dynamics along hillslope #6 in WS2 at quarterly time intervals during 1996. Abscissa is meters from ridge (plan view); ordinate is fractional soil-moisture content. Solid circles are measured data; error bars are 1 standard deviation. Hollow circles are simulated results.

interaction, the model uses a modified Rutter canopy interception module to account for interception loss, a Penman–Monteith representation of evapotranspiration, and a Feddes function to account for soil-moisture tension control over actual evapotranspiration. Calibration was conducted at the watershed scale using information from historical canopy interception studies and from stream-flow records. Calibration was conducted over both wet and dry antecedent storms to arrive at an invariant parameter set to be used regardless of season. The calibrated model predictions were compared with measured soil moisture on two different hillslopes for two different timescales; namely individual storm and multiple years.

The model was found to be effective at capturing mean soil-moisture response both during storm events (Figure 8.4) and over multiannual periods (Figure 8.6). Modeled soil moisture along hillslope planes was always within ranges expected for sandy loam and sandy clay loam soils. Additionally, the modeling approach used here represents hillslope gradient changes, with soil-moisture gradients along hillslopes responding appropriately during storm events and for seasonal changes.

Examining the hillslope gradient responses for either the storm event or across seasonal changes, however, shows that the model generally does not always hit the extremes as well as might be hoped. For example, during the extreme drought that extended into early Nov 91, the modeled moisture gradient is slightly higher than was measured (Figure 8.4). Then during the wet winter of 1996 on hillslope plane 6, the model prediction is lower than measured.

Another problem present in the results is that variation in soil moisture along the hillslope is not always well represented by the model (Figures 8.4 and 8.6). Lateral variation in soil properties is not incorporated in the model in its present form, although such specification is possible. Correlations between topographic features and lateral soil property distributions may allow this factor to be accounted in some modeling applications (e.g., Barling et al. 1994). For Appalachians watersheds, however, this assumption may not be appropriate, as ridges in these deep-soiled, humid watersheds have been found to have significant clay development (Boyer et al. 1990, Yeakley et al. 1998).

A different problem present in the results is that in the upper hillslope plane (#6), a persistent near-saturated condition was present near the stream. This condition was not present on hillslope #12, but it may be due to differences in planform. Plane #6 has a fairly severe constriction in plan form that occurs at approximately 300 m from the ridge. This location coincides with the increase in soil moisture to near-saturation or saturation in the modeled results. This occurrence must be addressed by examining the assumptions of a saturated wedge near the stream and by specification of saturated hydraulic conductivity in those near-stream areas.

Currently, efforts are ongoing to address these issues. In a different project using this same modeling framework in a near-stream vegetation manipulation (Yeakley et al. 1994), several years of piezometer data for near-stream areas have been collected on two hillslope planes in a nearby headwater watershed (WS56). It is hoped that ongoing model calibration for that project will provide further direction on this issue. In an extension of the work discussed in this paper, hillslope TDR transects over a total of 1.2 km have been installed in four watersheds in the Coweeta Basin representing the range of aspects and elevations. In addition to over three years of monthly soil-moisture data already collected, extensive soil samples have been collected and are presently being analyzed for texture and organic matter distributions along these transects.

8.7 CONCLUSIONS

A method of modeling hillslope soil moisture was implemented and assessed for a multiyear period in a 12.3-ha watershed (WS2) at the Coweeta Hydrologic Laboratory in the southern Appalachian mountains. The modeling framework consisted of three modules: (1) contour-based terrain analysis (TAPES-C); (2) above-ground canopy interception; and (3) a hillslope hydrology model (IHDM4) using a 2D Richards equation of subsurface moisture dynamics. Terrain analysis was conducted to identify hillslope planes in the watershed. The canopy interception module was calibrated based on 14 years of hourly climate data and extensive interception and throughfall measurements in the Coweeta Basin. The hillslope hydrology module was calibrated for stream flow using both wet and dry antecedent storm events and for soil moisture using time domain reflectometry measurements of soil moisture on a lower hillslope plane in WS2, over a period spanning a severe autumn drought through winter precipitation recharge. The accuracy of this modeling approach was examined using monthly soil-moisture measurements from Oct 94 to Oct 96 on another, more extensive hillslope plane in the watershed.

The model predicted mean monthly soil moisture for the hillslope plane with errors ranging from 1% to 23% in dry months and up to 45% in wet months. During wet months, the model predicted soil moisture along the hillslope better near the ridge and stream than on the midslope. During dry months, the model predicted soil moisture more accurately on the midslope.

Two issues need to be addressed in future model development: incorporation of lateral variation in soil hydraulic properties and more accurate representation of soil-moisture content near seepage faces. Ongoing research is being conducted to address both of these issues in several watersheds in the Coweeta Basin.

CHAPTER 9

Stochastic Analysis of a Coupled Surface/Subsurface Hydrologic Model

Gregory M. Pohll and John J. Warwick

9.1   INTRODUCTION

9.1.1   Statement of the Problem

The United States Department of Energy (DOE) is currently assessing subsidence craters (surface subsidence features that are created following an underground nuclear test) at the Nevada Test Site (NTS) for the purpose of low-level nuclear waste storage and to quantify their impact on the possibility of migration of radiowaste. Little is known about the connectivity between the surface water that collects in these craters and the groundwater. It is hypothesized that the surface alteration that forms following subsurface nuclear testing can enhance infiltration and radionuclide migration and may allow for the transport of nuclear byproducts from low-level waste or blast products to the groundwater (Tyler et al. 1992). Once in the groundwater, these nuclear contaminants may become highly mobile and nearly impossible to remediate. Assuming some degree of connectivity, a general understanding of the relationship between the surface water and the unsaturated subsurface moisture flow is required to predict large-scale radionuclide transport.

The focus of this chapter is to develop a more complete approach to determine past recharge fluxes and to develop a better understanding of the physics of flow from the surface to the vadose zone. No coupled computer models exist that can route the movement of surface water along a topographic surface into a zone of converging flow, compute the volumetric mass balance of the ponded water, compute the subsequent infiltration beneath this zone, and determine the extent of the vertical and horizontal moisture movement. Many models can compute the individual components within the hydrologic cycle. All available models were evaluated and then the most appropriate

Terrain Analysis: Principles and Applications, Edited by John P. Wilson and John C. Gallant.
ISBN 0-471-32188-5   © 2000 John Wiley & Sons, Inc.

ones were combined into a single, fully coupled surface/subsurface model that could accurately simulate all components of moisture movement.

9.1.2 Review of Existing Coupled Modeling Approaches

There are three distinct surface water routing approaches. The following model descriptions are not meant to be exhaustive, but are a description of the various approaches taken toward surface/subsurface modeling. The first type is characterized as physically based analytical and numerical models, such as ANSWERS (Beasley et al. 1980) and KINEROS (Goodrich 1990). These models use the kinematic wave approximation to the continuity equation, with all flow being routed over a set of user-defined planes. Infiltration is calculated using a curve number approach or an analytical approximation to the Richards equation, and there is no tracking of subsurface fluid movement.

The second type of coupled surface/subsurface flow models utilizes a grid structure such as that described by the Paniconi and Wood (1993) or SHE (Bathurst 1986) models. Paniconi and Wood (1993) developed a catchment-scale numerical model based on the three-dimensional transient Richards equation describing fluid flow in variably saturated porous media. The model takes advantage of gridded digital elevation databases by calculating various topographic parameters directly from the digital elevation model (DEM). A finite element mesh generator was developed to take advantage of the regular grid structure of the DEM data. The generator determines the catchment boundary, then discretizes a catchment into hexahedral elements. The element generator develops the element connectivities and initializes the system of storage for the various matrices used by the hydrologic model. The hydrologic simulator takes the information from the grid generator and routes surface overland flow and subsurface flow via iterative techniques. The Paniconi and Wood (1993) model uses a simple time-delay function to route the surface flow to the downslope stream elements, while the SHE model routes surface flow via the kinematic wave approximation.

The third approach is that of Grayson et al. (1992a), who developed a contour-based (rather than a grid-based) surface/subsurface model. The model simulates the surface hydrologic components of flow and predicts the areas and rates of infiltration. Currently, the model does not simulate subsurface moisture movement beyond surface infiltration. The suite of models incorporates four programs: (1) DIGITIZ, which permits contour lines to be line-digitized to create a vector DEM, (2) PREPROC, which is a preprocessing program that transforms the vector DEM into a north–south-oriented coordinate system, (3) TAPES-C, which subdivides a catchment into elements and calculates a variety of topographic attributes for each element, and (4) THALES, a hydrologic modeling program that calculates areas and rates of infiltration. The contouring method of partitioning interconnected elements allows the two-dimensional hydrologic equations to be reduced to a series of coupled one-dimensional equations. Figure 9.1 shows how TAPES-C partitions a partial catchment into elements (streamlines) based on the contour information. Nodal points on the network correspond to the midpoints on the upslope and downslope contour lines bounding each element. The kinematic

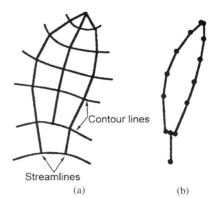

Figure 9.1. (a) A hypothetical hillslope segment "subdivided" into elements using TAPES-C; (b) its equivalent one-dimensional network (adapted from Moore and Grayson 1991).

wave approximation of governing flow equations are solved at each nodal point in the network.

For this study SWMS_2D (Simunek et al. 1994), a numerical, two-dimensional, axisymmetric, finite element, variably saturated model, was used in two capacities. First, it was used as a stand-alone model to simulate subsurface moisture movement given static surface boundary conditions. Second, it was dynamically coupled with THALES such that THALES would provide the surface boundary conditions (infiltration flux) for the SWMS_2D model.

Coupled hydrologic models require more input parameters and computation time and therefore add significant cost to the modeling project. To determine the benefits of a more complex model, the coupled model is compared with a more traditional vadose zone model, which imposes static surface boundary conditions.

9.1.3 Stochastic Model

Significant radionuclide contamination beneath the subsidence craters makes extensive field characterization of the site extremely difficult. Therefore, a mathematical modeling approach is investigated within a stochastic framework to quantify the uncertainty of the model predictions. Various stochastic models are investigated. Three different stochastic models, which include a dependence, independence, and partial dependence of saturated hydraulic conductivity (K), residual water content (θ_R), and van Genuchten pressure versus water content shape parameters α and β, are compared to determine how each affects simulated output uncertainty of moisture migration.

9.1.4 Study Site

The NTS is located 110 km northwest of Las Vegas, Nevada and encompasses 3500 km^2 in the southern part of the Basin and Range province (Black and Townsend) (Figure 9.2). The NTS has been the principal testing area for nuclear explosions in the United States. Since its inception in 1951 there have been 804 announced underground nuclear tests and 100 surface tests.

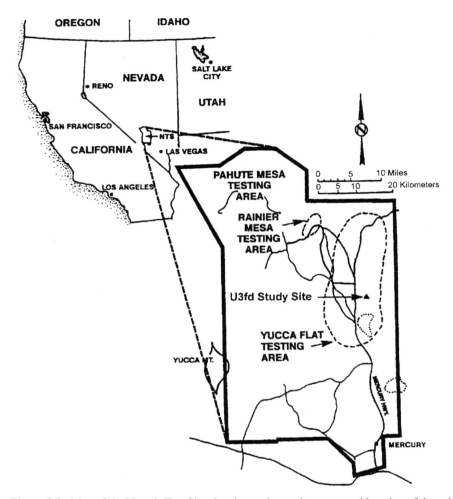

Figure 9.2. Map of the Nevada Test Site showing major testing areas and location of the subsidence crater U3fd (adapted from Tyler et al. 1992).

This investigation is focused on subsidence crater U3fd, which was formed by an underground explosion in 1971 (Figures 9.2 and 9.3). The site is located in the central portion of Yucca Flat within Area 3. The depth to the groundwater table was estimated at 500 m (Fenske and Carnahan 1975). The crater is underlain by coarse-textured soils, valley-fill alluvium, volcanics, and blast-induced fused rock glass units (Tyler et al. 1992). Crater U3fd is 180 m across its longest side and is approximately 13 m deep. The effective catchment area of the crater is approximately 25,000 m², of which only 20% is derived from outside of the crater's boundaries. A concrete pad, which was used for the nuclear device installation, exists at the cen-

Figure 9.3. Subsidence crater U3fd at the Nevada Test Site (looking east).

ter of the crater. The pad is moderately fractured and is simulated with a low permeability zone with a radius of 4.0 m.

9.2   DESCRIPTION OF NUMERICAL MODELS

9.2.1   Vadose Zone Model

The SWMS_2D (Simunek et al. 1994) model was chosen to simulate water flow in an axisymmetric, variably saturated flow domain. The governing flow equation is a modified form of the Richards equation:

$$\frac{1}{r}\frac{\partial}{\partial r}\left[rK(h)_r\left(\frac{\partial h}{\partial r}\right)\right] + \frac{\partial}{\partial z}\left[K(h)_z\left(\frac{\partial h}{\partial z} + 1\right)\right] = \frac{\partial \theta_v}{\partial t} - q \qquad (9.1)$$

where h is the pressure head (m), $K(h)_r$ and $K(h)_z$ are the relative hydraulic conductivity in the radial and vertical directions, respectively (m h^{-1}), r and z are spatial coordinates (m), t is time (h), θ_v is the volumetric water content (cm^{3}cm^{-3}), and q is the volumetric flow rate via sources or sinks per unit volume of porous medium (m^{3}m^{-3}h^{-1}). Given specified pressure, flux, or gradient-type boundary segments, SWMS_2D utilizes a Galerkin finite element method with linear basis functions to obtain a solution of the flow for Equation 9.1 (Neuman 1975, Zienkiewicz 1977, Pinder and Gray 1977). The axisymmetric option was employed to take advantage of the radially symmetric flow domain rather than utilizing a fully three-dimensional model.

A hysteresis algorithm in the soil-water retention curve is not accounted for in the vadose zone model because of the constant-flux upper boundary condition that would not impose wetting and drying effects on the system. The van Genuchten (1980) model is used:

$$\frac{\theta - \theta_r}{\theta_s - \theta_r} = \frac{1}{[1 + |\alpha h|^\beta]^\lambda} \tag{9.2}$$

where

$$\lambda = 1 - \frac{1}{\beta} \tag{9.3}$$

θ_s is the saturated volumetric water content (cm^{3}cm^{-3}), θ_r is the residual water content (cm^{3}cm^{-3}), α (m^{-1}), and β (dimensionless) are shape parameters.

It was assumed that four distinct hydrogeologic units exist (Figure 9.4): (1) the 442 m of alluvial material beneath the subsidence crater, (2) the fused rock/glass that formed at the bottom of the blast cavity, (3) the undisturbed tuff unit adjacent to the blast cavity, and (4) the collapsed tuff within the blast cavity, but above the fused section. For the relatively short-term simulations performed for this study, water infiltrated only into the surface alluvial unit.

The boundary conditions for the axysymetric flow domain were assumed constant throughout the entire simulation (Figure 9.4). The alignment of the left boundary with the axis of symmetry precludes no flow conditions. The right no-flow boundary was set such that moisture fronts would not come in direct contact. A specified pressure head boundary condition was applied to the lower boundary to coincide with the water table. The surface nodes were simulated with specified flux boundary conditions. Previous modeling studies concluded that the only significant infiltration flux occurred near the center of the crater. Therefore, a constant flux boundary condition was applied to a 6.0-m radius out from the center of the subsidence crater (Pohll et al. 1996). The flux rate was estimated at 1.4 m yr^{-1}, which was determined from the coupled modeling

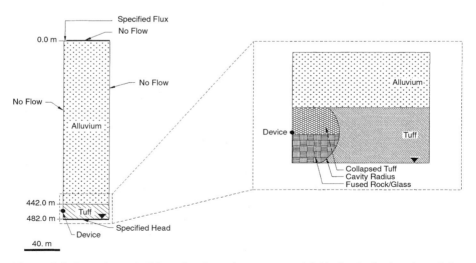

Figure 9.4. Boundary conditions for the vadose zone model (device is the location of the nuclear blast).

study of Pohll et al. (1996). A zero-flux condition was applied along the remaining surface boundary to represent the negligible flow in that region. The radial distance of the flux boundary was determined by trial and error calibration of the vadose zone model against the coupled model such that total surface infiltration was similar. The model domain was discretized onto a regular network of rectangles 2.0 m in width and 0.5 m in height. The 50 by 482-m model domain required a total of 24,100 nodes.

9.2.2 Coupled Surface/Subsurface Model

To determine the impact of using a more complete representation of surface flux, a coupled surface/subsurface model was developed using SWMS_2D, TAPES-C, and THALES. The coupling mechanisms are described below.

9.2.2.1 Surface Water Routing
The terrain-analysis module, TAPES-C (Moore and Grayson 1991), calculates the following topographic attributes that are required for the hydraulic model (THALES): (1) element area, (2) total upslope contributing area, (3) connectivity of upslope and downslope elements, (4) x, y, z coordinates of the element centroid, (5) x, y, z coordinates of the midpoint on the downslope contour bounding the element, (6) the average slope of the element orthogonal to the contour, (7) the widths of the element on the upslope and downslope contour lines bounding the element, (8) the flow distance across the element, and (9) aspect of the element. These geometric parameters are used by the surface routing module THALES to describe the series of one-dimensional flow tubes over which surface water flow is routed.

The topographic information required by TAPES-C was developed by surveying the subsidence crater and contributing area. The surveyed elevations were then converted to 0.5-m elevation contours via standard kriging techniques. A FORTRAN program was written to convert the vector elevation information to a format that could be read by TAPES-C. The elevation contours were then used by TAPES-C to develop the topographic attributes that are required for the THALES simulation.

THALES uses two alternative process-oriented hydrologic models to simulate runoff. The first simulates subsurface flow-saturation overland flow and the second simulates Hortonian overland flow. The former process is not important in arid regions, so only the latter is discussed here. Hortonian overland flow is routed between elements using the standard kinematic forms of the momentum equations. Brakensiek's (1967) four-point finite difference solution of both the kinematic overland flow and channel flow equations is used to route the surface flow between elements. The finite difference solution can be expressed as

$$A_4 + 2Q_4 \frac{\Delta t}{\Delta s_e} = A_1 - A_2 + A_3 + 2Q_2 \frac{\Delta t}{\Delta s_e} + 2iA_e \frac{\Delta t}{\Delta s_e} \tag{9.4}$$

where Q_i is the discharge (m^{3}h^{-1}), A_i is the cross-sectional area (m^2), A_e is the plan area of the element (m^2), i is the rainfall intensity (mm h^{-1}), Δt is the time interval (h), subscripts 1 and 3 refer to time t at positions s and $s + \Delta s_e$, respectively, and subscripts 2 and 4 refer to time $t + \Delta t$ at positions s and $s + \Delta s_e$, respectively. At any time and for any element the terms on the right-hand side of Equation 9.1 are known quantities; therefore, A_4 can be solved for using the Newton–Raphson method, since Q_4 can be expressed as a function of A_4 (Moore and Grayson 1991).

In the Hortonian overland flow model the infiltration rate, infiltration volume, and rainfall excess are computed for each element using a relationship based on the three-parameter Smith and Parlange model (Smith et al. 1993, Parlange et al. 1982). These infiltration fluxes become specified flux-type boundary conditions for the vadose zone model along the slope of the crater. The ponded infiltration near the crater bottom and subsequent vertical seepage beneath the ponded zone are determined entirely by the vadose zone model. The rainfall intensity is assumed to be uniform over all elements. The surface runoff can be modeled as sheet, rill, or channelized flow. Only sheet flow was employed for this modeling exercise. The solution scheme begins at the uppermost element, progressing along a contour and then down to the next element on the lower contour. The process is continued until the last element is reached, after which the process repeats for the next computational time step. At each element the stream lines represent no-flow boundaries; the upper contour is an inflow boundary and, in the coupled version, the downstream boundary is a no-flow boundary defined by the center of the crater or the radial extent of the ponding zone. The model calculates the surface flow, infiltration rate (up to the ponded zone), volumetric water content of the surface soil horizon, and rainfall excess for every element at every time step. An extensive discussion of the terrain-analysis, surface-routing, and infiltration algorithms can be found in Moore and Grayson (1991), Grayson et al. (1992a), and Smith et al. (1993), respectively.

9.2.2.2 *Ponded Zone Mass Balance*

The water routed via the kinematic wave approximation is then transferred to a mass balance algorithm, which computes the evolution of the ponded zone. Inflow via surface flow, evaporation, and infiltration are calculated to preserve the volumetric mass balance of the ponding zone. At each time step, surface inflow is tracked from all converging flow tubes. The site was surveyed along 10° and 5-m intervals to develop a high-resolution topographic database. Total surface flow is converted to a pond depth (at the crater bottom) based on a functional relationship between volume of stored water and pond depth at the crater center. This relationship was determined by developing a series of pond volumes and associated depths within a geographic information system that contained the high-resolution topographic information. The empirical form ($R^2 = 0.98$) for crater U3fd was

$$D = 0.001852 V^{(0.4581)} \qquad (9.5)$$

where D is the depth of water surface at the crater's lowest elevation (m) and V is the volume of water within the ponded zone (m^3).

The potential evaporation was calculated based on the mass-transfer approach developed by Langbein et al. (1951) for calculating evaporative fluxes from small stock ponds. It was assumed that evaporation was controlled by the windspeed and the vapor pressure difference between the water surface and atmosphere as follows:

$$E_0 = Nu(e_s - e_a) \tag{9.6}$$

where E_0 is the potential or evaporative flux from the water surface (m h^{-1}), u is the windspeed at two meters above the land surface (m h^{-1}), e_s is the vapor pressure of the water surface (mb), e_a is the vapor pressure of the air (mb), and N is a constant, known as the mass-transfer coefficient (mb^{-1}). The mass-transfer coefficient is determined by applying a linear regression between the product of the windspeed and the vapor pressure difference and the water level recession for the pond after linear recession was reached. The slope of the regression line is the mass-transfer coefficient and the intercept is the steady-state infiltration rate. The mass-transfer coefficient calculated for crater U3fd was 6.9×10^{-10} mb^{-1} ($R^2 = 0.99$), which is approximately 30–50% less than published values for ponds of equivalent areas (Dunne and Leopold 1978). The meteorological data used for this study were collected above the crater. One would expect decreased wind velocities at the crater bottom due to topographic effects, which is reflected by the low N value.

At each time step, the evaporative flux is calculated according to Equation 9.6 and subtracted from the pond depth. The pond depth is then converted to a volumetric measurement. At this point the vadose zone model is called to determine the pressure head distribution and infiltration beneath the ponding zone. The ponded infiltration is subtracted from the pond volume and then the program advances to the next time step.

9.2.2.3 *Coupling Vadose Zone Module*   The numerical model SWMS_2D (Simunek et al. 1994) was chosen to simulate the finite element approximation to the Richards equation (Equation 9.1). A hysteresis algorithm would be necessary to simulate the many wetting and drying cycles produced by the coupled model, but the purpose of this study is to compare the coupled model with the vadose zone model with a static upper boundary condition, which does not require a hysteresis algorithm. Therefore, hysteresis is not included in our discussion.

The dynamic coupling is performed by mapping the surface infiltration onto a two-dimensional axisymmetric finite element vadose zone network, as shown in Figure 9.5. At each time step the nonponded infiltration is calculated from the mean of all radially equidistant surface water routing nodes. The surface-water model utilizes the one-dimensional Smith and Parlange model to calculate infiltration. The infiltration flux is mapped to the equivalent surface node in the vadose zone model as a specified flux-type boundary condition. The surface nodes of the vadose zone model are automatically converted to a specified pressure-type boundary condition if the node becomes saturated or its pressure head becomes smaller than a specified equilibrium pressure between soil and atmospheric water vapor. As the surface becomes saturated due to ponding at the crater center, the surface water nodes are turned off and the ponded infiltration is handled by the vadose zone model. The boundary condition switches back to a specified flux type if the pond recedes below the particular surface element. The pressure head of the ponded nodes is calculated by taking the difference of the ponding depth and the nodes' relative elevation.

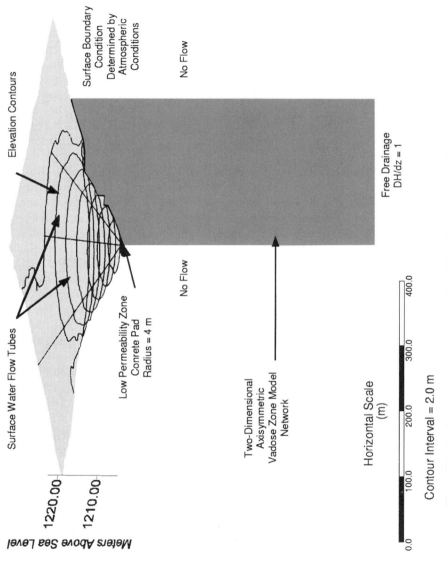

Figure 9.5. Surface flow tubes and subsurface finite element network for the coupled surface/subsurface model.

234

9.2.3 Stochastic Model

Four uncertain vadose zone input parameters are included in the stochastic analysis: the hydraulic conductivity (K), residual water content (θ_r), and the shape parameters (α and β) for the van Genuchten (1980) model of pressure head versus water content. It is assumed that (1) all four input parameters have an exponential semivariogram structure and the same correlation scale, and (2) the cross-correlation structure between variables is known and defined by a covariance matrix.

Carsel and Parrish (1988) compiled a soil database of over 10,000 soil samples to develop the cross-correlation structure (covariance matrix) for the four vadose zone model parameters according to soil textural classification. The class of transformed normal distributions, known as the Johnson system, was used to produce normally distributed data from the original distribution (Johnson and Kotz 1970). The Johnson transformations are of three main types: LN, lognormal; SB, log ratio; and SU, hyperbolic arcsine. They are defined as

$$\text{LN:} \quad Y = \ln(X) \tag{9.7}$$

$$\text{SB:} \quad Y = \ln\left[\frac{X - A}{B - A}\right] \tag{9.8}$$

$$\text{SU:} \quad Y = \sinh^{-1}(U) = \ln[U + (1 + U^2)^{1/2}] \tag{9.9}$$

where ln is the natural logarithm, X is the untransformed variable with limits of variation from A to B ($A < X < B$), and $U = (X - A)/(B - A)$. The normal (NO) Gaussian distribution was also used in those cases where no transformation was necessary to achieve normality.

To combine the Carsel and Parrish (1988) model with a specified spatial autocovariance structure, four spatially correlated random fields of independent standard normal deviates ($N(0, 1)$) were generated using currently available computer programs employing the turning bands method (Tompson et al. 1989). At each spatial location, the input parameter vector $\mathbf{y}$ (saturated hydraulic conductivity, residual water content, and van Genuchten's α and β) was generated by applying a linear Gaussian transform of the form

$$\mathbf{y} = \mu + \mathbf{T}'\mathbf{z} \tag{9.10}$$

where μ is the desired vector of means, $\mathbf{T}'$ is the transpose of an upper triangular covariance matrix derived from the textural classification, and $\mathbf{z}$ is the vector of independent standard normal deviates taken from the turning bands algorithm. The vector $\mathbf{y}$ is then inverse transformed into the original scaling of the hydraulic parameters. It can be easily shown that the linear transform given above yields a simple expression for the semivariogram of the transformed variables as the product of the variance and the original semivariogram. Therefore, this algorithm will generate spatially correlated random variables with a known semivariogram and joint probability distribution. Likewise, alteration of the mean vector does not influence the covariance or the autocovariance structure. Therefore, one can develop the mean vector from field data if available.

The result of one realization for a 100 × 100-m grid (grid spacing = 1 × 1 m) of the four vadose zone parameters is shown in Figure 9.6 for a loamy sand. The length and parameter value scales were omitted to emphasize relative correlation among

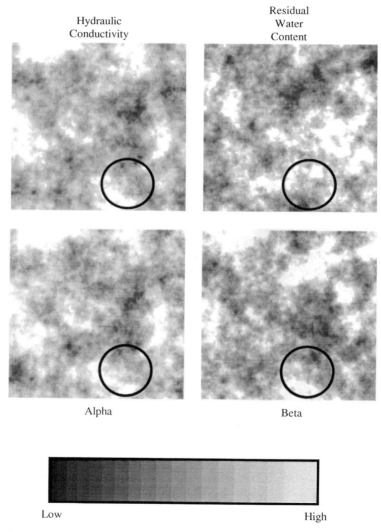

Figure 9.6. Synthetically generated transformed input parameter fields (K = hydraulic conductivity, θ_r = residual water content, and α and β = the van Genuchten shape parameters for the water content versus pressure head relationship) according to the algorithm presented above for a 100 × 100-m grid (grid spacing = 1 × 1 m; correlation scale = 10 × 10 m).

variables. An exponential autocovariance function with a correlation length of 10 m (in both the x and y directions) was used to generate the normal vector. Figure 9.6 suggests that the spatial autocovariance scaled properly. The covariance among the four hydraulic variables can be seen by focusing on zones of high (or low) conductivity and comparing them to other variables. As defined, the relationship is only partially dependent. For example, the negative correlation (-0.301) between β and θ_R is seen by comparing the circled areas within each plot. For these zones, θ_R has relatively larger values as compared to the β values.

9.3 SIMULATION RESULTS

9.3.1 Coupled Model Calibration

Two short-term deterministic simulations were performed for calibration and verification purposes. The first simulation represents an actual precipitation event that occurred during August–October, 1983. A total precipitation of 117 mm fell over a period of 15 days (United States Geological Survey–Water Resources Division [USGS-WRD], 1993, personal communication). This event was chosen as the calibration event because of the relatively large amount of precipitation, ponding depth, and overland flow contributions from the area adjacent to crater U3fd. The surface and subsurface initial conditions were set at a volumetric water content of $0.13 \text{ cm}^3 \text{ cm}^{-3}$ everywhere (except directly beneath the crater), which coincides with the water content found in the undisturbed borehole (U3fd-N2). The water content beneath the crater (within 10 m of the crater center) was set at $0.28 \text{ cm}^3 \text{ cm}^{-3}$ as determined by the shallow borehole (U3fd-N1) measurements. Figure 9.7 illustrates the simulated versus measured ponded water levels and

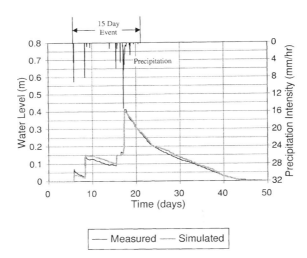

Figure 9.7. Measured versus simulated ponded water levels and associated precipitation intensities for the verification simulation (Aug–Oct 83).

the associated precipitation intensities for the calibration simulations. A root mean square error (RMSE) was used to evaluate the effectiveness of the model to simulate observed water levels:

$$\text{RMSE} = \left[\frac{1}{n} \sum_{i=1}^{n} (h_m - h_s)_i^2 \right]^{0.5} \tag{9.11}$$

where RMSE is the root mean square error of pond stage (m), n is the number of observed versus simulated comparisons that were made, and h_m and h_s are the measured and simulated stage at the crater center, respectively (m). The surface-water model matched the observed water levels with an RMSE of less than 0.04 m. Errors were largest for small stages because minor changes in pond volume result in significant water level fluctuations.

The verification event occurred during the months of January through April 1983 with a total precipitation of 142 mm falling over a period of 57 days (USGS-WRD, 1993, personal communication). All parameters used in the calibration simulation were used with the verification precipitation and meteorological time series. The simulated water levels were in general agreement with observed values, resulting in an RMSE of 0.12 m. The model tended to overestimate the peaks and underestimate infiltration (Figure 9.8).

9.3.2   Model Comparison (Vadose Zone Versus Coupled Model)

The standard Cartesian moments were used to compare the results of the vadose zone model and coupled model as

$$M_{ij} = \int_{-\infty}^{\infty}\int_{-\infty}^{\infty} A(x, z, t)x^i z^j \, dx \, dz \tag{9.12}$$

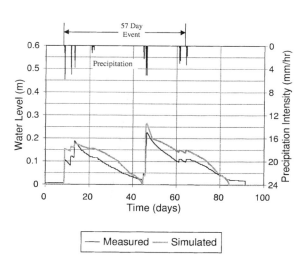

Figure 9.8. Measured versus simulated ponded water levels in the crater and associated precipitation intensities for the calibration simulation (Jan–Apr 83).

where $A(x, z, t)$ is the water content change from the initial condition, z represents the vertical direction, and x represents the lateral direction. The first moment in the lateral direction is normalized by M_{00} as

$$x_c = \frac{M_{10}}{M_{00}} \qquad (9.13)$$

The first moment in the vertical direction is normalized by M_{00} as

$$y_c = \frac{M_{01}}{M_{00}} \qquad (9.14)$$

The finite nature of the grid discretization did not warrant more than a visual fit of the first moments between the coupled and vadose zone model. Trial and error simulations led to an effective pond radius of 6.0 m. Figure 9.9 shows the simulated lateral (x_c) first moment of water content distribution, respectively, for the deterministic vadose zone and coupled models. The lateral x_c moment provides an indication of horizontal spreading of the moisture front. The coupled model shows a large amount of temporal variability associated with each precipitation event, while the vadose zone model shows very little temporal variability. The coupled model is representing the physical process of the pond extending laterally and then receding during dry periods. From the perspective of simulating the general movement of moisture in the subsurface, the vadose zone model simulates the general behavior of the moisture front.

Figure 9.10 shows the vertical first moments (y_c) of water content distribution for both models. The vertical moment (y_c) provides a measure of the movement of the moisture front vertically. Again the coupled model shows significant temporal variability with a general trend of increasing values, which represents the moisture front

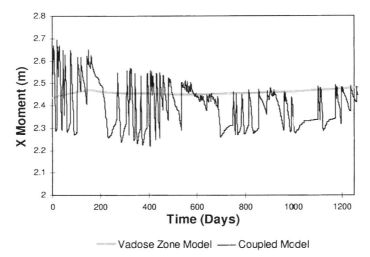

Figure 9.9. Mean lateral first moment (X_c direction) of moisture content distribution for the short-term vadose zone and coupled models.

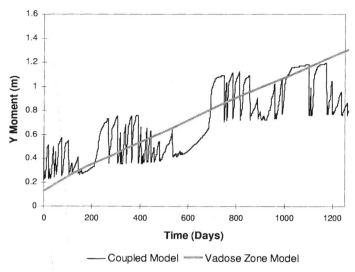

Figure 9.10. Mean vertical first moment (Y_c direction) of moisture content distribution for the short-term vadose zone and coupled models.

moving down into the soil profile. The increasing trend of the coupled model shows a step-type function, which suggests that only large precipitation events cause further movement of the moisture front. The vadose zone model with static surface boundary conditions shows a smooth upward trend in the vertical moment, indicating a constant downward movement. By the end of the simulation the vadose zone model overpredicts the downward position of the front, but in general provides a reasonable approximation of the moisture front behavior.

9.3.3 Long-Term Simulation Comparison of Stochastic Structures and the Vadose Zone Model

A series of vadose zone model simulations were performed to ascertain the impact of different stochastic assumptions on the wetting front behavior. Each case simulated 15 years of moisture movement since the inception of crater U3fd (1971–1986). Initial conditions were set at a uniform pressure potential of −50.0 m (mean water content 0.13 cm^3 cm^{-3}). For comparison purposes the vertical and lateral moments of moisture content were calculated from Equations 9.12–9.14. Three values (5.0, 10.0, and 20.0 m) for the horizontal correlation scale of the four input parameters (K, θ_R, α, and β) were used to determine the impact of horizontal correlation scale on the moisture front behavior. In addition, three types of stochastic dependence relationships were used. First, the four input fields were considered to be independent with zero correlation. Second, the four fields were assumed to be completely dependent. The third option utilized the stochastic structure discussed in the previous section, which has a partial dependence structure as defined by the known covariance matrix determined by textural classification. Twenty Monte Carlo simulations were performed for each case to provide stable summary statistics for the moment calculations.

The Pearson product–moment correlation matrix for a loamy sand is shown in Table 9.1 (Carsel and Parrish 1988). Table 9.2 shows the horizontal and vertical moisture content moments for the 15-year simulation. The horizontal correlation scale had little impact on either the mean vertical or the horizontal moisture moment. The increase in the standard deviation with increasing correlation scale suggests that Monte Carlo simulations with large horizontal correlation scales require more real-izations to stabilize the solution mean.

The stochastic structure for the input variables has an impact on the vertical and horizontal moments. The correlation matrix contains both positive and negative cor-relations, yet the output moments of moisture content fall very close to the depend-ent case. The similarity of the dependent and partially dependent cases may be due to the strong correlation between the hydraulic conductivity (K) and the van Genuchten shape parameters (α and β). The mean error (assuming that the partial dependence case is correct) for the independent and dependent solutions was calculated as

$$\text{MAE} = \frac{1}{n}\left[\sum_{1}^{n}|t_i - p_i|\right] \qquad (9.15)$$

where n is the number of samples, i is the time step, p_i is the partial dependence moment, and t_i is either the independent or dependent moment solution. The mean absolute error (MAE) is shown for the horizontal and vertical moments in Figures 9.11 and 9.12, respectively. The independent solution diverges from the other two solutions very quickly, but the dependent solution remains similar for nearly five years and diverges much less quickly than the independent solution. A dependent solution could be used in place of the partially dependent solution for short-term simulations because little is gained by adding covariance information for simulations of five years or less.

9.4 DISCUSSION AND CONCLUSIONS

9.4.1 Coupled Surface/Subsurface Model

To fully understand the physical linkage between the surface water and deep recharge a coupled surface/subsurface model was developed. This model incorporated the most recent advances in numerical modeling techniques. The coupled model system con-tains a terrain-analysis module to construct the geometric framework for a series of

TABLE 9.1   Sample Pearson Product–Moment Correlations

	Hydraulic Conductivity	Residual Water Content	Alpha
Residual water content	−0.359		
Alpha	0.986	−0.301	
Beta	0.730	−0.590	0.354

Source. Adapted from Carsel and Parish (1988).

TABLE 9.2   Horizontal (X_c) and Vertical (Y_c) Moisture Content Vertical Moments for the 15-year Simulation for Various Stochastic Structures and Correlation Scales

Stochastic Type	Correlation Scale (m)					
	5		10		20	
	Mean	SD	Mean	SD	Mean	SD
Horizontal Moment (X_c) for 15-year Simulation						
Independent	4.34	0.22	4.30	0.22	4.33	0.36
Dependent	3.94	0.03	3.95	0.02	3.93	0.02
Partially dependent	4.01	0.04	4.02	0.09	4.02	0.07
Vertical Moment for (Y_c) 15-year Simulation						
Independent	22.00	1.57	22.08	1.48	22.00	1.96
Dependent	25.29	0.47	25.52	0.67	25.64	0.75
Partially dependent	24.40	0.54	24.52	0.80	24.34	0.91

Note. SD, standard deviation.

one-dimensional flow tubes used by the surface water model. The surface-water model simulates overland flow using the kinematic wave equation following precipitation events. A mass balance algorithm was developed to calculate the evolution of the ponding within a single crater (U3fd). The mass balance algorithm converted the surface overland flow to a ponding depth and calculated evaporative losses using the mass-transfer approach. A numerical Richards equation solver was dynamically coupled to the surface-water model to calculate the moisture migration beneath the surface. The

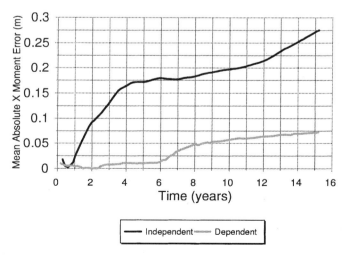

Figure 9.11. Mean absolute error of the lateral moisture moments (X_c) between the partially dependent simulation and the dependent and independent solutions.

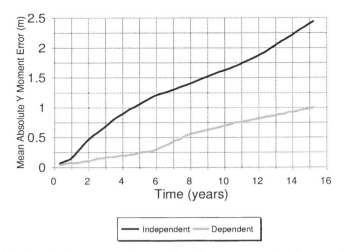

Figure 9.12. Mean absolute error of the vertical moisture moments (Y_c) between the partially dependent simulation and the dependent and independent solutions.

coupled model system was able to reach a good agreement between observed and simulated water levels following a precipitation event. The calibrated simulation matched the observed water levels with an RMSE of less than 0.04 m. The verification simulation had a slightly larger RMSE of 0.12 m. In general, the coupled model was able to simulate the hydrologic processes of interest and reach a reasonable level of agreement between measured and simulated pond levels.

9.4.2   Coupled Surface/Subsurface Versus Vadose Zone Model

A comparison was made between the coupled surface/subsurface model and a vadose zone model with static surface boundary conditions. The coupled model simulates the progression of the pond formation and the temporal variation in the lateral extent of infiltration. Although this added complexity provides a more realistic model of infiltration, once the moisture infiltrates into the subsurface, the lateral movement becomes damped. The coupled model showed significant temporal variability, yet the vadose model was able to match the "effective" behavior. Therefore, little benefit is gained with the simulation of the lateral moisture migration.

A comparison of the vertical moisture content first moment showed differences in the temporal movement of water between each of the models. The coupled model showed that the moisture front tended to advance in response to large precipitation events. The vadose zone model could not reproduce these effects with static surface boundary conditions and slightly overpredicted the vertical movement of the moisture front.

The numerical modeler needs to decide in advance the objectives of the modeling study due the added complexities and significant number of input parameters required by the coupled model. If the objective is to determine the temporal distribution of infiltration flux then the coupled model should be used. If the objective is to determine deep moisture migration, then a more simplified approach is justified, since the moisture movement becomes damped with time. Therefore, if a more simplified approach can be taken to determine the "effective" surface flux, then a standard vadose zone model can effectively simulate the movement of moisture.

9.4.3   Stochastic Model

The horizontal correlation scale had little impact on either the mean vertical or the horizontal moisture moment. Therefore, for a loamy sand texture little information is gained in collecting horizontal correlation scale information. One can use literature values to estimate the horizontal correlation scale and still have confidence in simulation results. The standard deviation in the moisture content moments did increase with increasing correlation scale, which indicates that more realizations are necessary to stabilize the solution mean with increased horizontal correlation scales.

For the statistical parameters used in this study, the dependent structure yielded results similar to the partially dependent case for simulation times less than five years. Thereafter, the inclusion of the joint distribution among input parameters becomes critical. This is especially true for deep vadose studies, as is the case beneath subsidence crater U3fd. Therefore, short-term simulations do not require the inclusion of the covariance between input parameters, yet long-term simulations require more stochastic information.

CHAPTER 10

The Role of Terrain Analysis in Soil Mapping

Neil J. McKenzie, Paul E. Gessler, Philip J. Ryan, and Deborah A. O'Connell

10.1 THE POTENTIAL OF TERRAIN ANALYSIS

Good-quality information from soil and land resource surveys is necessary for wise natural resource management. Only a few industrialized countries with intensive land use have a detailed and complete survey coverage. There is a pressing need to improve the survey coverage in many other countries but there are doubts as to whether conventional survey methods provide the most appropriate information. Terrain analysis can assist in providing a more useful survey coverage.

In conventional surveys, point observations of soils are extended to broader regions using qualitative and complex mental models of relationships with more readily observed landscape features: Qualitative terrain-related variables figure prominently as predictors, particularly at the local scale. The mental models are rarely stated and users of surveys find it difficult to separate evidence from interpretation (Austin and McKenzie 1988, Hudson 1992, Hewitt 1993). Accuracy and precision of mapping are seldom provided and soil variation is often portrayed as being discontinuous, with map units having sharp boundaries. In reality, soils may possess sharp boundaries in some areas, but it is more common for them to exhibit gradual variation. Conventional surveys also provide maps of preclassified soil types, with minimal information on patterns of variation within polygons, rather than estimates of the primary soil properties. The latter have greater utility for making interpretations relating to land use (McKenzie 1991).

The advent of digital terrain analysis and allied technologies has created an opportunity to develop a more scientifically based method of soil survey that overcomes several of the limitations of conventional survey. Terrain analysis has the potential to improve soil survey in three main areas. It can be used to

Terrain Analysis: Principles and Applications, Edited by John P. Wilson and John C. Gallant.
ISBN 0-471-32188-5 © 2000 John Wiley & Sons, Inc.

- Generate high-resolution environmental information of direct use in land evaluation (slope, net radiation, etc.)
- Create explicit environmental stratifications for survey design
- Provide quantitative spatial predictions of individual soil properties

Examples of the use of terrain analysis in these areas are given below along with an evaluation of impediments to routine application. However, a brief review of pedologic theory is a necessary precursor to appreciate the complexity of relationships between soil and terrain.

10.2   THEORIES OF PEDOGENESIS AND MODELING FOR PREDICTION

A knowledge of basic physical processes and their interactions is commonly held to be a prerequisite for making predictions about natural phenomena. The sheer complexity and great range of spatial and temporal scales over which soil-forming processes operate in landscapes makes the development of process-based models for spatial prediction an almost impossible task in routine soil survey, although there are some notable exceptions (e.g., Dietrich et al. 1995). Because of this complexity, simplifying theory is necessary and reliance has to be placed on approximate local models of pedogenesis with varying levels of empiricism.

The task of soil survey is to construct models that have a predictive capacity across complete regions. This objective differs from most studies of soil formation, which aim to provide models of pedogenesis (Butler 1964) and may provide a good understanding at a few locations, but fail to provide an operational tool for spatial prediction. The soil science literature is replete with soil formation models (mechanistic models for *understanding*) but contains very few operational models for survey (functional models for *prediction*)—most of the latter are still within the minds of soil surveyors (Hudson 1992, Hoosbeek and Bryant 1992, Hewitt 1993). Of course, operational models for spatial prediction should be more powerful when they are based on a good knowledge of local pedogenesis. It is therefore useful to evaluate the role of terrain analysis in modeling pedogenesis before assessing its potential role in soil survey.

10.2.1   The Functional Factorial Approach

Pedology was established by Dokuchaev and his colleagues in the youthful glaciated landscapes of the Russian steppes late last century. A key principle was the recognition that soil profiles had a functional relationship with their environment, with the so-called mature soil being in equilibrium with it.

Jenny (1941) accepted this concept and defined five factors of soil formation that were later termed state factors. Jenny's verbal model was presented in quasi-mathematical form with the general statement

$$s = f(cl, o, r, t, p, \dots) \tag{10.1}$$

where s is a soil property, cl = climate, o = organisms, r = relief, t = time, p = parent material, and the ellipsis refers to other unstated factors. Jenny realized the difficulty of attempting to solve the general equation. To overcome this dilemma, he derived a set of canonical functions for solving the equation one factor at a time. The functions were the climo-function, bio-function, topo-function, litho-function, and chrono-function. To solve each, the particular factor is allowed to vary and the dependency of one or more soil properties is determined, the other factors being constant. Terrain variables appear most commonly in topo-functions and climo-functions (Jenny 1980, Birkeland 1984, 1990).

Jenny's model encouraged pedologists to adopt an experimental approach to field studies where the sites selected for investigation are those where several factors are in a sense controlled. However, soil surveyors face situations in which information is needed on a set of soil properties that are dependent to some extent on several environmental variables and predictions cannot be restricted to a few ideal locations.

The nature of canonical functions has been well documented, with chronosequence and toposequence studies receiving the most attention (Yaalon 1975, Birkeland 1984, 1990). As noted earlier, the purpose of much of this work has been to develop models that provide explanations of soil formation rather than operational models that can be used to predict soil distribution for practical land evaluation. However, many studies have shown close, although often complex, relationships between soils and landforms (e.g. Milne 1935, Walker et al. 1968, Gerrard 1981, 1990).

It is widely accepted that the functional factorial approach provides the only principle of universal value in soil mapping—through the hypothesis that "the soil profile is the integrated expression of parent material, climate, topography, living organisms and the age of the landscape" (Dent and Young 1981). However, this general conceptual framework has an apparent simplicity that belies the complexity of most landscapes. An appreciation of this complexity is essential for the appropriate use of terrain analysis in soil and related geomorphological applications.

10.2.2 Contemporary Views

A range of theories competing with that of Jenny (1941, 1980) have been proposed (Crocker 1952, Runge 1973, Yaalon 1975, Birkeland 1984, Johnson et al. 1990, Paton et al. 1995), but there is general agreement that the five state factors provide a useful high level but qualitative classification of the complex subsidiary variables and processes controlling soil formation (Hoosbeek and Bryant 1992). We will refer to these subsidiary variables and processes as vectors of pedogenesis after Johnson et al. (1990).

Johnson et al. (1990) recognized exogenous and endogenous vectors of pedogenesis. Exogenous vectors include the general environmental factors of temperature, wetness, parent material, and topography, which can be used as predictors in soil survey—some can be readily predicted using terrain analysis. A range of specific agents also act as exogenous vectors, including disturbance, episodic erosion and deposition, and biomechanical impacts. Biomechanical impacts caused by earthworms, ants, and termites have a substantial effect on soil development in southeastern Australia, although because of their inherent spatial and temporal variability they have received inadequate attention to date (Paton et al. 1995). Biomechanical impacts are difficult to use as predictive variables in soil survey and any relationships with terrain may well be fortuitous.

Endogenous vectors are similarly difficult to use for prediction. These vectors evolve within the soil system and may create feedback mechanisms. For example, the formation of clay in the lower part of the profile may reduce permeability and cause a subsequent accumulation of salts. These acquired internal conditions, or pedologic accessions, may allow a soil to develop in a way that is largely independent of the external environment (Johnson and Hole 1994).

A summary of the most influential exogenous and endogenous vectors of pedogenesis in southeastern Australia is provided in Table 10.1 with comments on the degree to which they can be used as predictors in soil survey. Clearly, there will be some landscapes in which soil distribution will have a complex or poor relationship with landform because key vectors of pedogenesis are uncorrelated with terrain variables. A further complication to the prediction of soil distribution is caused by environmental change.

10.2.3 Environmental Change in Ancient Landscapes

The vectors of pedogenesis are always changing and, as a consequence, all soils bear the imprint of a sequence of environmental conditions. This is a critical issue in Australian landscapes, particularly those inland from the Great Escarpment (Ollier 1982, 1991) where some remnant areas have been exposed for millions of years. Major environmental fluctuations have occurred during this period and soils persist to now occupy environments markedly different from those in which they were formed.

For example, strongly weathered sesquioxidic soils (Alfisols, Ultisols, and Oxisols (Soil Survey Staff 1994)) were formed under humid and warm climates during the Late Cretaceous and Tertiary, and these soils now occupy extensive semiarid areas across Australia (Churchward and Gunn 1983). Similarly, eolian processes active during arid phases of the Pleistocene are the key control on soil distribution in the riverine landscapes of southeastern Australia (Butler and Churchward 1983). Both the strongly weathered Tertiary and eolian Pleistocene landscapes may still exhibit clear relationships between terrain attributes and soils (Hubble and Isbell 1983), but there is always the possibility that functional relationships between soils and terrain have been complicated or obliterated by imprints from successive sets of environmental conditions.

The impact of past environments is illustrated most graphically in areas where landscape inversion has occurred (Milnes et al. 1985, Ollier et al. 1988, Ollier 1991). Complete or partial landscape inversion occurs when materials (iron oxides, silica, gravels, etc.) accumulate in low-lying positions and become indurated or form a layer that is more resistant to erosion than the surrounding areas (Figure 10.1). The

TABLE 10.1 Vectors of Pedogenesis of Significance in Southeastern Australia and Potential Environmental Predictors

	Vector of Pedogenesis	Potential Process-Based Survey Predictors	Comments
Exogenous	Temperature	Climate surfaces	Past climates may have an overriding significance
	Wetness	Terrain attributes, climate surfaces	
	Parent material	Geology maps and geophysical remote sensing	
	Biomechanical impacts	No efficient predictors	Local scale ephemeral processes
	Disturbance (fire, tree fall)	No efficient predictors	
	Eolian accession	No efficient predictors	Often a function of past environments
	Erosion and deposition	Terrain attributes	Episodic events may dominate
Endogenous	Argillic horizons	Possibly ground penetrating radar	Radar methods have not been fully tested
	Oxic horizons	No efficient predictors	Empirical correlations with terrain are possible
	Natric horizons	No efficient predictors	Empirical correlations with terrain are possible
	Ferricrete, silcrete, calcrete	Geophysical remote sensing	Usually remnants of past environments
	Nutrient status	Possibly geophysical remote sensing	Further research is required

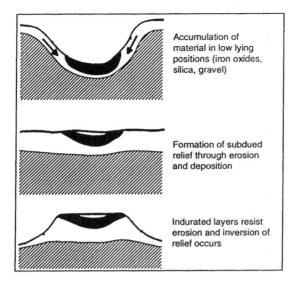

Accumulation of
material in low lying
positions (iron oxides,
silica, gravel)

Formation of subdued
relief through erosion
and deposition

Indurated layers resist
erosion and inversion of
relief occurs

Figure 10.1. A simplified schematic of landscape inversion. The process may occur over a period of hundreds of thousands, or more commonly, millions of years.

surrounding areas eventually erode and become the new low-lying areas. The difficulty for predicting soils from terrain in these landscapes is that features commonly encountered low in the landscape (e.g., soils with indicators of waterlogging or alluvial deposits) are found in elevated settings. However, there may still be strong relationships between soils and terrain that can be used for predictive purposes, especially where the regolith or surficial geology can be categorized in terms of both lithology and stability. This can be achieved using good geomorphological knowledge supported by modern gamma radiometric remote sensing (Cook et al. 1996b, Minty 1997). As with most landscapes, separate relationships between terrain and soil variables are to be expected in each regolith unit.

10.2.4   The Role of Landform

Soil formation is complex enough in young landscapes and even more so in ancient landscapes that have experienced a range of environmental conditions. This complexity is the norm across large areas of the continental plate centers that were free from glaciation during the Pleistocene. It may be expected that soils will have unpredictable relationships with terrain variables in old landscapes. There are too few comprehensive studies using quantitative methods of terrain analysis to reach general conclusions at this stage. However, it is our contention that landform has a major role in predicting soil properties in old landscapes, but its significance will be highly dependent on the materials and environmental history of each region and the ability to categorize and subdivide regolith. Furthermore, predictive relationships will range from those with a clear underlying set of processes to those with empirical correlations.

Good examples of process-based relationships between terrain and soil are found in hilly terrains and at higher latitudes where radiation, and hence evaporation, are strongly related to the geometry of the landsurface. Sites with lower radiation are wetter, have higher levels of organic matter, and support soils with different properties than do adjacent sites with higher radiation (Jenny 1980, Birkeland 1984). In these geomorphic environments, hillslope processes (degradation and accumulation) have strong relationships to terrain attributes such as the topographic wetness index and stream power (Moore et al. 1988a, Chapter 4).

Semiempirical correlations with landform are more widespread. For example, topographic variables such as specific catchment area (A_s), plan curvature, and topographic wetness index have been used to predict soil variation on hillslopes (Moore et al. 1993a, Gessler et al. 1995, Odeh et al. 1994). O'Connell (1997) has also found characteristics of the dispersal area to have predictive value. The processes that give rise to these patterns may have been a combination of water movement, sediment transport, and solute transport. While the precise set of processes cannot be tested in any simple way, they are all related to gravitational forces and conspire to create soil patterns correlated with terrain.

Empirical correlations are useful for prediction if nothing else. The distribution of organisms such as ants or earthworms, for example, may exhibit relationships with terrain that have no clear process explanation, but they may have a major impact on soil properties. Empirical correlations will have limited applicability and will rarely be useful beyond the immediate environment in which they have been developed.

Much research has been done on the spatial variability of soil properties (Burrough 1993), particularly in relation to variation within mapping units. These units are often defined using a combination of terrain and environmental variables (vegetation, aerial photograph tone, etc.). There are fewer studies on relationships between terrain variables per se and soil properties. Some of these studies are summarized in Table 10.2, and two recent studies are presented in Chapters 11 and 12.

10.3 EXAMPLES OF THE USE OF TERRAIN ANALYSIS IN AUSTRALIAN SOIL SURVEY RESEARCH

The following examples are drawn from recent studies exploring quantitative methods for the spatial prediction of soil properties. The Wagga Study (Gessler et al. 1995, Gessler 1996) investigated the use of terrain analysis in small agricultural watersheds. The results were promising and the methods for sampling and predicting soil distribution using terrain analysis were then applied in a realistic survey setting across a much larger area with limited financial resources in the Bago-Maragle Study (McKenzie and Ryan 1999). This study involved the quantitative survey of 50,000 ha of forested land where soil information was needed to assess the productivity of hardwood forests and determine the long-term impact of forest management. The Bago-Maragle Study area (E 148° 15′, S 35° 45′) is adjacent to the Snowy Mountains in southern New South Wales, while the Wagga Study area (E 147° 15′, S 35° 15′) is approximately 100 km to the northwest on more undulating slopes that grade into extensive and flat riverine plains. The Wagga Study area is comprised of older landscapes and evidence of past climates is more apparent, with relict aeolian features being notable (Beattie 1972).

TABLE 10.2 Summary of Selected Studies Predicting the Spatial Distribution of Soil Properties Using Terrain Analysis

Authors	Response Variable(s)	Explanatory Variables	Terrain Analysis Methods	Predictive Model	Location
Walker et al. (1968)	Field morphology	Elevation, slope, profile and plan curvature, distance from summit	Manual analysis of topographic profiles	Multiple regression	Iowa, USA
Furley (1968)	pH, organic carbon, total N, particle size	Slope with sampling restricted to sites with plan curvature ≈ 0	Manual analysis of topographic profiles	Linear regression	Oxford, UK
Furley (1974a,b, 1976)	Field and laboratory properties	Slope, position	Manual analysis of topographic profiles	Linear regression and graphical methods	Belize
Crosson and Protz (1973)	Color, pH, organic carbon, clay	Distance from apex, elevation, slope	Manual analysis of topographic profiles using aerial photographs	Multiple regression	Ontario, Canada
Anderson and Furley (1975)	Field and laboratory properties	Slope, position	Manual analysis of topographic profiles	Principal component analysis and graphical methods	Southern England
Malo et al. (1974)	Field and laboratory properties	Distance from summit	Manual analysis of topographic profiles	Multiple regression	North Dakota, USA
Roy et al. (1980)	Field morphology and particle size	Slope length, maximum slope, mean slope, convexity, toeslope concavity	Manual analysis of topographic profiles	Canonical correlation	Queensland, Australia
Munnik et al. (1984)	Soil depth	Slope, position, curvature	Manual analysis of topographic profiles	Contingency table analysis	Transvaal, South Africa
Kachanoski et al. (1985)	A horizon thickness, density and mass	Spectral estimates of land	Digital analysis of surface	Spectral analysis microtopography	Northern Great Plains, Canada
Gerrard (1990)	pH, particle size, field morphology	Slope, position, distance from slope base	Manual analysis of topographic profiles	Graphical methods and geostatistics	Dartmoor, Wye Forest and Kent, UK

Reference	Soil property	Terrain attributes	Method	Statistical method	Location
McKenzie and Austin (1993)	Field and laboratory properties (mostly chemical)	Relative relief, landform units, stratigraphic units	Manual analysis of topographic profiles	Generalized Linear Models	NSW, Australia
Moore et al. (1993a)	A horizon thickness, organic matter, pH, P, particle size	Slope, wetness index, stream power index, aspect, profile curvature	Digital elevation model and TAPESG	Multiple regression	Colorado, USA
Brubaker et al. (1994)	20 laboratory properties	Landscape position	Manual analysis of topographic profiles	Multiple regression	Nebraska, USA
Odeh et al. (1994)	Field morphology	Slope, profile and plan curvatures, upslope distance and area	Digital elevation model	Linear regression and co-kriging	Mt Lofty Ranges, South Australia
Gessler et al. (1995)	Field morphology	Plan curvature, TWI, upslope mean curvature	Digital elevation model and TAPESG	Generalized Linear Models	NSW, Australia
Boer et al. (1996)	Soil depth	Slope, aspect, profile curvature, specific catchment area, wetness index, LS factor	Digital elevation model with GRASS and PC-Raster	Principal components analysis, maximum likelihood classification with cross validation	Southeast Spain
McKenzie and Ryan (1999)	Soil depth, total P, total C	TWI, slope, relative elevation, temperature, gamma radiometrics, relief, downslope slope, plan curvature, Prescott Index, dispersal area	Digital elevation model and TAPESG	Generalized Linear Models and classification and regression trees	NSW, Australia

10.3.1   Improved Environmental Information

As noted earlier, terrain analysis can be very useful in soil survey and land evaluation regardless of the strength of relationships between soils and landforms. This is because it provides a much better basis for visualizing other data layers as well as generating sets of data that have value in their own right (e.g., slope and solar radiation).

10.3.1.1 Visualization   Views of shaded relief and drapes of coverages over digital elevation models (DEM) highlight relationships with landform. They provide a three-dimensional perspective difficult to attain using other methods. Three views of a portion of the Bago-Maragle study area are presented in Figure 10.2 (see color insert). Gamma radiometric and other remotely sensed images, in particular, are more readily interpreted when draped over a high-resolution digital elevation model. In Figure 10.2c the upper slopes emit a larger gamma radiometric signal and correspond with areas having younger soils and more exposed substrate. Low signals correspond with more weathered and deeper soils on the lower slopes or with areas of basalt on the elevated summit surface. The advent of dynamic viewing of digital landscapes offers a completely new way of examining land resource data.

10.3.1.2 Simple Terrain Characterization   Topographic variables such as slope and aspect can be used directly for land evaluation. For example, slope mapping has long been used in Australia for regulatory purposes in soil conservation. Klingebiel et al. (1987) provide an example in which slope, aspect, and elevation from a DEM were used to make a detailed soil survey. The quality of routinely available digital elevation models is at a stage where manual methods of slope mapping using aerial photographs can be replaced. However, slope, being a first derivative of elevation, is very sensitive to the resolution of the digital elevation data (Hammer et al. 1995, Gessler 1996, Chapters 2 and 5) and the means by which the DEM was derived.

10.3.1.3 Prediction of Microclimate   A more sophisticated application of terrain analysis for soil survey and land evaluation is the prediction of climate. In the Bago-Maragle study area, broad trends in rainfall and temperature are correlated with elevation. The ESOCLIM computer program (Hutchinson 1989a) uses long-term climate records and relationships with elevation to generate predictive surfaces of monthly climatic variables, including rainfall and mean temperature. Climatic variation at a more local scale in the area is caused by aspect differences, which can be modeled using the SRAD program (Moore 1992, Wilson and Gallant 2001, Chapter 4) to provide estimates of net solar radiation.

A high-resolution DEM with a grid size of 25 m was developed for the Bago-Maragle study area. The DEM was developed from digital contours, stream lines, and spot heights from the eight 1:25,000 topographic map sheets covering the area. The DEM was generated using the ANUDEM computer program (Hutchinson 1989b, Chapter 2).

There are many ways to integrate rainfall, temperature, and radiation data to provide indices of local climate. Simple indices of the water balance are among the most useful because they integrate the biologically important effects of rainfall and evaporation. They give an indication of the intensity of leaching and provide a relative measure of potential biological productivity. A useful estimate of the water balance is provided by the Prescott Index (Prescott 1948). It is calculated on a monthly basis and an annual average can be used to compare sites. The Prescott Index (PI) is

$$PI = \frac{0.445P}{E^{0.75}} \tag{10.2}$$

where P = mean monthly (annual/12) rainfall (mm) and E = mean monthly (annual/12) potential evaporation (mm).

As noted above, monthly rainfall (P) was derived from the ESOCLIM surface. E was estimated using the Priestly–Taylor Model (Priestly and Taylor 1972):

$$LE = \alpha \, \frac{s}{s + \gamma} \, (R_n + S) \tag{10.3}$$

where L = latent heat of vaporization of water, E = potential evaporation, α = a constant (1.26), s = slope of the saturation vapor pressure curve at the mean wet bulb temperature, γ = psychrometric constant, R_n = net radiation, and S = soil heat flux. S can be neglected for periods greater than 24 h and is a minor component of the energy balance when averaged over a year. The term $s/(s + \gamma)$ is tabulated as a function of temperature by Slatyer and McIroy (1961) and is well approximated by the following regression equation:

$$\frac{s}{s + \gamma} = 0.3943 + 0.01691T - 0.0001349T^2 \qquad R^2 = 0.99938 \tag{10.4}$$

with T being obtained from the ESOCLIM surface.

R_n was generated using the SRAD program where the units of R_n are Wm^{-2} (i.e., $Js^{-1}m^{-2}$). A conversion factor is required to estimate E in mm day^{-1}. Assuming that $L = 580$ cal g^{-1} (2427×10^6 J m^{-3}) and expressing R_n on a daily rather than a per second basis, the relationship for estimating E is

$$E = (0.01768 + 0.0007585T - 0.00000605T^2)R_n \tag{10.5}$$

Surfaces of average monthly evaporation and rainfall calculated on an annual basis were generated for the study area and used to derive the Prescott Index (Figure 10.2). The Prescott Index is useful because it provides a simple estimate of microclimate across the study area at a resolution that has not been available for survey purposes to date. In this instance it shows the distribution of cool, moist areas on the northern uplands of the study area. Contrasting local scale moisture gradients are also clear, particularly in the more dissected areas. The Prescott Index provides a cheap estimate of wetness suitable for predicting soil properties (McKenzie and Ryan 1999).

10.3.2   Explicit Survey Design

Sampling has been a weak feature of soil survey, although regionalized variable theory now provides a sound basis for intensive surveys (Webster 1977, Webster and Oliver 1990). There has been minimal development of explicit sampling schemes for medium- and low-intensity surveys where time, access, and resources are serious constraints. In these surveys, grid-based sampling schemes are inefficient. Terrain analysis can be used to generate stratified random sampling schemes in a way that has not been possible in conventional survey. In the Bago-Maragle survey, a design-based two-stage random sampling scheme was implemented using geology, landform, and climatic variables for stratification.

10.3.2.1 Geology   Generalized geology maps at a cartographic scale of 1:100,000 of the Bago-Maragle study area were digitized. The geology of the area was remapped using conventional field methods supported by airborne gamma radiometric remote sensing (Bierwirth 1996) as part of the study. Ninety-five percent of the Bago-Maragle area is covered by four rock types—granodiorite (Sgg), adamellite (Dga), basalt (Tb), and meta-sediments (Os)—and these formed an obvious first stratifying variable.

10.3.2.2 Landform   Local landform has a major impact on soils by controlling water and sediment movement. The steady-state topographic wetness index (TWI, Equation 4.40) is a useful integrative topographic variable that is a guide to water and sediment movement in particular landscapes. It quantifies the position of a site in the landscape and has proven useful for predicting soil properties (e.g., Moore et al. 1993a, Gessler et al. 1995). TWI uses the specific catchment area, A_s, expressed as square meters per unit width orthogonal to the flow direction, and β, the slope angle, as input variables. TWI was calculated using the TAPES-G and WET computer programs (Moore 1992, Chapters 3 and 4). The depressionless DEM option and DEMON algorithm for flow accumulation (Costa-Cabral and Burges 1994) were used to calculate A_s. As noted earlier, the cell size was 25 m. A_s was limited to a maximum value ($<100,000$ m^2) to prevent the generation of very large values along major streams. A small value (0.01) was added to grid cells with zero slopes to avoid a denominator of zero. Slopes above 300% were set to that value.

10.3.2.3 Climate   The Prescott Index (Equation 10.2, Figure 10.3) provides an appropriate basis for stratifying climate across the study area.

10.3.2.4 Exclusion Rules, Classification, and Randomization   Exclusion rules were applied to avoid sites of limited value, such as cleared land, roads, lakes, and stream beds. The geology, climate (PI), and landform (TWI) surfaces were then classified and used to generate the digital stratification as follows. Density functions of TWI and PI were used to calculate quantiles on an equal area basis. Each of the approximately 4×10^6 cells was then classified into one of three quantiles for PI and one of four quantiles for TWI. This produced 12 discrete environments for each geo-

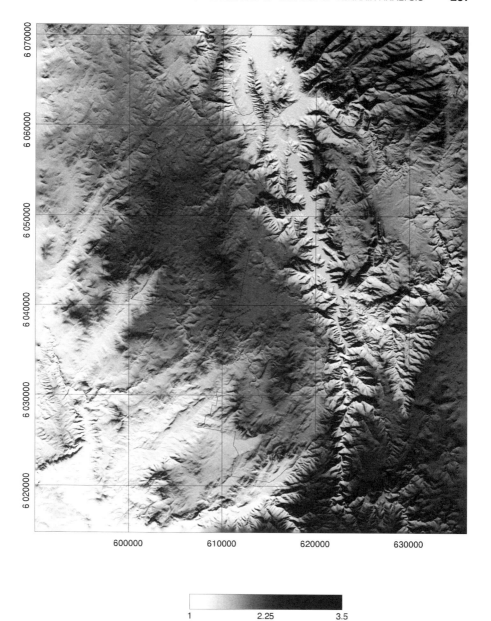

Figure 10.3. Prescott Index for the area covered by the Yarrangobilly 1:100,000 topographic map sheet (~40 × 60 km). The Bago-Maragle study area is outlined.

logical class (Table 10.3). Figure 10.4 (see color insert) shows classified TWI, PI, and the combined 12-class coverages for a 5 × 5-km area in the adamellite (Dga) rock type.

Discrete environments or patches were then selected randomly. Patches smaller than one hectare were excluded because they are difficult to locate in the field. Sites (individual cells) within each patch were then randomly selected. Replicate sites were selected in each of the 12 environments within each parent material class. The granodiorite class had an extra replicate because of its significance for forest production.

The field survey of the 144 selected sample sites involved detailed measurements of soil, vegetation, and fauna. These data were used for quantitative spatial prediction of key land qualities, including erodibility, nutrient status, and the soil-water regime (McKenzie and Ryan 1999).

The quantitative survey design has several positive features. Field sampling is unbiased and a consistency is forced upon the surveyor. The design ensures a complete sampling of the environmental variation, which is believed (at least as an initial hypothesis) to be affecting soil formation across the region at a range of scales. The design is explicit and therefore can be readily tested and modified or repeated elsewhere. The design could be improved by minimizing travel times by ignoring remote sites, but this may introduce bias. Additional sites of perceived pedological importance (swamps, terraces, etc.) can also be opportunistically sampled if required.

TABLE 10.3   Classification of the Granodiorite Geological Unit Within the Bago-Maragle Survey Area Using Landform (TWI) and Climate (PI)

Landform: Topographic Wetness Index	Climate: Prescott Index		
	PI < 2.0	2.0 ≤ PI < 2.2	PI ≥ 2.2
TWI < 6.17	*Class 1*: Dry sites (low rainfall and/or high evaporation) with divergent landforms (ridges and upper slopes)	*Class 2*	*Class 3*: Wet sites (high rainfall and/or low evaporation) with divergent landforms (ridges and upper slopes)
6.17 ≤ TWI < 6.84	*Class 4*	*Class 5*	*Class 6*
6.84 ≤ TWI < 7.68	*Class 7*	*Class 8*	*Class 9*
TWI ≥ 7.68	*Class 10*: Dry sites (low rainfall and/or high evaporation) with convergent landforms (lower slopes, flats and depressions)	*Class 11*	*Class 12*: Wet sites (high rainfall and/or low evaporation) with convergent landforms (lower slopes, flats and depressions)

Note. Written descriptions are given for the four extreme classes.

10.3.3   Quantitative Spatial Prediction

Terrain-analysis variables, in conjunction with those from other sources (geology, remote sensing, etc.), can be used to spatially extend point observations of individual soil properties using statistical models. The explanatory variables must be easier to obtain than soil variables, otherwise intensive sampling of soil variables in assocation with an interpolation or surface fitting procedure would be more efficient for spatial prediction.

Gessler et al. (1995), Gessler (1996), McKenzie and Austin (1993) and McKenzie and Ryan (1999) describe the approach and details of the statistical procedures. An example of quantitative spatial prediction of solum depth (depth of A and B horizons) is presented in Figure 10.5. Solum depth gives an indication of water storage capacity, nutrient reserves, and sediment transport through erosion and deposition. A generalized linear model was used to generate the surface in Figure 10.5. The model was derived from 73 observations across an area approximately ten times that shown in Figure 10.5. Two terrain variables are statistically significant predictors: the steady-state topographic wetness index and the sediment transport index (STRIN). Moore and Wilson (1992) defined STRIN and their definition is presented in Equation 4.1.

The third explanatory variable is the potassium concentration derived from high-resolution airborne gamma-ray spectrometric data collected on 400-m flight lines. Coverage of airborne radiometrics is expanding rapidly in Australia. The data are acquired primarily for mineral exploration but have great potential for land resource survey. The statistical model predicting solum depth accounts for 75% of the deviance, which is high by soil survey standards. Deviance is equivalent to variance in this instance because the generalized linear model has an identity link function and normal error distribution (McCullagh and Nelder 1989).

An alternative to regression models of the form presented above is provided by classification and regression trees (Breiman et al. 1984, Clark and Pregibon 1992). Tree-based models are fitted by successively splitting a dataset into increasingly homogenous subsets. The response variable can be either a factor (classification trees) or a continuous variable (regression trees) and the explanatory variables can be of either type. Output from the model-fitting process is a decision tree, and a measure of goodness of fit is provided by the reduction in deviance. Clark and Pregibon (1992) note that compared to classical regression methods, tree-based models are sometimes easier to interpret and discuss when a mix of continuous and factor variables are used as predictors. Missing values are dealt with in a more satisfactory manner and nonadditive behavior is captured more effectively.

The most significant disadvantage of tree-based methods is the lack of a well-accepted procedure for statistical inference. However, the reduction in deviance is a useful measure of goodness of fit. The set of data shown in Figure 10.5 has been reanalyzed using a regression tree and the results are presented in Figures 10.6 and 10.7. The model uses several extra terrain variables, including elevation, upslope mean profile curvature, and tangential curvature. Upslope mean profile curvature is the average profile curvature within the contributing catchment of a unit width orthogonal to the flow direction. Sites with a large negative upslope mean profile cur-

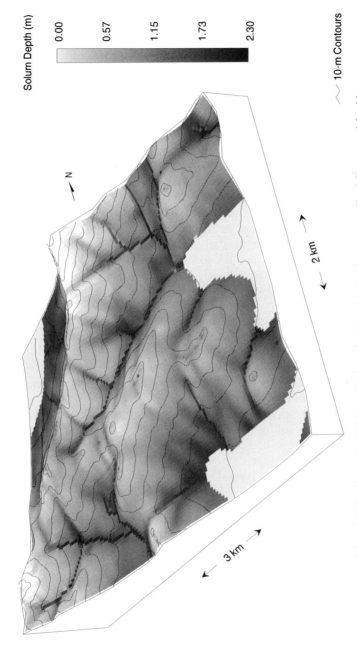

Figure 10.5. Gessler's (1996) prediction of solum depth using a generalized linear model with solum depth = −71.3 + 35.4 TWI − 0.81 potassium − 0.105 STRIN. The model accounts for a 75% reduction in deviance and is based on 73 observations. See text for a description of the explanatory variables.

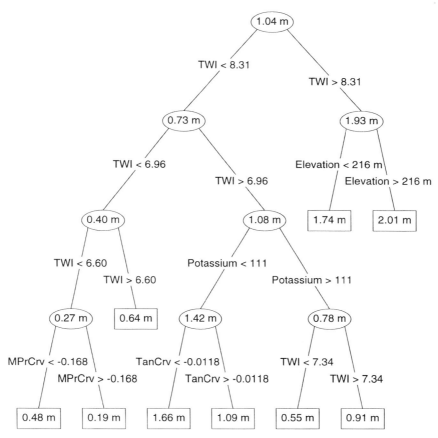

Figure 10.6. A regression tree for predicting solum depth (after Gessler 1996). The predicted solum depth is presented in the ovals and rectangles for the intermediate and terminal nodes, respectively. The model accounts for a 90% reduction in deviance and is based on 73 observations. See text for a description of the explanatory variables.

vature have a contributing catchment dominated by convex hillslopes. Tangential curvature is plan curvature multiplied by slope and provides a measure of local flow convergence and divergence (Gallant and Wilson 1996, Chapter 3). The tree-based model accounts for a much greater proportion of the deviance (90%) than the regression model and uses five variables. However, this is not strictly equivalent to five degrees of freedom because some variables are used more than once in the tree.

An advantage of generalized linear models and classification and regression trees is the ability to use explanatory variables that are continuous (e.g., TWI) or nominal (e.g., rock type). As a consequence, predictions are more realistic because they por-

Solum Depth (m)

0.00

0.57

1.15

1.73

2.30

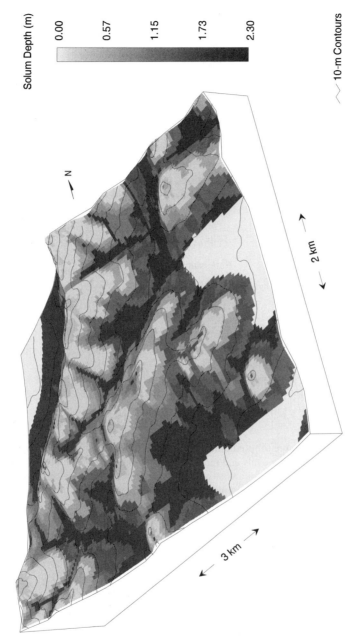

N

2 km

3 km

10-m Contours

Figure 10.7. Solum depth predicted using the regression tree presented in Figure 10.6 (after Gessler 1996).

tray soil variation as being gradual or discontinuous. The response variable of a generalized linear model can be continuous or binary. Gessler et al. (1995) and McKenzie and Austin (1993) provide examples of logistic regression models where the probability of a binary variable is predicted. Classification trees (Breiman et al. 1984, Clark and Pregibon 1992) have the further advantage of being able to predict multinomial response variables.

Quantitative statistical models allow spatial prediction of individual soil properties (not soil types) with a specified accuracy and precision. This is a considerable advantage because individual soil properties often display contrasting scales of variation (Beckett and Webster 1971). The models also allow predictions at local scales in a way that is not possible with conventional polygon-based approaches. This aspect is discussed below.

10.4 FACTORS AFFECTING THE UTILITY OF TERRAIN ANALYSIS

10.4.1 Landscape Complexity

Previous sections have highlighted the potentially complex nature of relationships between soils and terrain in many parts of Australia. McKenzie and Austin (1993), for example, found relationships between landform and soils to be more strongly expressed in younger alluvial units than in older landscape units. They suggested that the older units had been reworked several times, had experienced possible eolian accessions, and had undergone strong weathering during the last 100,000 years to effectively obliterate relationships between soils and the contemporary landform.

In some landscapes, relationships between soils and the geometry of the underlying weathering front may be more significant than surface morphometry. McKenzie and Austin (1993) also found the presence of impeding layers at shallow depths (<1.5 m) to be a strong predictor of soil properties, and the presence of geologic structures such as dikes and sills can control hillslope hydrology and soil patterns (e.g., Engel et al. 1987).

Finally, there will be circumstances in which soil variation occurs without relation to readily observable environmental variables. In these cases, detailed sampling is unavoidable and some form of interpolation or surface fitting is required to generate spatial predictions. The degree to which landscape complexity limits prediction will be clear only when quantitative studies using high-quality environmental analysis have been undertaken in a broad range of landscapes. This is a long-term challenge for pedology.

10.4.2 Issues of Scale

The scales at which processes of soil formation operate should ideally guide the scales used for measurement of soil and terrain variables. In practice this is not often possible because of logistic constraints. There are many issues relating to scale but we will restrict our comments to some recent observations on the scale of soil-forming processes and the resolution of digital elevation models.

A feature of the predictive relationships developed in the Wagga Study has been the explanatory power of contextual and local terrain variables. Contextual variables such as the steady-state topographic wetness index or upslope mean curvature (Gessler et al. 1995) are useful predictors and they indicate that processes at the hillslope scale are present and give rise to useful predictive relationships. Plan curvature has been a consistently useful local terrain variable. In our work it has usually been calculated across an area occupied by a three by three grid (5625 m^2) and indicates that more local scale processes also exercise control on soil patterns.

The local and hillslope scales of variation evident in the statistical models of (Gessler et al. 1995) are consistent with field experience. The significant development provided by digital terrain analysis using high-resolution digital elevation models is the capacity to predict soil variation at a more local scale. Conventional soil surveys cannot readily portray the ubiquitous and substantial short-range variation of soils (Beckett and Webster 1971). Webster (1977) concluded that conventional surveys could account for approximately half the total variance for physical and mechanical properties of soils and to less than a tenth for some chemical properties. The statistical models presented by Gessler et al. (1995) and Gessler (1996) account in some instances for a much greater proportion of the total variance.

Clearly, the appropriate resolution of a digital elevation model will depend on the scale of the processes controlling soil formation and this will be strongly landscape dependent. There are few guidelines at present, although Gessler (1996) found that a grid cell resolution of finer than 40 m is necessary in the catchments of the Wagga Study. At coarser resolutions, terrain variables behave erratically and quickly lose their predictive capacity. McKenzie and Austin (1993), working in extremely low-relief alluvial landscapes, found good predictive relationships between two-dimensional landform profiles and soil patterns, but noted that elevation data with a resolution of around 0.2 m were necessary.

10.4.3 Technology

Advances in computing technology are making sophisticated terrain analysis easier to perform, but it is still within the domain of research groups. The generation of predictive models such as those displayed in Figure 10.5 requires a good level of statistical and computing expertise that is not commonly found in land resource survey agencies. In Australia, several state survey agencies have started to incorporate terrain analysis into their routine survey program on an experimental basis. The ready availability of good-resolution digital elevation data with a grid resolution of better than 25 m across large areas is encouraging other agencies to use quantitative methods of terrain analysis, but it will be several years before their use is commonplace. An effective way to speed the transition would be to include terrain-analysis specialists in survey teams.

10.4.4 Quantitative Versus Intuitive Mental Models

Explicit models for spatial prediction conform to a more scientific approach to soil survey, but they probably do not utilize all of the predictive capacity in the intuitive mental models used in conventional survey. For example, an experienced pedologist usually draws on relationships between soils and environmental variables gained in other landscapes when mapping a new area. It is difficult to include this type of information in the explicit models presented here, although the use of an expert-system approach such as that described by Cook et al. (1996a) has great promise.

Another facet of conventional practice difficult to incorporate into a more quantitative system is the large amount of continuous observation that goes into intuitive mental models. Our quantitative models use data from observations at selected sites; however, many impressions of relationships between soil and more readily observed environmental variables are obtained as the survey area is traversed (e.g., opportunistic observations of road cuttings and other exposures). The degree to which the continuous observations improve an intuitive model has not been investigated. They may conceivably introduce bias and reduce the quality of spatial prediction. It is worth noting that continuous observations influence the form of quantitative analysis to some extent by assisting in the formulation of predictive models.

10.5 CONCLUSIONS

Terrain analysis is of value to soil survey and land evaluation, even in landscapes where soil–landform relationships are weak or complex, because it can be used to generate useful environmental information and provide a basis for visualization. The full potential of terrain analysis in soil survey will be realized only when it is integrated with field programs with a strong emphasis on geomorphic and pedologic processes. This is particularly important in ancient landscapes like those discussed in this chapter.

Statistical models for spatial prediction will range from those with a strong process rationale to those that are strongly empirical—The dictum "all models are wrong but some are useful" applies. In most cases, other environmental explanatory variables apart from those generated from terrain analysis will be essential (e.g., radiometric remote sensing). The statistical models are explicit and can be critically evaluated and revised where necessary. They have the capacity to mimic natural variation whether it be continuous or discontinuous or both, and predictions are provided of primary soil variables rather than soil types, which are classified entities.

The use of quantitative terrain variables as predictors of soil distribution is in its infancy and evidence of its worth, along with methodological refinements, will have to be accumulated in a broad range of landscapes. Subjective terrain analysis has been central to soil survey since its inception. The forms of terrain analysis presented here and in the next two chapters are a quantitative analog that aims to overcome the deficiencies of these conventional methods.

CHAPTER 11

Automated Landform Classification Methods for Soil-Landscape Studies

Stephen J. Ventura and Barbara J. Irvin

11.1   INTRODUCTION

In the past, qualitative and semiquantitative methods of inquiry dominated soil survey and geoecological research due to a lack of investigative techniques (Hudson 1990, Rohdenburg 1989). Now, high-speed computers, new statistical tools, and quantitative models provide unprecedented opportunities for analyses of environmental systems and problems. As discussed in the previous chapter, relationships between soils and terrain can be studied with more quantitative methods. Numerous slope profiles with multiple soil attributes can be handled where previously only a single profile was used. Whole drainage basins can be considered and the variability within them studied.

Geographic information systems (GIS) provide the means to manage these large data sets and to analyze and visualize the spatial relationships among elements. These systems can access databases of soil profile information and link them to landform information such as surficial geology, hydrography, elevation, slope, and aspect. New tools are becoming available to the realm of soil science that provide different methods and techniques for simulating earth surface processes. These include digital elevation models, numerical methods for classification, and geostatistical methods for interpolation. The integration of these tools within an information system provides methods to analyze and display data about soils and landforms. The use of these tools also allows for insight into sampling density, collection patterns and methods, and interpolation techniques. Field surveys can be planned to efficiently collect soil samples to adequately cover an area of interest.

The research presented here is part of a larger project aimed at developing a new approach for three-dimensional modeling of the soil–landscape characteristics based on a study area located in the Driftless Area of southwestern Wisconsin. The portion of the study described here is focused on the derivation of landform attributes and

Terrain Analysis: Principles and Applications, Edited by John P. Wilson and John C. Gallant.
ISBN 0-471-32188-5   © 2000 John Wiley & Sons, Inc.

classification of identifiable landform elements using numerical classification methods. This classification was also used to discern the relationships between soil characteristics and landform elements in the study area. As McSweeney et al. (1994) point out, an important assumption of this work is that soil and geomorphic patterns co-vary and are linked to process. Therefore, we ask

- Is the landform classification an accurate representation of the landscape?
- Does the landform classification provide for a better description than previous methods?
- Does the integration of GIS capabilities with numerical classification tools and the resulting analyses sufficiently explain the landscape processes?

The overall research study was designed to provide insight into landscape evolution in the Driftless Area and a means to test and consolidate new techniques within a single research effort. Since much of soil mapping is based on qualitative assessment of soil properties and landforms, these quantitative tools may enhance current methods of soil survey by providing new ways to visualize the soil–landscape system, by quantifying landform elements currently photo-interpreted, and by integrating data that are otherwise only part of a surveyor's intuitive model. We start out with a brief review of how landforms have been classified in modern soil surveys and soil–landscape studies.

11.2 ROLE OF LANDFORM CLASSIFICATION IN MODERN SOIL SURVEY AND SOIL–LANDSCAPE STUDIES

Landscape position influences soil development, forming the patterns of soils across landscapes. Hudson (1990) noted that there are discontinuities that allow soils to be mapped along natural terrain features, although soils are considered to be a continuum across the landscape. Soil surveyors use the relationship of soil characteristics to landscape position as part of an intuitive model when creating soil maps. This model is based on soil–landscape units, which are similar to landforms but are more narrowly defined using one or more soil forming-factors (Hudson 1990). For example, a slope could be divided into north-facing and south-facing soil–landscape units because of the impact of temperature and moisture on soil formation. The basic assumption of the soil–landscape model is that fairly homogenous soils form on each type of soil–landscape unit. Soil–landscape units have a predictable spatial relationship and are repeated again and again within given areas; stable landscapes produce a higher covariance between the soil and the landscape unit (Hudson 1990). Trained mappers can delineate these units, which serve as the basis for field observations.

The landscape essentially forms a continuous surface and any schemes of subdivision are necessarily somewhat arbitrary (Gerrard 1990). Similarly, delineating the soil into units or describing its qualities using a small number of properties contrasts with its continuous, complex nature. The following soil–landscape models provide a representative sample of the leading ways that researchers have viewed the division of landscapes into identifiable sections. Some are based primarily on soil-forming processes, some on landforms and landscape elements, and some combine these concepts.

Since Milne (1935) first introduced the catena, several models of soil–landscape systems have evolved. All share the basic conclusions of Milne—that there exists a relationship between soil attributes and landscape position. Since water movement is one of the most important driving forces in soil genesis and landscape evolution, the location of flow paths and water distribution on slopes is important for predicting pedogenic variability (Hall and Olson 1991). Most approaches to modeling land-forms attempt to delineate flow lines and highlight those areas of low and high flow and good and poor drainage. Existing models divide landscapes into landform units based primarily on these areas and on hillslope processes.

Ruhe and Walker (1968) delineated geomorphic units using downslope variations in the landscape and correlated them to soil properties (Nizeyimana and Bicki 1992). They describe hillslopes using gradient, slope length, and slope width. A slope is rectilinear if the gradient is constant per unit of length and is curvilinear if the gradient changes with length (Ruhe and Walker 1968). Curvilinear slopes can be either convex or concave downslope (profile view). Hillslopes are divided into five components (Figure 11.1). These terms are probably the most widely used for the description of hillslope positions (Hall and Olson 1991). Ruhe and Walker (1968) further identified geomorphic units of headslope, noseslope, and sideslope, which add divergent, convergent, and linear possibilities for flow lines (Figure 11.2), corresponding to curvature in planimetric view.

Troeh (1964) combined vertical and horizontal slope curvatures and developed four

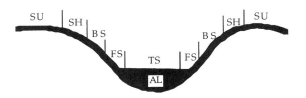

SU	Summit	An upland surface with an inclination which differs distinctly from the hillslope which ascends to it.
SH	Shoulder	The convex component between the summit and the backslope.
BS	Backslope	The typically linear, inclined part of the hillslope.
FS	Footslope	The concave part of the hillslope that welds the linear segments to lower terrain and is both an erosional and depositional surface.
TS	Toeslope	The region which extends away from the base of the hillslope and is composed of depositional debris.
AL	Alluvium	Unconsolidated depositional debris

Figure 11.1. Hillslope profile between two interfluves (Ruhe and Walker 1968, Ruhe 1969). Adapted with permission from Ruhe, R. V. (1969). Quarternary Landscapes in Iowa. Ames, Iowa: Iowa State University Press. Copyright © 1969 by Iowa State University Press.

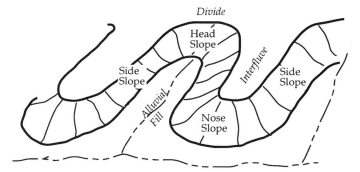

Figure 11.2. The geomorphic units (Ruhe and Walker 1968). Adapted with permission from Ruhe, R. V. (1969). *Quarternary Landscapes in Iowa.* Ames, Iowa: Iowa State University Press. Copyright © 1969 by Iowa State University Press.

basic convex–concave combinations for predicting soil drainage class. Arguing that the two-dimensional catena is valid only where flow lines run straight and parallel from watershed to stream channel, Huggett (1975) added flow lines to Troeh's combinations (Figure 11.3). Huggett (1975) proposed the basic unit of the soil system to be a three-dimensional body of soil known as a soil landscape system or a "valley basin," which is roughly the equivalent of the geomorphologic erosional drainage basin. The soil landscape is constructed of three-dimensional open systems whose boundaries are defined as drainage divides, the surface of the land, and the weathering front at the base of the soil profile.

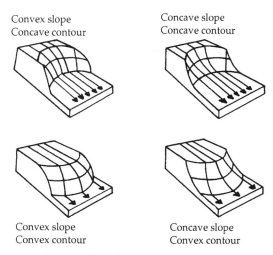

Figure 11.3. Basic slope shapes with surface flow lines. Reprinted from *Geoderma,* Volume 13, Huggett, A. J., *Soil Landscape Systems: A Model of Soil Genesis,* pp. 1–22. Copyright © 1975, with permission from Elsevier Science.

Conacher and Dalrymple (1977) proposed a landscape model in which the basic geomorphic unit is the land surface catena, extending from the center of the interfluve to the valley bottom and from the base of the soil to the soil–air interface. The model combines landform with contemporary pedogeomorphologic processes (movement of water and soil materials). Conacher and Dalrymple (1977) described nine landsurface units, not all of which occur within all landscape catenas (Table 11.1). The nine-unit model may not be applicable to all landscapes, though it may lend itself most to mature, relatively stable temperate landscapes (Gerrard 1992).

Where across-slope curvature is present, the flow lines are seldom parallel downslope, but rather diverge and converge along the contours. In general, hydraulic conductivity decreases with depth; water traveling downward in a profile will be unable to enter the less permeable layers and will be deflected laterally. Thus, material and soil solution throughflow vary with both profile (downslope) and plan (across-slope) curvature. This flux contains dissolved and suspended materials, which it moves from the upper reaches of the valley basin to lower parts (Birkeland 1984). This movement may result in eluviation (removal of soil particles) in the upper zone of the basin and illuviation (deposition) along the lower reaches.

Pennock et al. (1987) wanted to identify different moisture regimes as related to hillslope position, since moisture levels affect both pedogenic development and crop

TABLE 11.1   Landsurface Units Defined by Conacher and Dalrymple (1977)

1. Interfluve	An interfluve with predominant pedogeomorphic processes caused primarily by vertical (up and down) soil-water movements; 0°–1° gradient
2. Seepage slope	Upland area where responses to mechanical and chemical eluviation by lateral subsurface soil-water movements predominate
3. Convex creep slope	Convex slope element where soil creep is the predominant process, producing lateral movement of soil materials
4. Fall face	Areas with gradients greater than 45° characterized by the processes of fall and rockslide
5. Transportational midslope	Inclined surfaces with 1°–45° gradients and responses to transport of large amounts of material downslope by flow, slump, slide, erosion, and cultivation
6. Colluvial footslope	Concave areas with responses to colluvial redeposition from upslope
7. Alluvial toeslope	Areas with responses to redeposition from upvalley alluvial materials; 0°–4° gradient
8. Channel wall	A channel wall distinguished by lateral corrosion by stream action
9. Channel bed	A stream channel bed with transportation of material downvalley by stream action as the predominant process

yields. Using combinations of gradient, plan curvature, and profile curvature, they defined seven distinct landform elements (Figure 11.4). The elements correspond to Ruhe and Walker's (1960) hillslope classes: convex shoulders, concave footslopes, linear backslopes, and gently sloping summits and toeslopes. Summits and toeslopes are classified as level elements and are distinguished by their position relative to other elements. Agreeing with Huggett (1975) that across-slope flows are important, Pennock et al. (1987) use plan curvature to further subdivide shoulders, backslopes, and footslopes into convergent (concave) and divergent (convex) elements (Figure 11.5). The results of their study indicate that moisture content relates to elements in the sequence shoulders < backslopes < footslopes and divergent elements < convergent elements.

Speight (1974) identified broad-scale landform patterns and more localized landform elements. Landform patterns are generally on the order of 300 m in radius and are differentiated on the basis of altitude, landform elements, toposequences (catenas), network development, and relief. In later work, Speight (1990) identified about 40 types of landform patterns such as floodplain, dunefield, and hills. He described landform elements, which make up landform patterns, as being on the order of 20 m in radius and commonly ordered along a catena. They were described using slope, morphological type, dimensions, mode of geomorphological activity, and geomorphological agent. Slopes were described not only by percent but also by the relative inclination of the

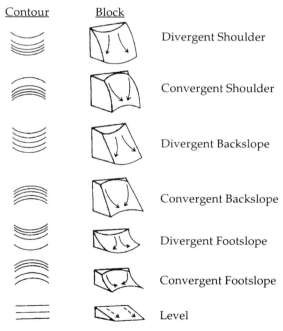

Figure 11.4. Landform element contour and block diagrams. Reprinted from *Geoderma*, Volume 40, Pennock, D. J., Zeborth, B. J., and E. DeJong, Landform Classification and Soil Distribution in Hummocky Terrain, Saskatchewan, Canada, pp. 297–315. Copyright © 1987, with permission from Elsevier Science.

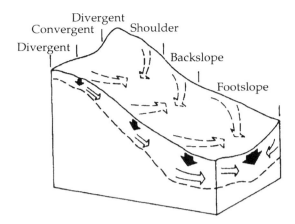

Figure 11.5. Summary diagram of vertical, surface, and through flow of water. Reprinted from *Geoderma*, Volume 40, Pennock, D. J., Zeborth, B. J., and E. DeJong, Landform Classification and Soil Distribution in Hummocky Terrain, Saskatchewan, Canada, pp. 297–315. Copyright © 1987, with permission from Elsevier Science.

slope elements. A waxing slope occurred between a gentler upslope element and a steeper downslope element; a waning slope occurred between a steeper upslope and a gentler downslope (Figure 11.6). He identified 10 morphological types: crest, hillock, ridge, simple slope, upper slope, midslope, lower slope, flat, open depression, and closed depression. Speight described the genesis of the landform element in terms of the mode of activity of a geomorphological agent. An activity, such as erosion, may result from different agents such as wind, creep, or landslide.

Early soil–landscape researchers defined landforms in the field (and from air photo stereopairs) using easily measured or observed attributes, such as aspect, slope, and observed slope shape (concave, straight, or convex). Given the difficulty of observing the covariance of more than a few soil properties, expressions of soil landscape processes (e.g., soil mapping units derived from landform interpretation) were based on assumptions of homogeneity and landform correlation that were not always well supported.

With the increasing use of computers for this type of research, landform descriptions are trending toward more quantifiable assessments that include terrain derivatives computed from digital elevation models instead of those interpreted from photographs, the analysis of larger numbers of spatially explicit soil properties, and the use of quantitative instead of cognitive expressions of the relation between the two. As I. D. Moore et al. (1991) pointed out, most efforts to link soil properties to landscape position until recently used broad classes (head slopes, linear slopes, footslopes) to characterize the landform elements instead of using specific topographic attributes (slope gradient, aspect, curvature). More recent studies have shown a trend toward the use of specific topographic attributes. The use of specific attributes may provide for more explicit linkage to underlying processes that influence soil–landscape patterns. For example, Pennock et al. (1992) studied spatial patterns and controls on denitrification in southern Saskatchewan using attributes derived from a digital elevation model (DEM) to assign

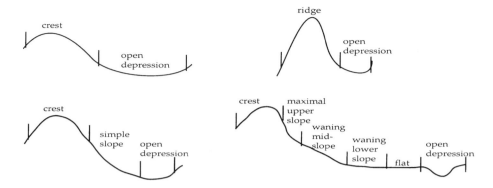

Figure 11.6. Examples of landform elements (Speight 1990). Reproduced by permission of CSIRO Australia.

sites to traditional landform elements (shoulder, footslope, level-convex, and level-concave). Moore et al. (1993b) used a DEM to derive terrain attributes but did not assign sites to landform elements; rather they correlated the individual terrain attributes to specific soil properties.

The following sections describe a soil–landscape study that is intermediate between these approaches. We used specific terrain attributes to partition the landscape into geomorphological units that could be assigned to named landform elements. However, data about specific properties of and variability around individual locations (pixels) in the DEM were retained so that this information could be used in analysis of soil property–terrain characteristic relations.

11.3 PLEASANT VALLEY STUDY

The research described in this chapter was performed as part of a project relating soil attributes to landforms in Pleasant Valley, located in western Dane County, Wisconsin. The study area falls within the "Driftless Area," an unglaciated portion of Wisconsin. Soils evidence points to considerable landscape instability during the Quaternary. A variety of processes have shaped the soils, including mass wasting, in situ weathering, loess deposition, and anthropogenically accelerated erosion. The 49-ha (120-acre) study area contains the five geomorphic units of Ruhe and Walker (1968): summit, shoulder, steep backslope, footslope, and alluvial toeslope.

To correlate soils with landforms, soils were sampled along a 50-m grid, resulting in 224 sample sites. Sampling was performed to the depth of the solum, or to 200 cm in the deeper alluvium. Attributes were measured in the field and samples were taken for laboratory analysis at selected depths along a sample core, representing what appeared to be horizons or at least distinct units of soils. Attributes recorded for each horizon were color, structure, boundary morphology, consistence, coarse fragments, and special features such as mottles. Lab analyses determined texture, pH, and organic matter content. Values for attributes were splined vertically using the technique of Ponce-Hernandez et al. (1986), allowing for estimates of values at any depth down the profile. Slater et al. (1994) discussed the soil data and performed a continuous classification of the soils of Pleasant Valley. He used fuzzy classification to assign membership in classes for each horizon, which were then vertically splined to give a continuous representation of the profile. Results were compared with traditional horizon designations for the area.

In Pleasant Valley, the prevailing landform elements related to soil-forming processes are the summit, backslopes, alluvial fans, and valley. Water movement (e.g., flow pathways) and incident solar radiation are also important. These features needed to be identified to examine the soil–landform relationships in the study area. A DEM for the area was created using a 10-m horizontal resolution and 1-ft vertical resolution (Figure 11.7). Savory (1992) concluded that this resolution was appropriate for the scale of the landforms in Pleasant Valley in similar work, delineating shoulders and footslopes using a moving window and a Laplacian of Gaussian edge detection operator. For the research reported here, the DEM was used to calculate primary and secondary attributes for the landform classification process. It was also used to create a two-dimensional (2D) surface for graphical display purposes within the GIS.

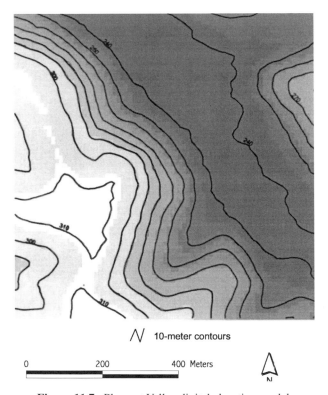

∧⁄ 10-meter contours

0   200   400 Meters

Figure 11.7. Pleasant Valley digital elevation model.

Two techniques were explored for the purpose of defining landform elements: an ISODATA unsupervised classification and a continuous, or fuzzy, classification. ISODATA creates a "crisp" classification of data points in which each data point is classified into a statistical cluster, which is assigned to a single information class. In continuous classification, each data point can have partial membership in several classes.

Clustering procedures are used to describe multivariate data in terms of clusters or groups of data points that possess strong internal similarities (Duda and Hart 1973). These techniques are often used for digital image interpretation (satellite images or scanned photographs) and can be either supervised or unsupervised. Supervised classification requires the operator to define the cluster characteristics, while unsupervised techniques rely on the classification process itself to define the clusters. The user is left to interpret the results and make conclusions about the resulting groupings (Peltzer 1992). Unsupervised classification techniques produce natural groupings of data in attribute space and can be used to gain insight into the structure of the data and may yield new subclasses of data or produce results that depart from the expected. When used with natural resource data, comparison of the classification with the environment may facilitate the identification of natural units (Ward et al. 1992).

Classification methods may identify a number and composition of classes that do not correspond to preconceived notions of the makeup of the landscape. Classes may be produced that do not fall within classic landform boundaries such as those of Ruhe and Walker (1968). Depending on the proposed use of the classification, classes may be combined to aid in visualization or analysis of the landforms. In this way, information classes are created from the statistical classes of the classification.

11.4 METHODS

The data used for the classifications were derived from a DEM that was stored and manipulated using ARC/INFO Rev. 7.0.2 (Environmental Systems Research Institute). The unsupervised classification was performed within the ARC/INFO GRID module primarily using the ISOCLUSTER and MLCLASSIFY routines. The continuous classification was performed using MacFuzzy, a Macintosh-based program utilizing fuzzy k-means with extragrades (Ward et al. 1992).

Six topographic attributes were used for each of the classification methods: elevation, slope, profile curvature, tangential curvature, steady-state topographic wetness index (TWI), and incident solar radiation. Each of these attributes related to soilforming processes was expected to give either algorithm the capacity to differentiate areas of the DEM where soil development varied, with the exception of elevation. The inclusion of elevation proved necessary as a means to consistently separate flat ridges from valleys.

Slope and both curvatures were calculated using TAPES-G, a program designed primarily to calculate hydrological factors from a DEM. TWI was calculated in the ARC/INFO GRID module using output from TAPES-G. Incident solar radiation was calculated using SOLARFLUX (Hetrick et al. 1993a, b, Rich et al. 1994), an Arc Macro Language (AML) routine written to run within ARC/INFO GRID.

11.4.1 Calculation of Topographic Attributes from a DEM

11.4.1.1 Primary Attributes Primary attributes such as slope, aspect, plan curvature, and profile curvature can be computed from DEMs as discussed in Chapter 3. TAPES-G was used to derive the primary attributes slope, aspect, and curvature. Slope (Figure 11.8) was calculated using the finite-difference approach (Equation 3.9). Aspect, as such, was not included as an attribute here, but is an important component in the calculation of incident solar radiation. Profile curvature (Figure 11.9a) measures the rate of change in the direction of maximum slope. Plan curvature is the rate of change transverse to the direction of maximum slope and is measured in the horizontal plane of the contour lines. Mitasova and Hofierka (1993) argue that a more appropriate measure for studying flow convergence and divergence is tangential curvature (Figure 11.9b), which is the curvature in the normal plane in the direction perpendicular to the gradient (or tangent to the contour line). Tangent curvature, by its definition, can be expected to provide a better representation of the collection or dispersal of water flowing over a surface perpendicular to the gradient.

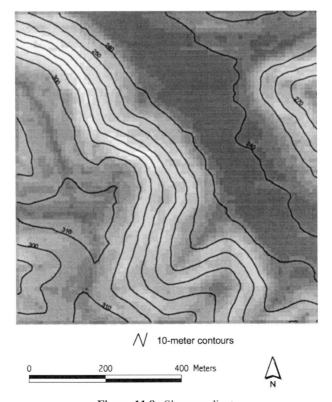

$\bigwedge$ 10-meter contours

0   200   400 Meters $\bigtriangleup$ N

Figure 11.8. Slope gradient.

(a)

(b)

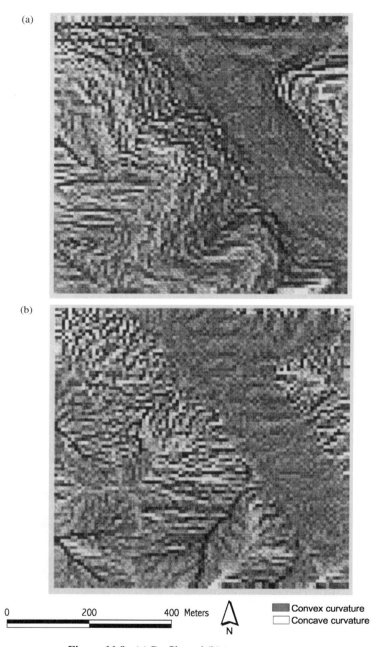

0   200   400 Meters

N

■ Convex curvature
□ Concave curvature

Figure 11.9. (a) Profile and (b) tangent curvature.

11.4.1.2 Steady-State Topographic Wetness Index   The steady-state topographic wetness index (Equation 4.40) was used to represent the spatial distribution of water flow and water stagnation across the study area. The derivation of TWI assumes that the gradient of the piezometric head is parallel to the gradient of the surface topography and steady-state conditions apply. The value of the index increases along with increases in specific catchment area and decreases in slope gradient. It is highest in valleys (high specific catchment area and low slope) and lowest on steep hillslopes (Figure 11.10a).

The upslope area draining into each pixel may be estimated several ways. The most common approach is the *D8* (deterministic-eight node) algorithm, which assigns all of the flow out of a cell to the nearest neighbor in the direction of steepest descent. Use of this algorithm tends to produce parallel flow lines on fairly planar slopes and it cannot model flow divergence on convex slopes (Moore 1996). The *FD8/FRho8* method used here allows flow from each cell to all its downslope neighbors on a slope-weighted basis. It distributes flow more evenly over planar or convex slopes but may underestimate the degree of flow convergence on hillslopes.

TAPES-G was used to produce a depressionless DEM from which it derived the slope and upslope drainage area for each pixel. The specific catchment area was then calculated as the drainage area divided by the width of the pixels (10 m). This specific catchment area and slope were used as input to ARC/INFO GRID for the calculation of TWI.

The area of the original DEM did not extend to the edges of the drainage basins of Pleasant Valley; therefore, if it had been used for calculation of the TWI, the upslope catchment area would have been underestimated, particularly for those pixels close to the edge of the DEM. For this reason, the area of the DEM was expanded to the ridges defining the drainage basins. Additional points were digitized along ridges and drainage lines from a USGS 7.5′ topographic map. These points were then combined with points sampled from the original DEM and then interpolated back to a 10-m pixel grid. The resulting DEM closely approximated the original and was used to calculate the upslope catchment areas. This expanded DEM was only used to calculate TWI, since the original DEM was suitable for calculating the other attributes.

11.4.1.3 Incident Solar Radiation   Aspect could have been used as an input for the classification methods, but its major effect on pedogenesis is its influence on soil temperature and weathering from frost and ice expansion. A better descriptor to use is an estimate of incident solar radiation (ISR) reaching each pixel. Spatial and temporal variation in site-specific solar insolation is predictable from basic geometric principles (Rich et al. 1994).

SOLARFLUX is a GIS-based program developed for modeling solar radiation interception using a DEM. It calculates solar insolation using surface orientation, solar angle (azimuth and zenith), shadowing produced by topographic features, and atmospheric conditions (Hetrick et al. 1993a, b, Rich et al. 1994).

SOLARFLUX calculates ISR by integrating direct, diffuse, and reflected radiation components over a specified time period. Direct radiation is transmitted unimpeded

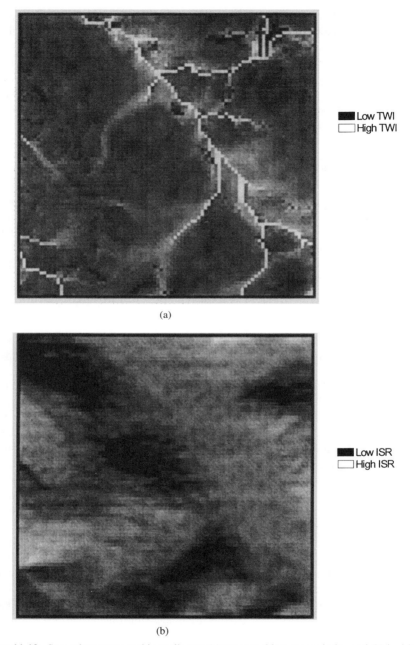

(a)

(b)

Figure 11.10. Secondary topographic attributes: (a) topographic wetness index and (b) incident solar radiation.

along the path between the sun and the earth and varies with sun angle and topographic shading (Hetrick et al. 1993a,b). Diffuse radiation results from atmospheric scattering of light rays, while reflected radiation consists of direct and diffuse radiation reflected off of surrounding terrain features (Rich et al. 1994). Total instantaneous insolation is the sum of direct, diffuse, and reflected radiation components (Gates 1980):

$$I = S_0 \tau^m \cos i + S_0 (0.271 + 0.294\tau^m) \cos^2\left(\frac{\theta}{2}\right) \sin a$$

$$+ rS_0 \sin a (0.271 - 0.706\tau^m) \sin^2\left(\frac{\theta}{2}\right) \quad (11.1)$$

where S_0 is the solar constant, τ is the atmospheric transmittance coefficient, m is air mass ratio, i is the angle of incidence between the normal to the surface and the solar rays, θ is the surface slope, a is the solar elevation angle, and r is the reflected radiation coefficient. This value is integrated to calculate total daily, monthly, seasonal, or annual insolation.

Atmospheric transmittance accounts for scattering of solar radiation with values ranging from 0 to 1, which vary according to location and elevation. In the absence of empirical data, Gates (1980) recommends an average value of 0.6 for clear-sky conditions. This recommended value was used for two reasons. First, since the study area is small, there is no discernible change in transmittance effects across the area. Second, both classification methods are looking at relative differences between pixels, so the absolute value of ISR reaching any individual pixel is unimportant to the classification.

Topographic shading (i.e., shading of a pixel due to a nearby hill) affects the direct insolation component; it changes with solar angle throughout the day. The importance of topographic shading increases with surface complexity and at times may be more important than surface orientation (Hetrick et al. 1993a, b).

For this project, daily insolation was calculated for four dates, one each season. These quantities were then summed to estimate relative annual ISR (Figure 11.10b).

11.4.2 ISODATA Unsupervised Classification

One of the most widely used unsupervised clustering algorithms is the ISODATA (Iterative Self-Organizing Data Analysis Technique) procedure (Tou and Gonzalez 1974). This algorithm is commonly used for satellite image classification. Spectral reflectances from multiple wavebands (equivalent to attributes) are used to determine classes. ISODATA incorporates an iterative process in which the user-defined number of clusters, k, are assigned arbitrary cluster means in multidimensional attribute space. All of the data points are then assigned to these clusters and new means are recalculated for every class. Using these new means, the data are then reclassified to the nearest cluster in attribute space and the cluster means are recalculated. This process is repeated until the migration between clusters is less than a threshold set by the user or program from one iteration to the next. The resulting classification must be analyzed by the user, whose knowledge of the phenomenon in question is integral to interpreting the significance of the various clusters.

The ISODATA classification of the data was accomplished using the ISOCLUS-TER and MLCLASSIFY functions in the ARC/INFO GRID module. ISOCLUSTER uses a modified iterative optimization clustering procedure, also known as the migrating means technique. The procedure must be run several times to produce different numbers of classes. The user must specify the number of classes and has the option of setting the maximum number of iterations, minimum class size, and sampling interval.

Elevation and the five derived attributes for each pixel were used as input to determine class compositions for Pleasant Valley. The ISOCLUSTER function returns a signature file, containing class means and covariance matrices, which are then used as input to a maximum likelihood classifier (MLCLASSIFY). The classifier uses the mean vector and covariance matrix of each class to compute the statistical probability that a grid cell belongs to the class. Each cell is assigned to the class to which it has the highest probability of being a member. The user can specify a reject fraction for the classification. The classifier calculates the level of certainty that the classification is correct for a cell. If it is not high enough, the cell is left unclassified.

To characterize output classes by mean vectors and a covariance matrix, the data for each class should be roughly Gaussian. Therefore, it is important to analyze the histograms of the input attributes. In some cases, a log transform is all that is needed to create a normal distribution of skewed data. Most of the DEM attributes exhibited adequate distributions and only the slope data were log transformed. However, the elevation data were bimodal—violating the assumption of a Gaussian distribution. To mitigate this problem the data set was split midslope (at the midpoint of the elevation range). As a result, each half of the scene had a seemingly normal distribution for this parameter. ISOCLUSTER and MLCLASSIFY were then run for each half of the data set and the results combined.

11.4.3   Continuous Classification Overview

Fuzzy set concepts were first presented by Zadeh (1965) to provide a mathematical method for dealing with continuous data and they have been used in various applications. Fuzzy logic has been used not only to describe vagueness in data (i.e., inaccuracies in measurement) but also in the semantics used to describe data. In this light, its incorporation into geographic information systems has been discussed (Burrough 1989, Wang et al. 1990, Kollias and Voliotis 1991, Lowell 1994). However, the use of fuzzy logic for representing vagueness should not be confused with its use as a classification tool. For this reason, McBratney and de Gruijter (1992) proposed the term *continuous classification* to better describe the method of grouping data using the fuzzy k-means technique. Though the term "fuzzy" implies vagueness, actually the method provides much more precise information than that provided by use of discrete classes.

Continuous classification techniques utilizing fuzzy k-means do not require data points to be exclusive members of one and only one class. Data points are allowed partial membership in many classes according to how close a point is to each class center in attribute space. The idea of fuzzy clustering based on fuzzy set theory was first proposed by Ruspini (1969). Others have refined the concept by developing additional algorithms

(Bezdek et al. 1981, McBratney and de Gruijter 1992). The most widely used fuzzy clustering method is probably fuzzy k-means, also referred to in some literature as fuzzy c-means (McBratney and de Gruijter 1992), with k or c referring to the number of clusters or classes. The number of classes may be determined by the application (e.g., comparing the fuzzy classes to existing classes on a one-to-one basis) or the user may determine the optimal number of classes with the help of performance indices.

Natural phenomena that are continuous in nature have been effectively described using continuous classification. These procedures have been used to classify natural resource phenomenon such as climate data (McBratney and Moore 1985), geologic data (Bezdek et al. 1984), and soils data (McBratney and de Gruijter 1992, Odeh et al. 1992, Slater 1994). Continuous classification is accomplished through an iterative process in which individual data points are allocated class memberships using simple statistical measures of distance in attribute space (Ward et al. 1992). The membership value indicates how closely related an individual is to the centroid (attribute mean) of that class.

Some data may be poorly represented by any of the defined classes and, therefore, have a low membership in most or all of the classes. Their characteristics are different enough from other points to warrant not being classified. McBratney and de Gruijter (1992) term these points *extragrades* and place them in an outlier class.

Continuous classification is similar to the ISODATA method described previously in that both derive groupings solely from the data points without referencing existing classification systems. They both use iterative processes to determine class means by minimizing distances in multidimensional attribute space. However, they differ in their treatment of class memberships: The ISODATA method creates classes in which each data point is wholly a member of a class, while the continuous classification creates class memberships for each point. Membership values range between 0.0 and 1.0 and the sum of values for any point equals 1.0.

MacFuzzy, a Macintosh-based program (Ward et al. 1992) utilizing fuzzy k-means, was used to perform the continuous classification. It permits choices of algorithm (with or without extragrades), statistical distance metric, desired number of classes, k, and degree of fuzziness. It also provides performance indices to aid in the choice of the optimum number of classes. The program incorporates a fuzzy k-means clustering algorithm and the further extension of that algorithm to allow for extragrades by McBratney and de Gruijter (1992), as discussed below.

11.4.3.1 Fuzzy k-Means Algorithms

Conventional set theory constrains set membership to either 0 (no membership) or 1 (exclusive membership). Consider a set of n individuals (data points) classified into k classes; conventional classification gives membership functions $M = \mu_{ic} = 1$, where individual i belongs to class c, and $M = \mu_{ic} = 0$, where i is not a member of c. Three conditions ensure that conventional sets are exclusive and jointly exhaustive:

$$\sum_{c=1}^{k} \mu_{ic} = 1 \qquad 1 < i < n \qquad (11.2)$$

(The sum of memberships of an individual across all classes is 1)

$$\sum_{i=1}^{n} \mu_{ic} > 0 \qquad 1 < c < k \tag{11.3}$$

(Classes are nonempty—at least one individual belongs to each class)

$$\mu_{ic} \in \{0,1\} \qquad 1 < i < n; \quad 1 < c < k \tag{11.4}$$

(An individual either belongs to a class or does not belong at all)

Equation 11.4 establishes the difference between hard classes and fuzzy classes. Fuzzy set theory relaxes this condition of all or nothing membership so that individuals are allowed to be partial members of a class:

$$\mu_{ic} \in [0,1] \qquad 1 < i < n; \quad 1 < c < k \tag{11.5}$$

(An individual's membership in a class is in the 0 to 1 range, a continuous function from no to complete membership)

To determine the optimal fuzzy classification, an objective function is minimized. A generalized objective function for fuzzy sets uses fuzzy k-means to minimize the within-class sum of squares function, $J(M, C)$ (Bezdek et al. 1984):

$$J(M, C) = \sum_{i=1}^{n} \sum_{c=1}^{k} \mu_{ic}^{f} d_{ic}^{2} \tag{11.6}$$

where $C = (c_{cv})$ is a $k \times p$ matrix of class centers; p is the number of attributes; c_{cv} is the value of the center of class, $c,$ for variable, $v;$ d_{ic}^{2} is the square of the distance between the individual i and the class center c according to a selected metric; and f is a user-chosen fuzziness coefficient ranging from 1 to infinity. This function is minimized, constrained by the conditions summarized in Equations 11.2, 11.3, and 11.5. The fuzziness exponent, $f,$ determines the fuzziness or overlap between classes. Classes are discontinuous at $f = 1$; as f increases, fuzziness increases. At $f = 1$, minimization is achieved through an iterative process of assigning individuals to classes. If $f > 1$, the function can be minimized by Picard iteration (McBratney and de Gruijter 1992).

This fuzzy k-means analysis provides for representation of data points that fall between classes, called *intergrades*. However, extragrades, or points that fall far away from the centroids of the classes, are not well represented. To accommodate these points, McBratney and de Gruijter (1992) defined a new objective function:

$$J(M, C) = \alpha \sum_{i=1}^{n} \sum_{c=1}^{k} \mu_{ic}^{f} d_{ic}^{2} + (1 - a) \sum_{i=1}^{n} \mu_{i*}^{f} \sum_{c=1}^{k} d_{ic}^{-\beta} \tag{11.7}$$

Following the lead of McBratney and de Gruijter (1992), we estimated the value of the mean membership of the extragrade class, $a,$ as the average of each of the regular classes. The term μ_{i*}^{f} represents membership in the extragrade class. The extragrade exponent β determines the effect of distance from centroids on membership allocation between the normal and extragrade classes. A value of 2 for β is often used, but smaller values allow higher normal class membership and reduce membership in the extragrade class.

11.4.3.2 Distance-Dependent Measures

Distance functions, or metrics, are a measure of the similarity of a data point to a class; generally, the smaller the distance in attribute space, the greater the similarity. The metric indicates the method of measurement of that distance (Peltzer 1992).

The Euclidean distance metric gives equal weight to all measured variables and is insensitive to statistically dependent variables (Odeh et al. 1992). It is useful when attributes are independent and the clusters have the general shape of spherical clouds, but this method is not appropriate to use when attributes have widely varying average values and standard deviations, since large values in one attribute will prevail over smaller values in another (Ward et al. 1992). The diagonal metric compensates for distortions in the assumed spherical shape caused by disparities in variances among the measured variables (Odeh et al. 1992). It transforms the data set to one in which all attributes have equal variance. Likewise, the Mahalanobis metric compensates for differing variances but also accounts for statistically dependent measured variables (Odeh et al. 1992). Its use results in a transformation of data so that all attributes have zero mean and unit variance. Because all of the attributes used for classification were generated from the elevation data of the DEM, the data are not independent; therefore, the Mahalanobis metric is a better choice to account for their interdependencies and differing variances.

11.4.3.3 Performance Indices

Choosing the optimum number of classes or the degree of fuzziness can be an intuitive approach based on the user's knowledge of the data (McBratney and Moore 1985) or it can be based on the intended use, for example, comparison of fuzzy classes to a predetermined number of conventional classes.

The optimal degree of fuzziness is determined subjectively by the user. Several analyses need to be performed and the results evaluated. The fuzzy exponent determines the extent to which groups are compact and separated (Odeh et al. 1992). If it is too low, the classes are too discrete, meaning that the membership values are close to 0 or 1. If the exponent is too high, the classifier may fail to converge or the classes will not be clearly separated for analysis.

While there are formal functions to aid in the evaluation of the optimal number of clusters, the final decision on number of classes is left to the judgment of the user. The fuzzy performance index (FPI) and normalized classification entropy (NCE), discussed in depth by Odeh et al. (1992), reflect the degree of continuity and disorganization in the clustered data after classification. Since the minima of these functions indicate the optimum number of classes where there exists a balance between continuity and structure (McBratney and Moore 1985), these results are useful as a guide to the choice of number of classes.

11.5 RESULTS AND DISCUSSION

The fuzzy and ISODATA classifications were compared to three delineations of landscape divisions (two manual and one automated) and the published soil map units. Two soil scientists experienced in the area delineated landform elements using an airphoto stereo-pair (1:9600 scale, leaf-on) and a mirror stereoscope. These two delineations picked out features similar to those found in modern soil surveys published at scales in the 1:12,000 to 1:24,000 range, though with somewhat more spatial detail (Figure 11.11). The shoulder and footslope breaklines identified by Savory (1992) using an automated edge detection technique on the DEM were also useful for comparison purposes.

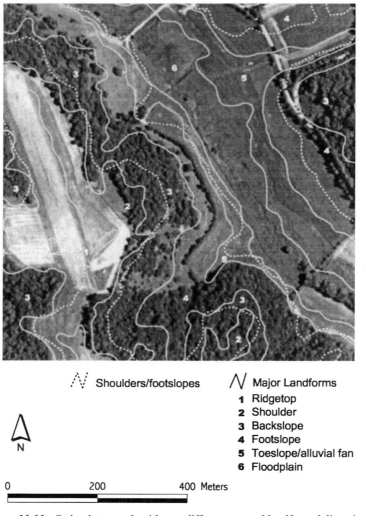

⋏⋎ Shoulders/footslopes   ⋀⋁ Major Landforms
                                 1 Ridgetop
                                 2 Shoulder
△   3 Backslope
N   4 Footslope
                                 5 Toeslope/alluvial fan
                                 6 Floodplain

0   200   400 Meters

Figure 11.11. Orthophotograph with two different manual landform delineations.

The ISODATA classification produced broad landform classes that were easily decipherable at eight classes; smaller, less pronounced (and possibly less interpretable) features were picked up as the number of classes increased. The classes roughly follow the automated breaklines and soil map units (Figure 11.12). When compared with the manual delineation, it can be seen that this automated method produced more detailed results. The classification also differentiated features based on aspect, so that north-facing slopes were in a different class from otherwise-similar

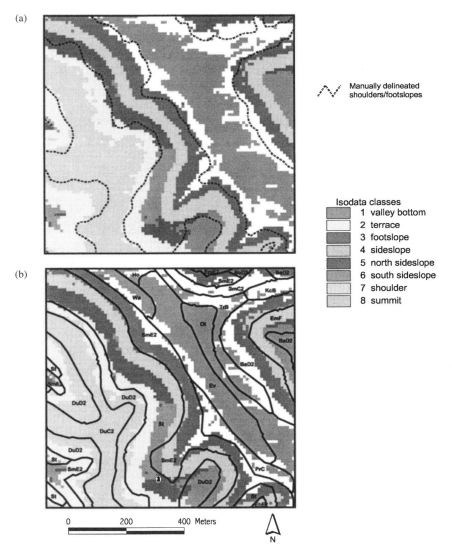

Figure 11.12. Isodata classification results comparisons: (a) 8 classes with manual shoulder and footslope breaklines and (b) 8 classes with USDA–NRCS soil map units.

south-facing slopes. Some discrepancy in boundaries existed, possibly due to the subjective judgment of the interpreters in manual delineation of airphotos.

Class centroid means of component terrain derivatives can be used to interpret similarities and differences between classes (Table 11.2). For instance, the alluvial fan or terrace area (class 2) was slightly higher in elevation and a little steeper than the valley bottom (class 1). Class 3 contained the footslope, which was somewhat steep and exhibited concave profile curvature. The steep sideslope of class 4 was a continuous band; however, at a higher number of classes, this class split into northern and southern exposures. Class 5 contained steep, north-facing slopes, while class 6 slopes had southern exposures. Class 7 exhibited the somewhat steep, convex downslope curvature of a shoulder. Class 8 contained the flat, convex ridgetop area of the summit.

The results of the continuous classification were visualized using a series of grids. An individual grid was produced for each class to show the distribution of the class memberships across the study area. A continuous sequence of tones represented the memberships across the color spectrum, with purple signifying no membership and red representing 95–100% membership.

A fuzzy exponent of $f = 1.25$ was tried first, but the classes were too distinct. An exponent of $f = 1.30$ created classes that were easily discernible but not overly differentiated. While the performance indices (FPI and NCE), in this case, did not clearly point to an optimal solution, minimums occurred at 9 and 14 classes (Figure 11.13). Nine classes was deemed to be insufficient because some of the less prominent, but important, features (i.e., the alluvial fan/terrace and footslope) were no longer separately classified. The results of 14 classes appeared to split some of the landforms needlessly (Figure 11.14), but these same landforms were split into separate classes even at nine classes. Thus, the 14-class solution was chosen for subsequent analyses.

This method also provided more detailed results than the manual delineation for the same reasons as the ISODATA results. Two or more classes often fell within one of the manually delineated landform units. Slight changes in elevation, differences in

TABLE 11.2   Class Means Generated with ISODATA Unsupervised Classification

Class ID	Landscape Unit	Elevation	Slope	Profile Curvature	Tangent Curvature	TWI	Incident Solar Radiation
1	Valley	238.4	3.1	−0.058	−0.033	10.80	2121
2	Terrace	241.7	9.3	−0.248	−0.028	9.42	2112
3	Footslope	249.2	23.8	−0.354	0.042	8.57	1994
4	Sideslope	261.9	33.1	−0.055	0.106	8.25	1841
5	North Sideslope	281.7	40.8	0.150	0.093	8.07	1609
6	South Sideslope	279.0	24.2	0.070	−0.010	8.35	2056
7	Shoulder	298.0	22.7	0.155	−0.088	8.42	1992
8	Summit	308.4	11.2	0.219	0.062	8.25	2092
	Mean	269.8	21.0	−0.015	0.018	8.77	1977

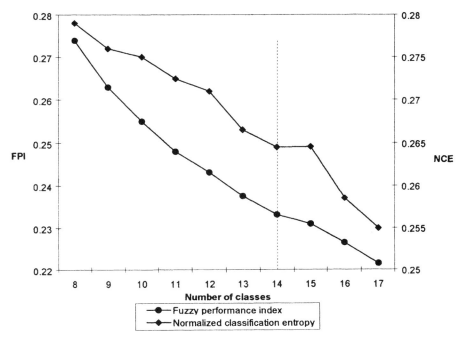

Figure 11.13. Continuous classification performance indices.

profile or tangent curvatures, or aspect changes too small to be picked up on the photographs all contributed to the disparity (see Figures 11.7–11.10 for examples).

Classes within similar traditional geomorphic units (i.e., backslope or valley floor) exhibit comparable characteristics but may differ on one or two attributes, such as the topographic wetness index or incident solar radiation. The differences between these classes may be significant to pedogenic processes (e.g., predominant vegetation or freeze–thaw cycles), resulting in differences in soil attributes within traditional landform units.

Analysis of the class centroids provided additional information about the differences between classes (Table 11.3). The three "summit" classes (A, H, and N) were quite similar except for curvatures. Another similar class (L) located near the summit was slightly steeper with flatter curvatures and higher incident solar radiation. The valley floor classes (C and M) were very similar except that M was slightly flatter and slightly more convex, resulting in a much higher TWI value. Classes B, G, and K were all steep, north-facing slopes with low incident solar radiation. Tangent and profile curvatures ranged from concave to convex, providing the differentiation between the classes. Classes E, F, and J were all steep slopes with southern exposures but widely varying TWI and curvature values. Class I appeared to be a footslope class containing south-facing, low-lying, convex areas with high TWI and solar values. Similarly, class D described low-lying, flatter, less convex areas suggesting a terrace class. The extragrade class contained the drainage area that corresponded to class 1 of the ISODATA

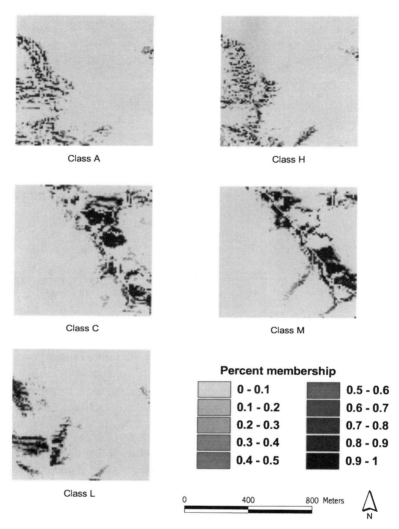

Figure 11.14. Continuous classification results for representative classes.

set. At higher numbers of classes, these cells would most likely constitute a separate class; at 14 classes, there were not enough cells to warrant a separate class.

The value of continuous classification is evident in those areas where class memberships are not clear-cut, such as the terrace area (D) or the footslope (I). Cells in these transition areas can be expected to have significant partial membership in two or more classes. Natural phenomena in these areas may have different patterns than those in areas with more distinct classes. In other areas, the variability between cells may be too great to warrant one class. For instance, the summit area is classified as a

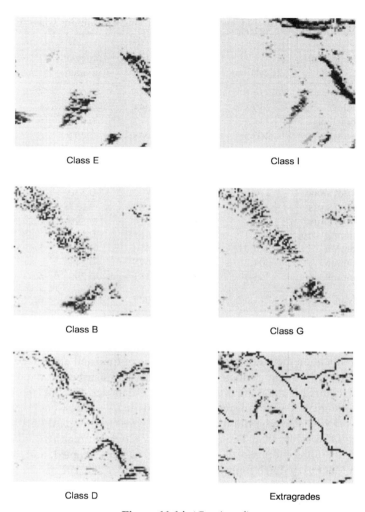

Class E

Class I

Class B

Class G

Class D

Extragrades

Figure 11.14 *(Continued)*

mixture of three classes (A, H, and N). Assigning partial memberships to these cells may increase the effectiveness of analyses by providing more information about the nature of the landform at each cell.

When the two classification techniques and the manual method were compared, all of the methods picked up significant features, such as the valley bottom, terraces, convex shoulder slopes, and the summit. As pointed out in the discussion of results, the automated methods seemed to have a slighter higher resolution of detail, which was due in part to the method of manual delineation and the use of attributes not discernible on the photographs for the automated methods.

TABLE 11.3   Centroids for 14 Classes Identified Using Continuous Classification

Class ID	Landscape unit	Elevation	Slope	Profile Curvature	Tangent Curvature	TWI	Incident Solar Radiation
M	Valley	239.6	4.7	−0.086	−0.047	8.88	2121
C	Valley	240.3	6.1	−0.044	−0.009	6.76	2110
I	Footslope	246.5	14.3	−0.312	−0.040	6.74	2189
D	Terrace	246.5	19.5	−1.048	−0.290	6.36	1925
G	North slope	259.8	32.0	−0.001	0.678	5.58	1762
F	South slope	267.9	30.4	0.148	1.223	5.80	2105
K	North slope	268.3	34.2	0.980	0.359	5.66	1708
B	North slope	270.0	35.8	−0.402	−0.530	5.85	1694
E	South slope	270.9	31.9	0.049	0.057	5.78	2212
J	South slope	275.0	29.1	−0.076	−0.870	6.28	2067
L	Summit slope	303.1	18.1	0.061	−0.182	5.85	2193
H	Summit	303.9	13.4	0.418	0.664	5.67	2038
A	Summit	304.2	14.7	0.675	−0.111	5.81	2002
N	Summit	305.4	13.5	−0.483	−0.290	5.93	2005
	Mean	271.5	21.3	−0.009	0.044	6.21	2009

The classification methods also varied in terms of time and effort required. Each method utilized the same attributes, but the Gaussian distribution requirement for ISODATA required some extra work to split the data set. Overall, the ISODATA method was simpler to perform since the data and the algorithm were both contained within ARC/INFO. The algorithm itself was very fast to run, partly because of the small size of the study area. The continuous classification required the data to be transferred to a different platform (UNIX to Macintosh), sampled, and then repeatedly run in MacFuzzy. Each run was several hours in length. The results were transferred back to ARC/INFO, where grids were generated for visualization.

Each classification method has its advantages and disadvantages. The unsupervised classification provides summary information about landforms in a relatively quick manner. Graphic renditions are easy to interpret but lack information about transition zones. Also, consideration must be given to the distribution of the data; a non-Gaussian distribution for an attribute may yield misleading results. The continuous classification provides much additional information on each point, which may be useful for analysis particularly in process-related research. However, it is a more time-consuming method and the results are not as easily visualized.

The unsupervised classification seems to be best suited to identifying landforms, or areas of rapid change, which may require more intensive sampling to obtain a complete picture of the soils in an area. Because the fuzzy classification provides more complete information about the landscape characteristics, it may be better suited for the prediction of soil properties in unsampled areas.

11.6   CONCLUSIONS

The use of numerical classification methods for identifying landform elements appears to be a promising technique. In the study area, the ISODATA unsupervised and continuous classifications identified both contiguous areas of the landscape that could be correlated to commonly used landform classes (summit, shoulder, slope, footslope, terrace, toeslope) and areas of high and low wetness and high and low incident solar radiation. Consequently, areas of high water flow were isolated and north-facing slopes were distinguished from south-facing slopes. The conditions and processes occurring in the Driftless Area also occur in many other soil–landscape systems, but these numerical techniques need to be verified in other landscapes and at various scales. It is possible that different attributes, such as vegetation, may be significant in other settings or that a change in scale may produce different results.

Unsupervised and fuzzy classification methods both produce results that can aid soil scientists and others interested in soil–landscape processes. A few caveats apply. Some knowledge of the landforms in question is necessary to evaluate the efficacy of the results of either method. Care should also be taken in the choice of attributes; they should reflect the nature of the landscape and the phenomena being studied.

The ISODATA classifier provided a relatively quick and easy way to delineate landform elements. This unsupervised classification algorithm is available within widely used raster GIS packages (e.g., ARC/INFO GRID, ERDAS Imagine). As discussed in more detail in the previous chapter, landform element designations can be overlaid on any digital map or orthophoto, which could prove useful in designing sampling schemes based on landform designations or in providing a first cut for soil unit delineation.

Continuous (or fuzzy) classification provides more information about the character and variability of the data than the ISODATA method, affording more insight into the nature of the data and making it more amenable to statistical analysis. This technique requires a conceptual shift to accommodate the new form and degree of information. Additional work is necessary to place this type of categorization in soil surveys or soil–landscape studies. More data manipulation is required to move between packages, since the algorithm for fuzzy classifications is not yet available in commercial GIS packages. It has only recently been incorporated into one commercial image processing package (MATLAB Image Processing Toolbox, The Math Works Inc, Natick, MA).

Geomorphometric techniques using DEMs hold great promise for the identification of landscape units for soil–landscape and other geomorphological studies. DEMs, while useful in various applications, are particularly well suited for hydrological analyses, such as flow routing and flow accumulation. Researchers in the past developed their own code to derive information from DEMs, but algorithms to calculate primary attributes, and even some secondary attributes, are now becoming available in commercial GIS and image-processing packages. Those wishing to calculate more complex secondary attributes may turn to software programs from the academic sector, such as TAPES-G. Often the programming code is available so that modifications can be made to customize the program. One drawback of the approach taken in this research is that a relatively high-resolution DEM was necessary for the Driftless Area landscape. Such DEMs are not universally available. However, "soft-copy" photogrammetry-based methods for automating the creation of digital orthophotographs are becoming commercially available, providing high-resolution DEMs as a by-product.

Landform classification using manual methods is a subjective process, different from person to person and day to day. Numerical classification methods used to identify landform elements need some subjective input, but are largely pragmatic and repeatable. An understanding of the landscape is still necessary to properly analyze the results of the classifications, but the techniques do bring a measure of objectivity to a process that has been largely intuitive and experiential.

Within a GIS, the use of DEMs, landform information, and soil attribute data provides for useful data analysis and visualization. In the study area, the numerical classifications corresponded to manual landform delineations, the existing soil map units, and a different automated approach, while providing more information about the nature of the landform elements. New insights can be gained from combining and viewing the data in ways not possible using manual approaches. The viewing environment of a GIS provided new ways to look at data, including perspective renditions with soil characteristics or terrain attributes "draped" over the surface to enhance visualization of three-dimensional processes.

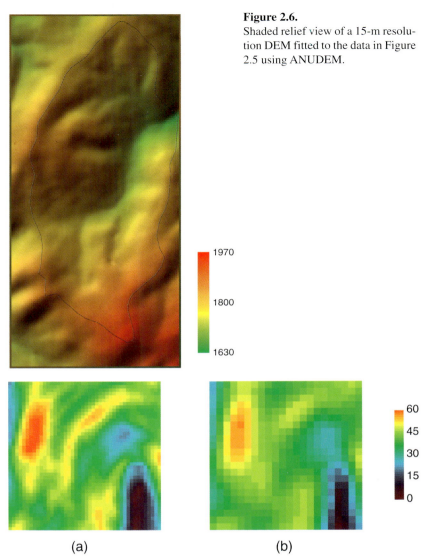

Figure 2.6.
Shaded relief view of a 15-m resolution DEM fitted to the data in Figure 2.5 using ANUDEM.

1970

1800

1630

(a)   (b)

60

45

30

15

0

Figure 2.9.
Derived slope for the subarea from (a) the 15-m DEM and (b) the 7.5-m DEM.

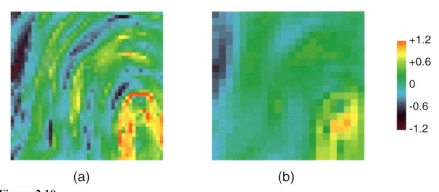

(a)   (b)

+1.2

+0.6

0

-0.6

-1.2

Figure 2.10.
Derived profile curvature for the subarea from (a) the 15-m DEM and (b) the 7.5-m DEM.

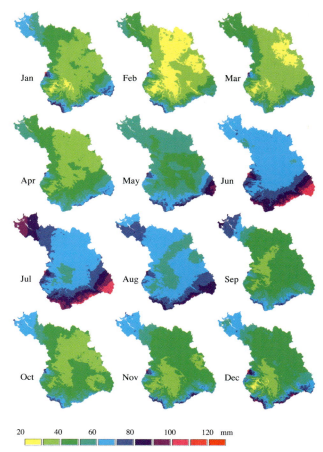

Figure 6.3.
Spatial interpolation of mean monthly precipitation in the Elbe drainage basin.

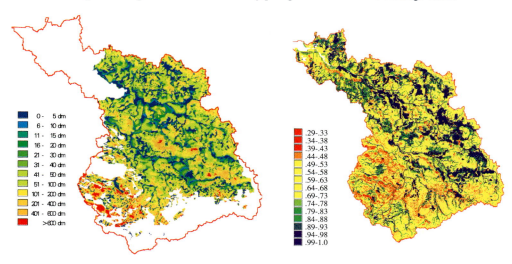

Figure 6.6.
Mean groundwater depths for eastern Germany based on a map of groundwater isohypses, 1:50,000 scale.

Figure 6.7.
Mean monthly soil wetness index for July (relative units, from 0 to 1) in the Elbe drainage basin.

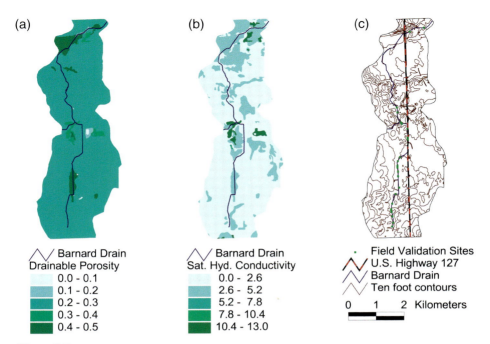

Figure 7.2.
The Barnard Drain subwatershed: (a) drainable soil porosity, (b) saturated hydraulic conductivity, (c) topography (10-ft contour interval), hydrography, highway route, and locations of field validation sites (wet spots and gullies).

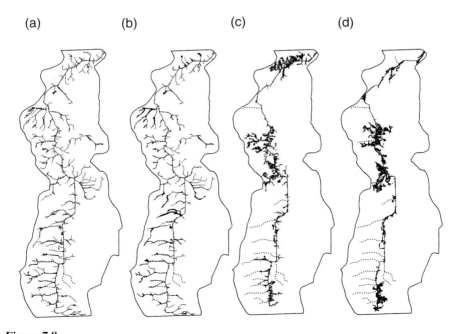

Figure 7.8.
Cost-distance buffers at the 5 percentile threshold with interpreted flow paths (dotted lines) overlaid for the (a) static, D8, (b) static, DEMON, (c) dynamic, uniform soil, and (d) dynamic, variable soil models.

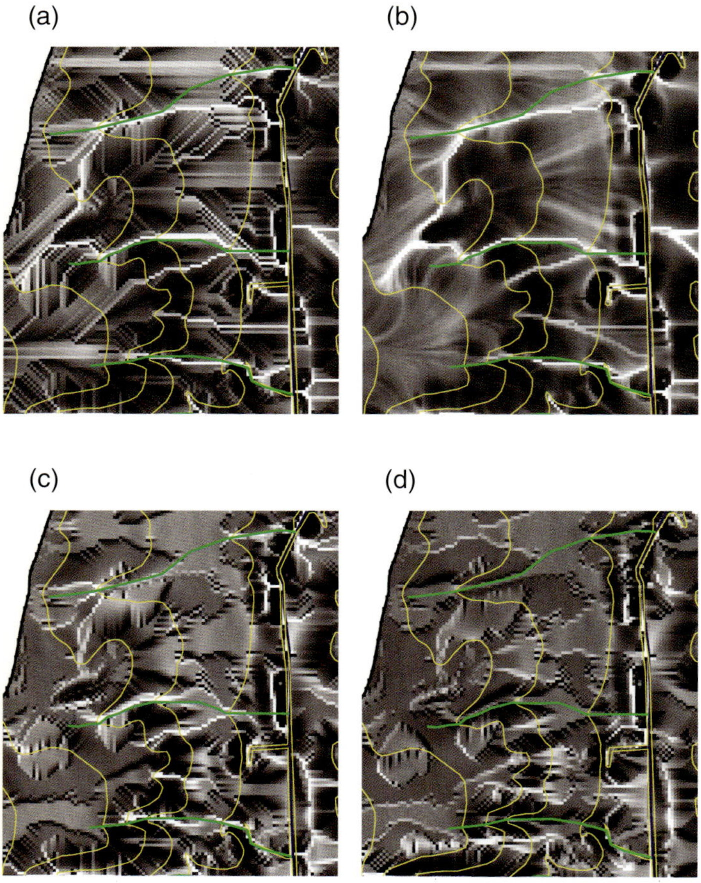

(a)   (b)

Figure 7.10.
Interpreted flow paths (green) and 10-ft contours (yellow) overlaid on topographic wetness index maps (linear gray stretch) for a portion of Barnard Drain produced by (a) static, D8, (b) static, DEMON, (c) dynamic, uniform soil, and (d) dynamic, variable soil models.

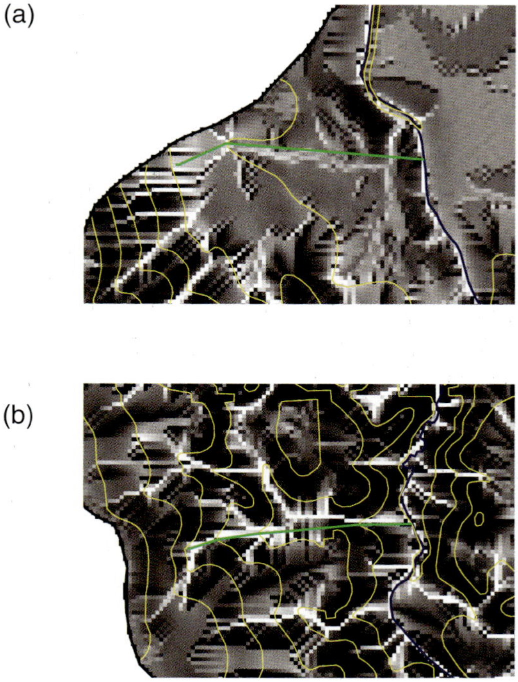

(c)   (d)

(a)

Figure 7.11.
Potential flow paths (strands of high topographic wetness) that (a) disperse due to flat topography and (b) diverge from flow paths manually delineated by watershed experts.

(b)

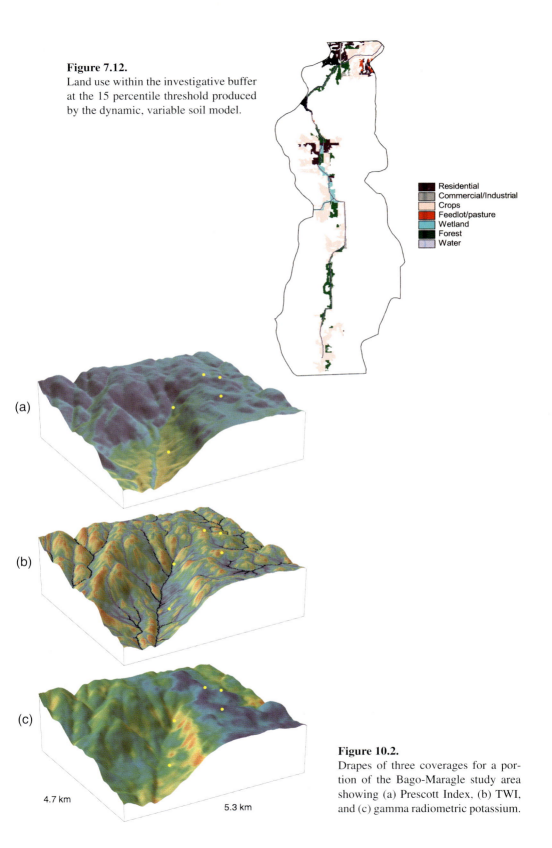

Figure 7.12.
Land use within the investigative buffer at the 15 percentile threshold produced by the dynamic, variable soil model.

Residential
Commercial/Industrial
Crops
Feedlot/pasture
Wetland
Forest
Water

(a)

(b)

(c)

4.7 km

5.3 km

Figure 10.2.
Drapes of three coverages for a portion of the Bago-Maragle study area showing (a) Prescott Index, (b) TWI, and (c) gamma radiometric potassium.

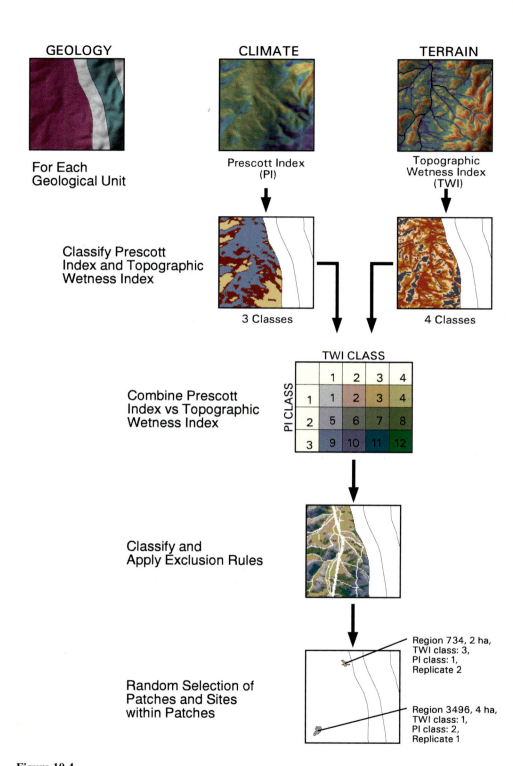

GEOLOGY

For Each
Geological Unit

CLIMATE

Prescott Index
(PI)

TERRAIN

Topographic
Wetness Index
(TWI)

Classify Prescott
Index and Topographic
Wetness Index

3 Classes

4 Classes

Combine Prescott
Index vs Topographic
Wetness Index

TWI CLASS

PI CLASS	1	2	3	4
1	1	2	3	4
2	5	6	7	8
3	9	10	11	12

Classify and
Apply Exclusion Rules

Random Selection of
Patches and Sites
within Patches

Region 734, 2 ha,
TWI class: 3,
PI class: 1,
Replicate 2

Region 3496, 4 ha,
TWI class: 1,
PI class: 2,
Replicate 1

Figure 10.4.
Sampling strategy for a 25-km^2 portion of the Bago-Maragle study area occupied by the adamellite geologic unit.

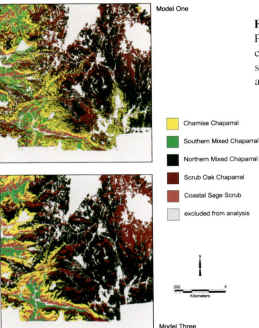

Model One

Figure 14.8.
Predicted map of chaparral vegetation classes (Table 14.2) for the subarea shown in Figure 14.1 from (a) model 1, and (b) model 3.

■ Chamise Chaparral

■ Southern Mixed Chaparral

■ Northern Mixed Chaparral

■ Scrub Oak Chaparral

■ Coastal Sage Scrub

□ excluded from analysis

Model Three

Figure 14.10.
Predicted map of riparian vegetation life-form classes (Table 3) overlain on the TM greenness image from Figure 14.5 (shown for all areas predicted to be nonriparian from the model) for the Viejas Mountain quadrangle (outlined in Figure 14.1). Also overlain are the vectors describing stream locations from the USGS DLG.

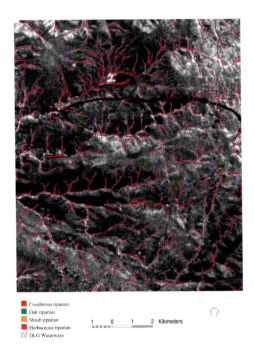

■ Coniferous riparian
■ Oak riparian
■ Shrub riparian
■ Herbaceous riparian
△ DLG Waterways

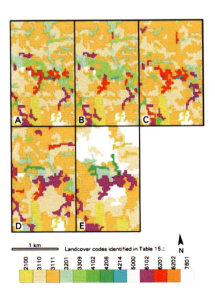

Figure 15.11.
Potential land cover maps for part of study area predicted with initial classification method and varying numbers of secondary topographic attributes and minimum map units as follows: (a) standard inputs using a 0.4-ha minimum map unit; (b) addition of quasi-dynamic topographic wetness index using a 0.4-ha minimum map unit; (c) addition of quasi-dynamic topographic wetness index and short-wave solar radiation using a 0.4-ha minimum map unit; (d) addition of both attributes using a 0.8-ha minimum map unit; and (e) addition of both attributes using a 2.0-ha minimum mapping unit.

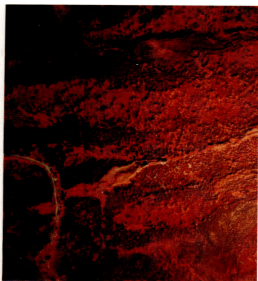

Figure 16.2.
Infrared aerial photo of part of the Rinker Lake Research Area.

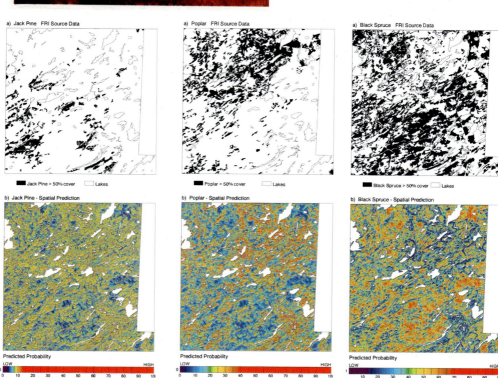

Figure 16.16.
Predicted environmental domain for jack pine (b) based on observed probabilities presented in Figure 16.10. (a) Species location data derived from the forest resource inventory maps used to calculate the observed probabilities.

Figure 16.17.
Predicted envronmental domain for trembling aspen (b) based on observed probabilities presented in Figure 16.11. (a) Species location data derived from the forest resource inventory maps used to calculate the observed probabilities.

Figure 16.18.
Predicted envionmental domain for black spruce (b) based on observed probabilities presented in Figure 16.12. (a) Species location data derived from the forest resource inventory maps used to calculate the observed probabilities.

CHAPTER 12

A Soil-Terrain Model for Estimating Spatial Patterns of Soil Organic Carbon

Jay C. Bell, David F. Grigal, and Peter C. Bates

12.1 INTRODUCTION

Rising concern over observed increases in atmospheric CO_2 levels and the potential effects on global climate have led to efforts to increase our understanding of the carbon cycle. The annual release of atmospheric carbon from anthropogenic sources, mainly fossil fuels, is estimated at 6 Pg yr^{-1}, of which 3.4 Pg yr^{-1} may be released to the atmosphere. Scientists have speculated that these releases are sufficient to trigger global climate changes. An understanding of the global carbon budget requires knowledge of potential sources and sinks of carbon, the potential rates of sequestration and release, and the environmental factors controlling these processes. Studies by Houghton (1995) suggest that C release from terrestrial ecosystems may account for an additional 0.3 to 2.5 Pg yr^{-1}. As such, a better understanding of the sinks and sources of carbon in terrestrial ecosystems is required. While estimation of carbon fluxes is the crucial question, flux estimates must be based on an inventory of carbon reserves. Total soil carbon based on global soil maps and assumed carbon levels for soil classes is estimated at 3293 Pg, with 1738 Pg as soil carbonates and 1555 Pg as soil organic carbon (SOC) (Eswaran et al. 1993). The soil is both a potential source and sink of atmospheric C, depending on the relative rates of C incorporation and decomposition by soil organisms. Accurate estimation of the net flux from terrestrial ecosystems is predicated from an understanding of soil-forming processes and the spatial variability of SOC within landscapes.

Spatial patterns of SOC storage in the landscape are important both from a carbon inventory perspective and for understanding the biophysical processes that may be affecting SOC fluxes. These patterns and processes will vary considerably among

---

Terrain Analysis: Principles and Applications, Edited by John P. Wilson and John C. Gallant.
ISBN 0-471-32188-5   © 2000 John Wiley & Sons, Inc.

different landscapes, limiting extrapolation across broad climatic–geomorphic regions. As such, region-specific studies are warranted at appropriate spatial scales to understand specific processes affecting C fluxes. Our contention is that the appropriate landscape unit for study is the hillslope for humid region landscapes because hydrologic processes active on hillslopes lead to considerable differences in soil characteristics, especially the depth and darkness of the A horizon, which are highly correlated with SOC (Thompson and Bell 1996). In soil science, the term "catena," meaning landscape chain, is used to describe these consistent hillslope relationships (Milne 1935).

Although the catena concept is firmly established in soil science, quantification of changes in SOC with hillslope position is lacking. This is especially true in more recently glaciated regions such as the lake states, where climate and a poorly developed drainage network have led to abundant peatlands, occupying 10–30% of most basins. Modeled scenarios of global change suggest that both forest tree distribution and the balance between precipitation and evaporation may change, affecting C storage. Quantification of the SOC stored in these immature landscapes, and the relationship of that storage to landscape features will all aid in predicting the magnitude of C response to various scenarios of global change.

The overall objective of our research is to better understand the mechanisms responsible for C sequestration at landscape scales. Although not our ultimate goal, the predictive mapping of SOC is a necessary first step in this understanding. While existing soil maps, such as those produced by the National Cooperative Soil Survey (NCSS) in the United States, are quite valuable, there are significant shortcomings because (1) some of these maps are not prepared at a sufficiently detailed spatial scale to discern the crucial spatial patterns, (2) detailed soil maps are not available for most regions of the globe, and (3) soil maps do not necessarily quantify the relationships between landscape features and soil characteristics. Making a sufficient number of direct observations of soil characteristics (usually with a hand auger) to map soil characteristics from observation to observation is not practical when mapping large geographic areas at fine scales (<1:50,000). A common method of mapping soils at fine scales is to formulate consistent relationships between easily observable landscape features and difficult to observe soil characteristics. As such, the soil surveyor develops a mental, qualitative model relating landscape features to soil characteristics. Consequently, a soil map in reality is a delineation of landscape features that are assumed to be related to spatial patterns of soil characteristics. These mental models are frequently difficult for soil surveyors to articulate and their application may not be consistent across the landscape.

An alternative approach that has been made possible through advancements in geographic information systems (GIS) and allied technologies is to explicitly define relationships between landscapes features and soil characteristics (Bell et al. 1992) and map the spatial distribution of soil characteristics using landscape feature information stored in a digital geographic database (Bell et al. 1994a). This approach uses empirical relationships quantified in a soil–landscape model that can be consistently applied across the landscape and easily updated as new information and better modeling techniques become available. Numerous researchers have investigated this approach in recent years (Moore et al. 1993b, Bell et al. 1994b, McKenzie and Austin 1993, Gessler et al. 1995, Thompson et al. 1997) to map spatial patterns of various soil characteristics. Terrain attributes derived from digital elevation models (DEMs) have been a significant component of all these soil–landscape modeling efforts, and two additional projects that utilized the TAPES software tools are described in Chapters 10 and 11.

Quantitative soil-landscape modeling approaches have considerable potential for mapping spatial patterns of SOC for many landscapes given the consistent relationships with terrain attributes and vegetative cover. Our specific objective for this study was to quantify the magnitude and spatial patterns of SOC storage in a glacial outwash landscape in the north-central United States using quantitative soil–landscape modeling techniques.

12.2 METHODS

12.2.1 Study Site

Our study site was the Cedar Creek Natural History Area (CCNHA), located 50 km north of Minneapolis–St. Paul, Minnesota, on the Anoka Sand Plain (45° 25′N, 93° 10′W) (Figure 12.1). This area is a Long-Term Ecological Research (LTER) site sponsored by the National Science Foundation and is typical of sandy outwash areas for much of the Great Lake States. The Anoka Sand Plain was formed during the retreat of the Grantsburg sublobe (~14,000 ybp) of the Des Moines ice lobe during the late Wisconsinan ice age (Grigal et al. 1974). Outwash sediments range from coarse gravels to fine sands with inclusions of lenses of finer soil material. Following the retreat of glacial ice during the Holocene and before vegetation was well established, winds sculpted much of this landscape, creating an undulating terrain consisting of sand dunes and closed depressional basins. These basins were typically formed as blowout features as sand was eroded by wind down to the water table surface. As such, many of the basins are shallow with relatively level bottoms that developed into peatlands. Other depressions were formed by stranded blocks of ice in the outwash creating ice-block depressions that are typically lakes. Soils are derived from deep (20-m) well-sorted glacial outwash of very uniform fine sand (>90% sand). The uplands are dominated by Udipsamments, lowland mineral soils are Endoaquolls, and depressions are predominantly organic soils (Saprists). The regional water table is at or near the soil surface in low-lying areas and only a few meters below the surface in the highest elevations. Nearly 40% of the CCNHA is occupied by wet mineral or organic soils so that the 2200-ha site is an intricate upland–wetland mosaic. Slopes are gentle, usually less than 15%, and local relief ranges up to about 5 m. Natural vegetation includes prairie, oak savanna, closed forests, shrub carrs, marshes, and bogs (Pierce 1954). Many of the present upland ecosystems developed following agricultural abandonment. The regional climate is continental with a mean annual temperature of 6°C and annual precipitation of 660 mm (Grigal et al. 1974). A few meters of horizontal distance often separate upland oak forests or old fields from adjacent wetlands.

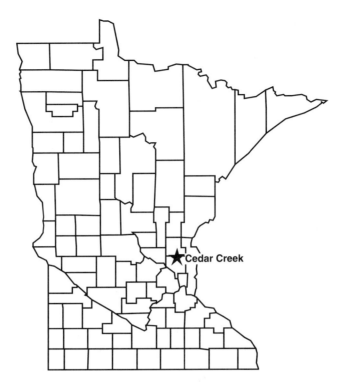

Figure 12.1. Location of the Cedar Creek Natural History Area.

12.2.2   Field Sampling

Samples and descriptions of soils were collected along 19 transects crossing topographic changes of the major vegetation types. We sampled at two levels of intensity; level #1 sites were at 25-m increments along the transects (total = 745). We recorded landscape position and topography (slope gradient, profile curvature, and aspect); vegetation type; tree basal area, diameter, and condition; tall shrub cover and height; low shrub and forb cover; forest floor and A-horizon thickness; and fiber type and depth of organic soil materials (if present). Geographic locations of all sampling points were determined using a differential global positioning system (DGPS). Approximately one-third of the level #1 sites were randomly selected for level #2 sampling (total = 250). Of these, 198 were located in mineral soils and 52 in peatlands. Soil samples were collected for laboratory determinations of organic C content by sampling at depths of 0–25 and 26–100 cm with a bucket auger, and organic soils were sampled in 50-cm increments until impenetrable with a McCauley peat auger. Field observations from the level #1 sampling were used for other studies; this study utilized the data collected from the level #2 sampling. A 5% subset of soils samples were subsampled for bulk density from soil cores.

12.2.3 Laboratory Methods

Loss-on-ignition (LOI) was determined for all forest floor and soil samples by ashing overnight at 450°C in a muffle furnace. Approximately 30% of the samples were used to develop relationships between LOI, total organic C on a gravimetric basis (g kg^{-1}) (David 1988), and soil bulk density (kg m^{-3}) (Grigal et al. 1989). The relationships were used to calculate C and bulk density for the remaining samples. SOC on a volumetric basis was calculated to a 1-m depth for mineral soils and to the depth of sampling for organic soils.

12.2.4 Spatial Data

We developed a digital geographic database that included vegetation cover type, peatland locations from an order II soil survey, and a digital terrain model. The information was stored in a raster format using 10-m grid cells. Scales for input data ranged from 1:15,000 to 1:20,000. Vegetation cover type was interpreted from color infrared aerial photography onto a Mylar sheet. Vegetation class and ground control points were hand digitized from the Mylar overlay (Figure 12.2). Soil map unit delineations were digitized from a detailed (scale: 1:17,600) soil survey prepared for the

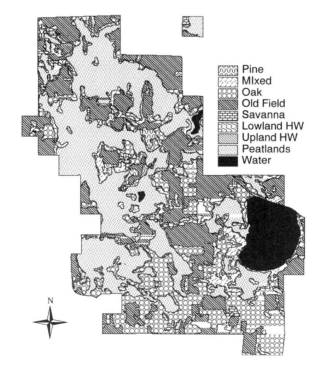

Figure 12.2. Vegetative cover map for the Cedar Creek Natural History Area interpreted from aerial photography.

CCNHA (Grigal et al. 1974) and surface topography was digitized from 1-ft (0.31-m) elevation contours derived from a ground survey of the CCNHA. Soil mapping units were reclassified to create a binary map of peatlands or mineral soil conditions (Figure 12.3). Common ground control points were identified on each data layer and geographic coordinates for these control points were collected in the field using DGPS. These ground control points were digitized from each data layer and were rectified to a common base using bilinear interpolation. Because topographic relief across the site was less than 5 m, terrain relief displacement errors were not a significant source of spatial error.

A 5-m (horizontal grid-cell size) digital elevation model (DEM) with a vertical resolution of 0.1 m was interpolated from the elevation contours using ANUDEM (Hutchinson 1989b), which implements a cubic-spline procedure with drainage enforcement algorithms (Figure 12.4). Digital maps of slope gradient, surface curvature, aspect, plan and profile curvature using TAPES-G, and both horizontal (distance to peatlands) and vertical proximity (elevation above peatlands) to peatlands and mineral soils (distance to mineral soils) (Bell et al. 1992) were calculated for

Figure 12.3. Peatlands reclassified from a detailed survey of the Cedar Creek Natural History Area (Grigal et al. 1974).

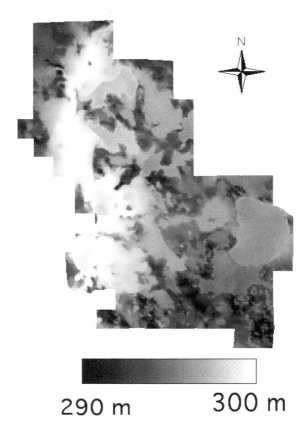

290 m   300 m

Figure 12.4. Digital elevation model for the Cedar Creek Natural History Area interpolated from 1-ft elevation contours.

each grid cell in the geographic database using the DEM and peatland location map. The transect sampling locations and digital geographic database were registered to standard geographic coordinates (universal transverse mercator) such that vegetation type, topographic attributes, and peatland versus mineral soil conditions were derived for each field sampling location based on geographic location.

12.2.5   Spatial Modeling

Estimates of SOC were partitioned by vegetation type. Eight vegetation types (pine, oak, upland mixed hardwoods, savanna, old field, lowland conifers, lowland hardwoods, and lowland nonforested) were recognized in both the geographic data base and at transect locations. The latter three types were primarily located in peatlands. Linear and exponential statistical models were developed to predict SOC. A stepwise procedure was used to identify the landscape attributes that were significantly related

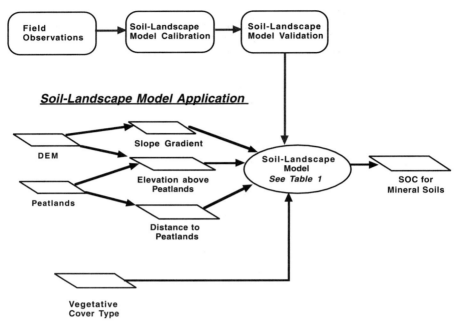

Figure 12.5. Outline of approach used to develop and apply the soil-landscape model at the Cedar Creek Natural History Area to estimate SOC for mineral soils.

($p \leq .05$) to SOC. Peatland depths cannot be ascertained from simple surface data (Swanson and Grigal 1989), and were estimated by functions related to distance to mineral soil and surface elevation. A 75% random subset of transect observations was used to calibrate statistical relationships between landscape attributes and SOC. The remaining observations were used to obtain an unbiased estimate of model performance (model validation). After the quantitative relationships were calibrated and validated, digital overlay techniques were used to define the landscape attributes for each grid cell in the geographic database, and the statistical models were applied to these attributes on a cell-by-cell basis to predict and map SOC (Figure 12.5).

12.3 RESULTS AND DISCUSSION

The SOC and LOI characteristics for the subset of samples were closely related, with $r^2 = .97$ for mineral soils with LOI < 14% dry weight, $r^2 = .83$ for mineral soils with LOI $\geq$ 14%, and $r^2 = .88$ for organic soil materials. As such, soil carbon concentrations estimated from LOI were assumed to be reliable for both mineral and organic soils.

12.3.1 Mineral Soils

SOC contents based on the point samples collected from the field indicated that the total SOC in the upper meter varied considerably by vegetative cover and was highest for lowland hardwoods and lowest for old, abandoned agricultural fields (Figure 12.6). We speculate that these differences in SOC are related to the chemical composition and relative rates of organic C incorporation into the soil. Additionally, many of the vegetative cover types are consistently found in unique landscape positions, often related to soil-moisture regime. As such, higher rates of SOC incorporation in lowland hardwood sites is also probably a function of wetter soils leading to periodic saturation and development of anaerobic soil conditions. Rates of organic matter decomposition are slower under anaerobic conditions and organic C generally accumulates in anaerobic soils.

While a portion of the variability of SOC in this landscape is explained by vegetative cover, a substantial amount of variability still exists within each vegetative cover type with respect to SOC (Figure 12.6). Coefficients of variation ranged from a low of 46% for old fields to 85% for savannas. Our hypothesis is that a combination of terrain attributes can further explain a significant ($\alpha < .05$) portion of the remaining variability and spatial patterns of SOC within each vegetative cover type. For upland (mineral soil) sites, SOC concentration varied significantly with terrain attributes, including slope gradient and horizontal distance and elevation above peatland (Table 12.1). The relationships derived from the nonlinear regression equations are congruent with the conceptual model for where and how soils with high organic C contents are formed within the Anoka Sand Plain. As previously discussed, the region is characterized by a shallow water table. Peatlands develop in continuously saturated basins, and the surface elevation of the peatland represents the local eleva-

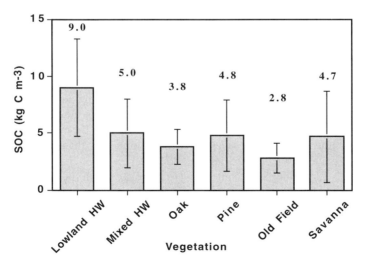

Figure 12.6. Mean and standard deviations for soil organic carbon storage under the six vegetative cover types at the Cedar Creek Natural History Area based on field samples.

TABLE 12.1 Significant Terrain Variables, Regression Equations Used to Predict SOC in Upper Meter, r^2 for Calibration Data (r_c^2) and Validation Data (r_v^2), and Number of Samples by Vegetative Cover Type at the Cedar Creek Natural History Area

Vegetative Cover	Terrain Variables	Regression Equation	r_c^2	r_v^2	n
Lowland hardwoods	Slope gradient (S)	$156.4 * (S + 1)^{-0.75}$	.93	.58	8
Mixed hardwoods	Elevation above peat (E_p)	$(0.43 * (E_p + 1)^{0.38})/$	.93	.66	13
	Slope gradient (S)	$0.0023 * (S + 1)^{0.34}$			
Oak	Elevation above peat (E_p)	$51 * (E_p + 1)^{-0.28}$	.95	.37	74
Pine	Distance to peat (D_p)	$67.8 * (D_p + 1)^{-0.21}$	.94	.70	9
Old field	Distance to peat (D_p)	$8.3 * (D_p + 1)^{0.23}$	.93	.31	59
Savanna	—	42 Mg ha^{-1}	—	—	35

tion of the water table. The highly permeable nature of the outwash sands of the mineral soils minimizes lateral movement of soil water, and the depth to the long-term water table can be approximated by taking the difference in elevation between the ground surface and the elevation of the nearest peatland. Again, as previously discussed, the duration and depth of soil saturation is a major determining factor in SOC accumulation due to the development of anaerobic conditions resulting in decreased rates of SOC decomposition over time.

This conceptual model agrees with the spatial pattern of soil properties found in the detailed soil survey (Grigal et al. 1974) and with detailed observations of soil properties and depth to groundwater along hillslopes as part of ongoing hydrologic monitoring studies at the CCNHA (Reuter and Bell, unpublished data). As such, slope gradient and the horizontal and vertical proximity variables describe these spatial relationships and could be used to explain a significant component of SOC variability within all vegetative cover types except for savannas. Correlations with the data used to calibrate the relationships between SOC and terrain attributes (r_c^2) ranged from .93 to .95, and correlations with the validation data (r_v^2) ranged from .31 to .70 among vegetation types (Table 12.1). These results support our hypothesis and we conclude that we can use these relationships to estimate spatial patterns of SOC storage in mineral soils. The level of correlation for the validation data sets is similar to results found in similar quantitative soil–landscape modeling studies (Thompson et al. 1997, Moore et al. 1993b, Bell et al. 1992). The lack of association between SOC and topographic attributes in savannas is congruent with the conceptual model. Savannas are found in upland landscapes where factors other than soil wetness are probably the main determinants of SOC variability. Likewise, old fields are primarily found in uplands and higher SOC contents were associated with field areas adjoining peatlands as would be expected.

12.3.2 Peatlands

While the regression equations in Table 12.1 can be used to estimate spatial patterns of SOC for the mineral soils at the CCNHA, the highest levels of SOC will be found in the peatlands. Based on our field sampling of the peatlands, we found a significant

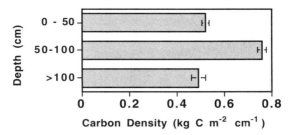

Figure 12.7. Carbon densities with depth for peatlands at the Cedar Creek Natural History Area.

($\alpha < .001$) difference in carbon density (kg C m^{-2} cm^{-1}) with depth from the surface because of differences in both C concentrations and bulk density (Figure 12.7). As such, our hypothesis is that horizontal spatial variability and patterns of carbon storage in peatlands are primarily a function of peatland depth. We found that peatland depth was best estimated by distance to nearest mineral soil and elevation by

$$P_d = -3415 + 2.121E_s + \ln(D_m) \qquad (12.1)$$

where P_d is the peatland depth, E_s is the surface elevation, and D_m is the distance to mineral soil. This relationship was significant ($\alpha < .05$) with $r_c^2 = .29$ (calibration data) and $r_v^2 = .39$ (validation data). A rough interpretation of these relationships suggests that peatland depth increases toward the center of the peatland (P_d increases with D_m). E_s accounts for differences in this relationship associated with increasing elevation. We speculate that this is related to a change in the structure and/or size in peatlands basins based on the proximity to Cedar Creek. If we consider the predictive relationships of both peatlands and minerals, the mean $r_c^2 = .83$ and the mean $r_v^2 = .50$.

12.3.3 Estimating Spatial Patterns of SOC

We estimated spatial patterns of SOC storage in the upper 1 m of the mineral soils by applying the regressions equations derived from our field sampling (Table 12.1) to the digital geographic coverages of mineral soils, vegetative cover type, slope gradient, elevation above peatland, and distance to peatland (Figure 12.5). Because a significant relationship was not found among terrain variables and SOC for savannas, we used the mean value of 42 Mg C ha^{-1} as an estimate for these areas. Estimates were made only in areas of mineral soils and the appropriate regression equation was used within each vegetative cover type on a cell-by-cell basis to create a predictive map of SOC storage in the upper 1 m of mineral soil (Figure 12.8). A summation of the total SOC stored at the CCNHA divided by the area of mineral soils yields a mean value of 4 kg C m^{-3}.

Similarly, by applying Equation 12.1 to the digital maps of peatland areas, distance to mineral soils, and surface elevation, we derived a predictive map of peatland

Figure 12.8. Predictive map of SOC for the mineral soils at the Cedar Creek Natural History Area created using soil-landscape modeling approach.

depth. By applying the carbon densities we determined for different peatland depths (Figure 12.7) and integrating over peatland depth, we derived a predictive map of SOC stored in peatlands (Figures 12.9 and 12.10). By summing the SOC stored in peatlands and dividing by the peatland area we arrived at a mean value of 131 kg C m^{-3} for the total soil depth and 62 kg C m^{-3} for the upper 1 m. When considering all of CCNHA, organic C storage is heavily weighted toward the peatlands (97%).

Reports of C storage in peatlands are rare. Gorham (1991) estimated global carbon storage in peatland soils, and his estimate (133 kg C m^{-2}) is nearly equal to our estimate for peatland soils at the CCNHA (131 kg C m^{-2}). In terms of C storage in uplands, Grigal and Ohmann (1992) reported mean C storage of five upland forest cover types in five geographic zones in a gradient across the lake states. The SOC estimates for upland sites at the CCNHA (4 kg C m^{-3}) are much lower than their range for SOC (7.0–15.0 kg m^{-3}). The lower estimates for the CCNHA reflect the

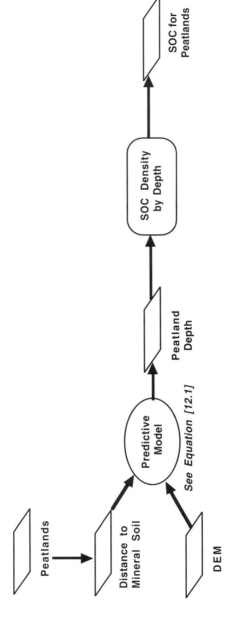

Figure 12.9. Outline of approach used to create predictive map of SOC for peatlands at the Cedar Creek Natural History Area.

307

Figure 12.10. Predictive map of SOC for the peatlands at the Cedar Creek Natural History Area.

influence of the coarse texture of the mineral soil. Although the pine types in the gradient study were also coarse-textured, their surfaces were somewhat finer-textured. In addition, the CCNHA has experienced more agricultural disturbance than have the sites from the gradient study. Finally, because the CCNHA is at the southwest extreme of lake states forests, precipitation is lower and temperatures are higher than for most of those forests, further reducing C storage (Jenny 1941).

Total SOC storage over all 2070 ha of the CCNHA (excluding lakes) is 1.08×10^{12} g. If C storage in peatlands is only considered to 1 m, as in mineral soil, C storage in soil is 0.55×10^{12} g. Our approach to estimating organic C storage in soil, which explained roughly half the spatial variation at the CCNHA, treats the landscape as a continuum rather than as a set of discrete map units. The SOC storage was based on terrain attributes derived from a continuous representation of the topographic surface. In contrast, a discrete approach can also be used to estimate soil C, based on the area and the estimated C density of each soil map unit derived from the detailed soil survey (Grigal et al. 1974). Using that approach, estimated soil C storage to 1 m at the CCNHA is 0.50×10^{12} g (Figure 12.11). The continuous approach is much more amenable to modeling changes in C storage with changes in climate on these landscapes and can be used in areas where soil survey data are not available.

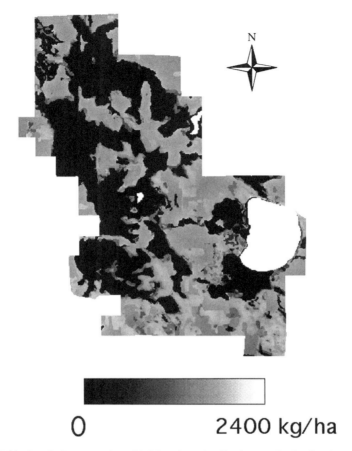

0   2400 kg/ha

Figure 12.11. Predictive map of total SOC (mineral soils plus peatlands) for the upper 1 m at the Cedar Creek Natural History Area.

12.4 CONCLUSIONS

We found significant differences in SOC among vegetative cover types at the CCNHA and were able to further refine estimates of both SOC storage and spatial patterns using terrain attributes derived from a digital elevation model. Significant relationships were found between a combination of slope gradient, elevation above peatlands, and distance to peatlands and SOC for all vegetative cover types except savannas. For peatlands, we found that carbon density varied significantly with peatland depth. As such, we estimated peatland depth from a combination of distance to mineral soil and surface elevation and then estimated SOC by integrating SOC densities over peatland depth. Mean SOC for mineral soils was estimated at 4 kg C m^{-3} and for peatlands at 131 kg C m^{-3}. As such, peatlands store over 97% of the SOC in the landscape at the

CCNHA. This wide disparity between mineral soils and peatlands and the spatial patterns of SOC storage in these environments are important considerations in studies involving terrestrial and atmospheric C fluxes in the lake states.

More work is needed to reduce the uncertainty currently associated with estimates of SOC pools in northern forests. The techniques that we used are applicable to broader areas, and could rapidly move our understanding of C storage in terrestrial ecosystems forward. Our approach is based on environmental correlation between soil properties and landscape variables and results in continuous, rather than discrete, maps of SOC, providing a more realistic representation of spatial patterns of variations within the soil continuum. Further work is also needed to ascertain the geographic scope of this soil–landscape model. Assuming that soil–landscape conditions at the CCNHA are representative of the Anoka Sand Plain, this soil–landscape model could conceivably be applied to this entire geomorphic region with some additional refinement.

CHAPTER 13

Shallow Landslide Delineation for Steep Forest Watersheds Based on Topographic Attributes and Probability Analysis

Jinfan Duan and Gordon E. Grant

13.1 INTRODUCTION

Mass movements triggered by rain and rain-on-snow events have been a major concern in forest management in many parts of the world (Sidle et al. 1985). Mass movements are dominant sources of sediment and affect the geometry and disturbance regimes of channel and riparian areas in steep forested lands (Swanson et al. 1987). Landslide-caused damage exceeds $1 billion annually in the United States (Schuster and Krizek 1978) and poses threats to life. Downstream effects of mass movements also affect water quality, water quantity, and aquatic habitat.

Landslides result from a combination of interacting factors that include topography; soil thickness, conductivity, and strength properties; rainfall intensity and duration; subsurface flow orientation; bedrock fracture flow; and vegetation surcharge and root strength (Montgomery and Dietrich 1994). These controlling factors are unevenly distributed in space and time, making quantitative assessment of landslide risk complex and difficult.

Mapping or delineating areas susceptible to landslides is essential for land-use activities and management decision-making in hilly and mountainous areas. Ideally, land-use activities, such as forest harvesting and road construction, should avoid vulnerable slope areas. Sites prone to mass failures can be identified by analytical and

Terrain Analysis: Principles and Applications, Edited by John P. Wilson and John C. Gallant.
ISBN 0-471-32188-5   © 2000 John Wiley & Sons, Inc.

empirical methods. A variety of approaches have been used in landslide mapping and can be classified into five categories: on-ground monitoring, remote sensing, factor overlay, statistical models, and geotechnical process models (Schuster and Krizek 1978). Each of these methods has its value for certain applications and disadvantages for other objectives (Ward et al. 1982, Montgomery and Dietrich 1994).

A computer-based framework incorporating topographic, vegetation, and soil information with geotechnical models is a useful way to locate unstable areas (Ward 1976, Okimura and Ichikawa 1985, Wu and Sidle 1995, Montgomery and Dietrich 1994). The recent trend of using spatially explicit approaches is largely due to the recognition of the importance of accounting for spatial heterogeneity, and newly developed tools and technology, especially geographic information system (GIS) and related remotely sensed imagery, and digital terrain analysis, which make it possible to consider small-scale heterogeneity. Incorporating digital terrain models into landslide prediction explicitly accounts for the importance of both topographic form and landscape position in slope stability. Recent work has demonstrated that both form and position represent first-order controls on landslide initiation (Dietrich et al. 1986, 1993, 1995, Montgomery and Foufoula-Georgiou 1993, Montgomery and Dietrich 1994, Zhang and Montgomery 1994).

Other studies have used spatial terrain data in regional stability analyses in conjunction with process-oriented models. Ward and co-workers (Ward 1976, 1985, Ward et al. 1978) delineate possible shallow landslide areas based on the uncertainties of control variables, but ignored the strong topographic control of shallow landslides by flow accumulation and convergence. Wu and Sidle (1995) developed an event-based slope stability model for forest watersheds, but the model is computationally intensive. Montgomery and Dietrich (1994) coupled near-surface flow characteristics with a slope stability model, but assumed steady-state rainfall and uniform soil and vegetation properties. Later work using the same model (Dietrich et al. 1995) explicitly accounted for both spatial variation in soil depth and the influence of root strength on slope stability. Their work did not, however, address the influence of temporally varying precipitation duration and intensities on slope stability.

In this chapter, we expand on these previous efforts to present a quantitative approach for evaluating both spatial and temporal factors influencing shallow landslides, using a more probabilistic framework than has previously been used. Topographic attributes from a digital elevation model (DEM) are linked with a process-based geotechnical equilibrium model. In this model, the high variability of factors controlling landslide occurrence and temporal variability of subsurface saturation are treated using probability analyses. Our intent is to explore an alternative representation of the uncertainty and variability inherent in simulating climatic and topographic controls on landslide initiation. This representation includes a dynamic simulation of rainfall intensities and treats the spatial distribution of key soil parameters stochastically, using a Monte Carlo simulation approach.

The following sections describe the study area, the theory underlying the model, model parameterization, and results of testing it using data on observed landslides in a 64-km² watershed in western Oregon. We conclude with a discussion that considers the advantages and disadvantages of this approach with respect to other models that predict landslide location and frequency.

13.2 STUDY AREA DESCRIPTION

We tested the model against landslide inventory data from the H. J. Andrews Experimental Forest (hereafter HJA) in the western Cascade Range of Oregon. The HJA includes the 64-km² drainage basin of Lookout Creek, a tributary of the Blue and McKenzie rivers (Figure 13.1). Elevations range from 410 to 1630 m.

Lower elevations of the HJA are underlain mainly by Oligocene and lower Miocene volcanic rocks composed of tuffs, ash flows, and stream deposits. In higher areas, bedrock is composed of andesite lava flows of Miocene age and of younger High Cascade rocks. Stream erosion, a variety of types of landsliding, and glaciations have created a deeply dissected, steep landscape. Soils developed in these parent materials are mainly Inceptisols with local areas of Alfisols and Spodosols. Field estimation of soil depth over the Andrews Forest from soil pits found a pattern of thick soils in the upper two-thirds of the basin, which is underlain by deep-seated earth flows, with thinner soils in the lower elevation zone. Thick alluvium borders many of the lower-elevation streams (Dyrness 1969) (Figure 13.2). Soil depth is quite

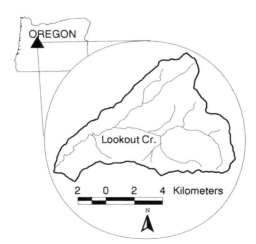

Figure 13.1. Location of the H. J. Andrews Experimental Forest, comprising the Lookout Creek watershed. Shaded areas show gauged watersheds with long-term stream flow and climate observations.

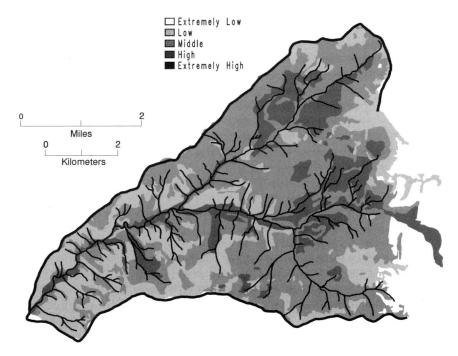

Figure 13.2. Soil depth in the H. J. Andrews Experimental Forest.

variable locally and, because of the extensive mass movement history of the area, does not always conform to topography.

The maritime climate has wet, mild winters and dry, cool summers. At the primary meteorological station at 430 m elevation, mean monthly temperature ranges from near 1°C in January to 18°C in July. Average annual precipitation varies with elevation from about 2300 mm at lower elevations to over 3500 mm at upper elevations, falling mainly from November through March. Rain predominates at low elevations; snow is more common at higher elevations. Highest stream flow occurs generally from November through February during warm rain-on-snow events.

When it was established in 1948, the Andrews Forest was covered with old-growth forest. Douglas fir, western hemlock, and western red cedar dominate lower elevation forests. Upper elevation forests contain noble fir, Pacific silver fir, Douglas fir, and western hemlock (Figure 13.3). Before timber cutting began in 1950, about 65% of the Andrews Forest was in old-growth forest (400–500 years old) and the remainder was largely in stands developed after wildfires in the mid 1800s to early

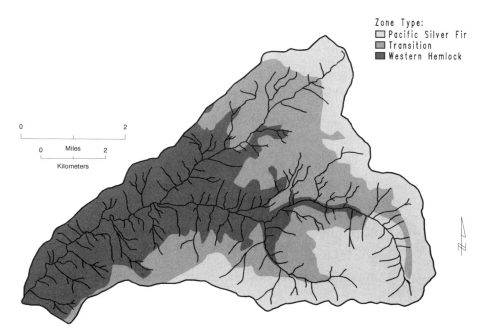

Figure 13.3. Vegetation zones in the H. J. Andrews Experimental Forest.

1900s. Clear-cutting and shelterwood cuttings over about 30% of the Andrews Forest have created young plantation forests varying in composition, stocking level, and age (Figure 13.4). Old-growth forest stands with dominant trees over 400 years old still cover about 40% of the total area. Mature stands (100 to 140 years old) originating from wildfire cover about 20%. Wildfire was the primary disturbance in the natural forest; windthrow, landslides, sites of concentrated root rot infection, and lateral stream channel erosion were secondary disturbances.

13.3 METHODS AND DATA SOURCES

To delineate landslide-prone areas, the infinite slope model was coupled with a probabilistic analysis of precipitation and landscape parameters. This approach is discussed in the following sections.

Figure 13.4. Land-use history in the H. J. Andrews Experimental Forest.

13.3.1 Infinite Slope Model

The factor of safety (FS) is commonly used as a quantitative expression of the hazard index of landslide initialization. It is customarily expressed as the balance between resisting and driving forces such that

$$FS = \frac{\text{Resistance of the soil to failure (shear strength)}}{\text{Forces promoting failure (shear stress)}} \qquad (13.1)$$

In equilibrium analysis, an FS of unity is a critical condition and would indicate imminent failure, and an FS greater than one indicates a stable condition. The infinite slope method is a popular slope stability analysis tool because it is simple and applicable to many shallow landslides. Generally, the infinite slope model does not adequately predict deep-seated, rotational failures, but is appropriate for failures of a soil mantle that overlies a sloping drainage barrier that may be bedrock, or a less permeable and well-compacted soil layer (Hammond et al. 1992).

A variety of slope stability models characterize stability using factor of safety. The infinite slope model is widely used (e.g., Ward 1985, Wu and Sidle 1995, Montgomery and Dietrich 1994). It can be written as

$$\text{FS} = \frac{C + \Delta C_r + \{[(D - h)\gamma_m + h(\gamma_{sat} - \gamma_w)] + W_r\} \cos \beta \tan \phi}{[(D - h)\gamma_m + h\gamma_{sat} + W_r] \sin \beta} \qquad (13.2)$$

where C is soil cohesion; ΔC_r is root strength; W_r is the surcharge weight of vegetation; γ_m and γ_{sat} are the unit weight of soil at field moisture content and saturation, respectively; D is the soil depth normal to the slope; h is the normal saturated depth; β is the surface slope; and ϕ is the internal frictional angle.

13.3.2 Parameter Estimation from Probability Analysis

Hillslopes commonly have a variety of rock types, soils with different properties and thicknesses, and vegetation with different surcharges and root strengths. Deterministically quantifying all these controlling factors is impossible for regional hazard mapping. Quantifying error and determining error sources and uncertainties are difficult, if not impossible. Many model equations may be inappropriate, because they may only be used for defined boundary conditions at specific scales: Equation 13.2, for example, is not valid for deep-seated earthflows. Even if measurement errors are assumed negligible and back-calculation of parameters is well defined, point measurements (e.g., strength tests) cannot effectively describe the larger-scale heterogeneity of the measurable parameters, such as frictional angle and cohesion.

Because the parameters in Equation 13.2 are inherently uncertain, a probability of failure for each hillslope element must be estimated (Ward et al. 1982) rather than predicted deterministically. In fact, the predictions of landslide occurrence from simple deterministic models actually represent statements of probability (Montgomery and Dietrich 1994). Some parameters that are highly variable in space and time, including soil and root strength, soil depth, and saturated depth, can be considered as random variables or assumed to be uniform (as we do for all but saturated depth). Other parameters, such as slope angle and unit weight of soil, can be estimated with greater accuracy from terrain analysis and/or GIS software, or from field surveys. In some landscapes, factors such as soil depth are strongly correlated with topography (i.e., Dietrich et al. 1986, 1995); however, this assumption is less valid in our landscape because of extensive mass movement and glacial overprinting.

Two important assumptions simplify the problem. First we treat the highly variable parameters as independent, following Ward et al. (1982), while recognizing that some of the stochastic variables in the infinite slope model are correlated (Hammond et al. 1992). There is no good way to handle this dependency quantitatively: Typically, soil cohesion and friction angle are inversely related (correlation coefficient of

−0.2 to −0.85) (Cherubini et al. 1983), while friction angle and soil density are positively related (Hammond et al. 1992). Although Hammond et al. (1992) adopted methods using transformed variables to treat the correlations between soil cohesion and friction and between friction angle and soil density, further research is needed to evaluate the statistical soundness and general applicability of this approach.

The second assumption involves using prior assigned distributions for certain parameters. In this chapter, all variables except saturated depth are based on an assumed uniform distribution (Ward et al. 1978) (Table 13.1). Saturated depth is affected by various factors, such as precipitation, evapotranspiration, vertical recharge, and deep ground water seepage, that are controlled by soil and vegetation characteristics. Based on the concept of the time–area curve and hillslope unit hydrograph (Iida 1984), Barling et al. (1994) derived a quasi-dynamic relationship between recharge rate and the saturated depth as

$$h(t) = \frac{r}{K_s \tan \beta} \, a(t) \tag{13.3}$$

where $h(t)$ denotes the depth of saturated throughflow at time t; r is the constant recharge rate; $a(t)$ is the upslope drainage area divided by the downward width of the element in this case; and K_s is the saturated conductivity of soil.

The assumption of uniform recharge rate is only a simplification of the complex behavior of the soil–vegetation–atmosphere interface. If we further ignore canopy interception and evapotranspiration loss, r can be viewed as the precipitation mean intensity, I_m, which we define as an exponential distribution (e.g., Beven 1987):

$$f(I_m) = \frac{e^{-I_m/\lambda_P}}{\lambda_P} \tag{13.4}$$

where $f(I_m)$ denotes a probability density function and λ_P is mean storm intensity, in mm/h.

TABLE 13.1   Stochastically Treated Parameters and Their Assumed Probability Distributions

Random Variable	Probability Distribution
Soil depth	Uniform
Soil cohesion	Uniform
Internal friction angle	Uniform
Root strength	Uniform, mean dynamic change with age
Biomass surcharge	Uniform, mean dynamic change with age
Precipitation intensity	Exponential

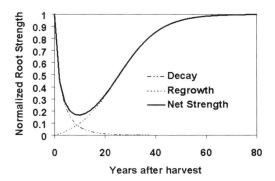

Figure 13.5. Root strength change after forest cutting. Only the mean is shown; variance is assumed constant for different years after cutting.

13.3.3 Effect of Vegetation

Vegetation influence on slope stability is also an important factor in steep forested watersheds, especially through root strength (Sidle 1992). Tree roots provide lateral support and vertically anchor and buttress the soil mass (Hammond et al. 1992). A sigmoid curve was used to model root cohesion during regrowth (Sidle 1992); an exponential curve was used to model the decay of residual root strength following tree harvesting (Sidle 1992) (Figure 13.5). The general shape of conceptual root strength regeneration is

$$C_r = C_{max} \left[(a_r + b_r e^{-k_r t})^{-1} + d_r \right] \tag{13.5}$$

and the root deterioration is

$$C_{rd} = C_{max} \, e^{-K_d t^{m_1}} \tag{13.6}$$

Total root strength is given as

$$\Delta C = C_r + C_{rd} \tag{13.7}$$

where a, b, d, m, and k are all parameters (Sidle 1991, 1992) and C_{max} is the maximum root strength.

Biomass surcharge is generally not a significant factor in landslide initiation (Hammond et al. 1992). We include it, however, because it is explicitly influenced by both timber harvest and tree regrowth and may affect the stability threshold at which sliding occurs on marginally stable sites. Different functions have been used for calculating biomass after harvesting. A function similar to the root regrowth Equation (13.5), as described by Sidle (1987), is used here (Figure 13.6):

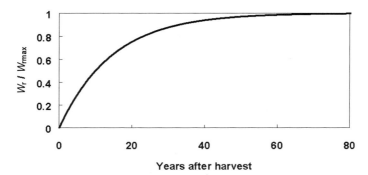

Figure 13.6. Biomass surcharge after forest cutting. Only the mean is shown; variance is assumed constant for different years after cutting.

$$W_r = W_{max} [(a_w + b_w e^{-k_w t})^{-1} + d_w] \tag{13.8}$$

where W_r is the biomass over the element at t years after cutting; W_{max} is the maximum weight trees can grow in unit area; and a_w, b_w, c_w, and d_w are constants. As indicated by Sidle (1992), the model assumes that tree surcharge is uniformly distributed across each element.

Based on the above equations characterizing the changing vegetation factors after harvesting, we obtain the means of the uniform distribution for onsite root cohesion and biomass surcharge. The variances of these two distributions are treated as constants equal to one-tenth of the normalized root cohesion and biomass.

13.3.4   Terrain Analysis and Climate Data Acquisition

The data and parameters used for this investigation were derived from or obtained by existing Level 1 DEM, aerial photograph and field surveys. The DEM data stored in the ARC/INFO GIS have a resolution of 30 × 30 m and the slope and topographic convergence index were computed for each pixel using internal GIS functions. Although considering flow as multidirectional out of each pixel is more sophisticated conceptually, the GIS tool we used incorporated only a single-direction algorithm. Because this algorithm accumulates drainage area only along cardinal directions and 45° to those directions, it inevitably results in the "striping" artifacts visible on Figures 13.8 and 13.9. Newer flow-routing algorithms, like those discussed in Chapters 3 and 5, may eliminate this problem.

The soil and vegetation characteristics and the land-use history were rasterized to the DEM resolution in ARC/INFO. For the seven soil types that occupy most of the area of HJA, parameters of soil strength and saturated conductivities were derived from field measurements (Table 13.2), and the average of the known seven types was used for the remaining unmeasured types. The error introduced by this method is limited because of the small area involved. Although there are three vegetation zones at HJA, we used the same vegetation parameters for all types in the random simulation

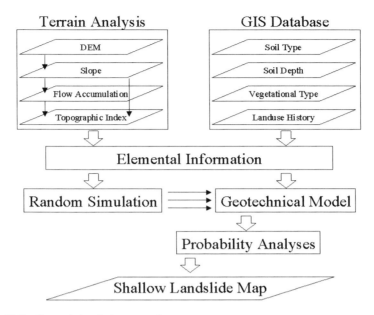

Figure 13.7. General simulation procedure.

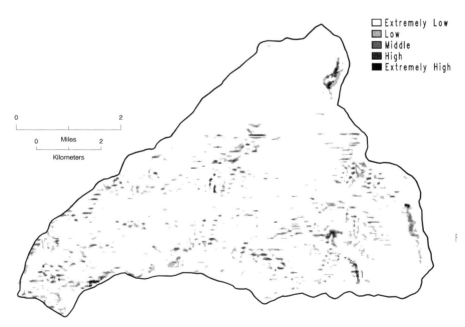

Figure 13.8. Potential shallow landslide area in the H. J. Andrews Experimental Forest under old-growth condition.

TABLE 13.2 Soil Parameters Used in Model Test

Soil Parameters	Flunky	Blue River	Limberlost	Budworm	Slipout	Frissell	Mckenzie	Others	Reference
Bulk density (g/cm³)	2.317	1.034	1.421	1.564	1.477	1.954	1.585	1.6002	Rothacher et al. 1967
Porosity (vol/vol)	0.29	0.63	0.52	0.5	0.48	0.4	0.53	0.486	Rothacher et al. 1967
Field capacity (vol/vol)	0.102	0.285	0.277	0.324	0.343	0.217	0.265	0.2852	Rothacher et al. 1967
Saturated conductivity (m/s)	0.004178	0.001567	0.002089	0.001567	0.000627	0.000627	0.000418	0.001044	Calibrated
Cohesion (kPa)					0–15				Hammond et al. 1992
Frictional angle (°)					30–45				Hammond et al. 1992

TABLE 13.3   Vegetation Parameters Used in Model Test

Parameters	Value Used	Reference
Root strength		Sidle 1991
C_{max}, maximum (kPa)	12.5	
k_d, coefficient in residual	0.402	
m_1, coefficient in residual	0.647	
a_1, coefficient in regrowth	0.952	
b_1, coefficient in regrowth	19.05	
d_r, coefficient in regrowth	−0.05	
k_r, coefficient in regrowth	0.12	
Biomass		Oral communication with Harmon 1995, Sidle 1992
W_{max}, maximum (kPa)	3	
a_w, coefficient	0.952	
b_w, coefficient	19.05	
c_w, coefficient	−0.05	
k_w, coefficient	0.12	

because the root strength and biomass surcharge values were similar among types (Table 13.3).

Storms were defined as periods with a mean precipitation intensity greater than 1 mm h^{-1}, separated by at least 12 h with precipitation less than 1 mm h^{-1} (Beven 1987). We analyzed precipitation records for the climate station that has the longest history of observation at HJA over a 36-year period (water year 1958 to 1993). A total of 1097 storms fitting the above criteria had an average intensity of 1.35 mm h^{-1}.

13.3.5   Monte Carlo Simulation and Probability Derivation

The simulation of probability of landslide occurrence relies on spatially distributed information of small elements. From terrain analysis and GIS, topographic attributes such as slope, flow accumulation and topographic convergence index can be derived from the DEM for each element. Overlaying the distributed elements with a GIS database helps with the derivation of other land surface or near-surface characteristics, such as soil, vegetation and land-use history (Figure 13.7).

For each element, 1000 random simulations were produced that represented the ranges in probable value of saturation, soil and root strength, and biomass, using the appropriate probability distributions (Tables 13.1 and 13.2). The variables in Table 13.1 were treated as random variables, which means we randomly simulated them, using the assumed probability distribution and the mean and variance calculated from the terrain and precipitation data. For each simulation set of these variables, a factor of safety was calculated based on the geotechnical model (Equations

13.2–13.4). Then the mean (μ_{FS}) and variance (σ_{FS}) of factor of safety from the Monte Carlo simulation for each element were computed. By assuming a normal distribution for the factor of safety (Ward et al. 1982), the probability of failure, defined as the probabilities of factor of safety less than and equal to 1 for each element, was calculated using the normal probability distribution:

$$P(\text{FS} \leq 1.0) = \int_{-\infty}^{1.0} \frac{1}{\sqrt{2\pi}\sigma_{FS}} \exp\left[-\frac{1}{2}\left(\frac{x - \mu_{FS}}{\sigma_{FS}}\right)^2\right] dx \qquad (13.9)$$

which can be solved numerically. Another way to construct the probability of failure is to use the ratio of the number of sets with computed FS less than or equal to one divided by total number of Monte Carlo simulation sets. The latter has the advantage of not assuming a distribution for factor of safety. Our tests demonstrated that when large numbers of simulation sets are computed, there is little difference in the resulting slide maps between these two methods.

The 1000 simulations for each element were constructed based on the random parameters of soil, vegetation, and precipitation characteristics, using Equations 13.3 and 13.4. The probability of failure was computed for each element using the simulation. Two conditions were considered: First, we analyzed probability of failure for old-growth conditions; second, we examined the probability of failure at 12 years post clear-cutting for the whole watershed, when root strength is usually at a minimum and the landscape is, therefore, most unstable following cutting.

We qualitatively define the probability of slope movement as extremely low ($P \leq 0.2$), low ($0.2 < P \leq 0.4$), moderate ($0.4 < P \leq 0.6$), high ($0.6 < P \leq 0.8$), and extremely high ($0.8 < P \leq 1.0$). This relative ranking of stability of hillslope elements does not have an absolute timescale attached to it. However, since the 36-year climate record used to generate the precipitation input incorporates events ranging up to approximately a 100-year return period, the implied timescale is of one to several centuries in length. This corresponds closely with the interpreted return period of sliding for topographic hollows with mature forest cover of several hundred years in this area (Swanson et al. 1987).

13.4   RESULTS AND DISCUSSION

The two modeled vegetation conditions yielded dramatically different patterns of landslide probabilities (Figures 13.8, 13.9). Under old-growth conditions most of the hillslopes are stable and only scattered areas have a high potential for shallow landslides (Figure 13.8). Under old-growth conditions, 98% of the area is stable; 92% of the areas is ranked extremely low in probability of failure (Figure 13.10). Only 1% of the uncut forest has a high probability of sliding and none is rated extremely unstable. Under clear-cut conditions, on the other hand, only 76% of the area is rated in a stable state; the area with high failure probability has increased to 11%, with 3% of the area ranked as extremely unstable under the recent clearcut condition. Under both

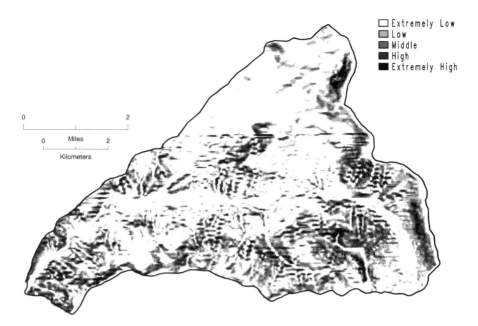

Figure 13.9. Potential shallow landslide area in the H. J. Andrews Experimental Forest under condition that the whole drainage is clear-cut.

scenarios, the proportion of area in each category declines exponentially with increasing landslide risk.

Historically, approximately 137 landslides occurred at the HJA (Figure 13.11) (F. Swanson, unpublished data on file at Pacific Northwest Research Station). Of these 137 slides, 71 (52%) are road-related and 35 (25%) are clear-cut-related, with another 31 (23%) in forest. This inventory clearly demonstrates the acceleration of

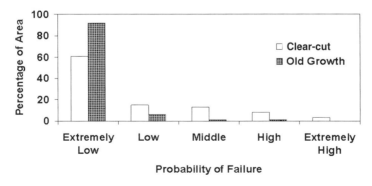

Figure 13.10. Percentage of areas under different slide potential categories for old growth and clear-cut.

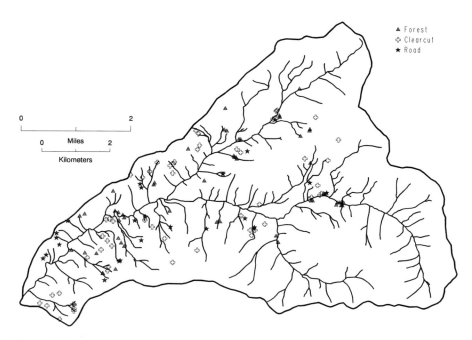

Figure 13.11. Observed landslide sites and associated land use conditions for 1950–1990.

landsliding due to road construction and cutting at HJA (Swanson and Dyrness 1975). Comparison of observed landslide sites with predicted unstable areas for current land-use conditions, incorporating existing vegetation cutting patterns (but not roads) (Figure 13.12), can be used to test model performance. Table 13.4 compares the modeled and observed landslide sites based on a point registration that treats any landslide site as a single point in the spatial database.

Overall, the model does not predict the observed sliding behavior very well. Many landslides occurred at elements mapped as low-probability areas. These discrepancies may be due to several factors, which point to the limitations of this model or landslide models in general for accurately predicting slide locations. A key limitation may be errors related to accurately pinpointing landslide sites as points in GIS coverages. Average areas for individual slides are approximately 200–300 m^2, assuming an average slide depth of 2 to 3 m (Swanson and Dyrness 1975); so that the slide area is less than the 900 m^2 (30 × 30 m) "window" generated with 30 m square-grid DEMs. Given the uncertainties attached to both accurately locating field-inventoried slides on the DEM as well as pinpointing them with the model, a more robust test might utilize a larger (e.g., 3 × 3 cell) window, associating slide risk with the highest probability recorded in any of the 9 cells. A related critical factor is the coarseness of the DEM data itself; 30-m data do not typically display the topographic detail that drives landsliding (Zhang and Montgomery 1994). The single-flow-direction algorithm used to calculate drainage area also imposes an artificial structure that may influence predictions in unknown ways.

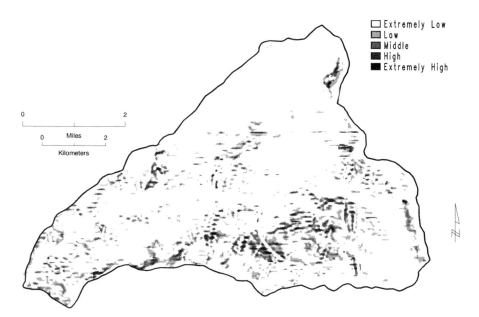

Figure 13.12. Potential shallow landslide area in the H. J. Andrews Experimental Forest under current stand conditions.

More importantly, this model does not incorporate geologic information, which plays an important role in the geographic pattern of shallow landslides. The model overestimates the landslide potential for high-elevation areas of the HJA, where competent andesite lava flows underlie the hillslopes. Soil developed from this rock type typically has a very high rock content and therefore does not build the high pore pressures necessary to trigger landslides. Another consideration is that many landslides occur during rain-on-snow events, which typically occur in lower to middle elevation areas. The model does not consider that high-elevation areas may have snowpacks that persist through storms (Swanson and Dyrness 1975). Nor does it explicitly consider the contribution of roads and associated altered drainage patterns to slope stability. Finally, 50% of the landslides observed within the 40-year inventory period occurred during two storms in the winter of 1964–65. The climatologic conditions necessary to trigger major episodes of landsliding are probably underrepresented by the probabilistic simulation of precipitation.

TABLE 13.4   Percentage of Observed Slide Sites and Delineated Probability for Different Land-Use Categories

	Extremely Low	Low	Middle	High	Extremely High
Road	60	21.5	13.8	3.1	1.5
Clear-cut	67.6	8.8	20.6	2.9	0
Forest	62.5	31.3	0	6.3	0

Other mass movements, such as earth flows, are present at HJA, in addition to shallow debris slides. The infinite slope model used in this project is valid only for shallow landslides and not for deep-seated earth flows. Other types of topographically controlled representations are needed to delineate deep slides.

Because of these factors, agreement between the model and inventory data is only fair. Future models for this area must explicitly consider the effects of geology on soil strength and hydraulic conductivity. A simple partitioning of the HJA basin by geology alone gave much better agreement with actual slide data (Swanson and Dyrness 1975). This raises the possibility that model performance might be improved if geologic variation were explicitly considered in the model, perhaps by defining ranges for soil cohesion based on geology. However, poor resolution of soil mapping due to dense forest cover contributes to uncertainties in defining soil and geological properties, however.

13.5   IMPLICATIONS FOR MANAGEMENT AND FUTURE MODELING

The relatively poor fit to the observed data, together with uncertainties as to how to interpret goodness-of-fit in the first place, raises interesting questions regarding how mixed dynamic and probabilistic models like this one, or other more deterministic models (i.e., Montgomery and Dietrich 1994), should be used in management applications. Use of topographically driven landslide models in land-use management is still rather new. Increasing concern about accelerated risks from landsliding following forest harvest, however, is prompting managers to look for more sophisticated tools to predict location and probability of failures. Understanding the strengths and limitations of these models is therefore crucial.

Both deterministic and probabilistic models must address the inherent uncertainty in predicting parameters controlling slide initiation; these parameters include both climatic and topographic variables. Measures of model success can include the spatial pattern of slides, the timing with respect to triggering storms, and the overall slide frequency integrated over multiple years or decades.

In general, the success rate for predicting slide *location,* as measured against observed slide locations, appears to be greater for the more deterministic models. Even simple models that incorporate the strong topographic controls on landslide location (i.e., convergent topography and hillslope gradient) appear to predict the overall spatial pattern of slides rather well, where sliding is primarily controlled by topography (Montgomery and Dietrich 1994, Dietrich et al. 1995). These models have been less well-tested in terrain such as the H. J. Andrews, where deep-seated earth flows introduce other factors controlling sliding (i.e., long-term mass movement history) that are not well-represented by topography. The Monte Carlo approach described here does less well, perhaps because the key variables influencing slide initiation that were simulated, such as soil and root strength, are not truly random but are topographically controlled as well. Land-use decisions requiring maps that broadly predict landslide risk and slide-prone sites would do well to rely on the more deterministic models, as long as underlying assumptions of mechanisms responsible for slide initiation were not violated.

Predicting slide frequencies, either at an event (i.e., individual storm) or longer-term (i.e., multiyear) timescale, however, may require better representation of climatic forcing functions, since landslide initiation may be due as much to transient precipitation intensities as to average or mean precipitation amounts. The theoretical distribution of precipitation events utilized here is a first attempt to explicitly incorporate a more detailed hydrology into landslide models. The increased value of this approach in terms of predictive power, however, is hard to demonstrate because all the other sources of uncertainty and error obscure any incremental gains. Future work should focus on testing model predictions of landslide frequencies at both event and decadal timescales.

13.6 CONCLUSIONS

The model described here explicitly considers the spatial heterogeneity of landscape characteristics, such as soil and root strength, using Monte Carlo simulation. Precipitation events, the most influential variable on landslide frequency and timing, are also stochastically treated based on a theoretical exponential distribution. The advantage of the probabilistic approach is that it may offer a less biased way of distributing inherent uncertainties across the landscape. Whether this necessarily results in an increase in ability to predict either timing or location of slides remains to be seen.

As expected, this topographically explicit method is strongly influenced by terrain-analysis methods (e.g., grid-based versus contour-based methods, single flow direction versus multiple flow direction algorithms). Different terrain-analysis methods produce distributed elements of different sizes and shapes that reflect different topographic attributes, such as slope, aspect, and drainage area, hence landslide probability. Future research should also explore the influences of different terrain-analyses methods, scale, and DEM data quality on landslide delineation.

CHAPTER 14

Terrain Variables Used for Predictive Mapping of Vegetation Communities in Southern California

Janet Franklin, Paul McCullough, and Curtis Gray

14.1   INTRODUCTION

Distributions of plant species on the landscape are related to soil-moisture balance in the semiarid Mediterranean-type climate of southern California. Moisture supply, subsurface soil moisture, soil properties related to moisture-holding capacity, and soil-moisture demand (evapotranspiration) have been shown to be related to topographic variables. Those variables include elevation, aspect, slope, curvature, hillslope position and drainage basin position, and more complex topographically derived indices such as potential solar insolation and soil-water balance (Chapters 1, 3, and 4 and references therein).

In this chapter we discuss the use of terrain variables for predictive vegetation mapping (reviewed by Franklin 1995) of chaparral and riparian vegetation in southern California. Before describing our case studies, we review the following issues related to ecological applications of terrain models:

- Degree to which variance in species distributions can be explained with site variables
- Problems with ecological field data used in conjunction with digital geographical databases—scale of observations and locational accuracy
- Problems with digital terrain models—resolution and nonsystematic and systematic errors
- Choice of model forms, based on theoretical and empirical considerations

Terrain Analysis: Principles and Applications, Edited by John P. Wilson and John C. Gallant.
ISBN 0-471-32188-5   © 2000 John Wiley & Sons, Inc.

14.1.1 Species Distributions and Environmental Gradients

The relationship between plant species distributions and environmental site variables is frequently studied using gradient analysis (Whittaker 1951) and gradient modeling (Kessell 1996). While gradient analysis is based on the continuum model of plant species distributions (species are distributed along environmental gradients with a Gaussian response function), ecologists continue to debate the relative merits of the continuum versus community model for vegetation composition (Austin and Smith 1989, Collins et al. 1993). In this chapter we use the species assemblage or association, rather than the individual species, as the categorical variable to be mapped. This is partly because it is a generalization conventionally used in mapping, and partly because these assemblages, which we refer to as vegetation classes, can be observed on the landscape on the basis of the co-occurrence of certain species (Austin and Smith 1989), and identified by clustering techniques commonly used in vegetation analysis.

Studies showing relationships between environmental site characteristics and plant species distributions suggest that models predicting vegetation distributions based on mapped environmental variables (especially those topographically related variables described in Chapters 3 and 4) might be a powerful mapping technique. However, disturbance factors and succession, and biotic factors (competition) are also very important in determining the presence of a species at a site. This is especially true for both chaparral and riparian vegetation, characterized by disturbance that occurs with relatively high frequency and magnitude (fire and flood, respectively). In fact, it has been asserted that attempts to explain chaparral species distributions based solely on edaphic and topographic site variables have "produced weak correlations" (Keeley and Keeley 1988, 167).

Therefore, a model that considers site variables only and not disturbance history might be expected to explain only a portion of the variance in plant species distributions. If disturbance history was known it could be used in an empirical model, by including a fire history map for example, or succession could be simulated explicitly (Malanson et al. 1992). However, even this type of model could not account for the many stochastic and biotic factors affecting the realization of a plant species distribution at a point in time. Thus, there are fundamental limits to predictive mapping of biotic distributions (cf. Morrison et al. 1992, 260).

14.1.2 Ecological Field Data in Geographical Databases

In the case studies described below, we have used field measurements of vegetation composition, collected in the course of previous studies, to calibrate or "train" our models—to serve as measurements of the response variable. Field observations of this type are expensive to collect (Austin and Heyligers 1989, Lees and Ritman 1991, Mackey 1994b), and may be valuable data sources, especially when machine-learning models are used (below), which require large numbers of observations. However, several challenges can arise when using field observations with explanatory variables estimated from digital geographical databases, especially when the field data were collected for other purposes. The first problem is locating plots accurately and precisely with reference to the geographical database. For example, the chaparral and riparian vegetation data used in this study had been collected for other purposes by the U.S. Forest Service, and plot locations were marked as points or lines on USGS 7.5' topographic quadrangle maps with unknown locational accuracy. Only in the last few years have geographers, foresters, and field biologists started using portable, low-cost global positioning systems (GPS) to precisely georeference the locations of field measurements.

The second issue concerns the scale of observations. Vegetation composition may vary over short distances, and field observations may represent areas smaller than the spatial resolution of the geographical database. Further, when environmental site variables such as slope and curvature are measured in the field, they might also represent a smaller, or unknown, area when compared to the same terrain attributes estimated from a digital terrain model. The two estimates may be poorly correlated (Davis and Goetz 1990, Fels 1994, Giles and Franklin 1996). For example, in our case study the chaparral plots provided by the U.S. Forest Service were ~0.03 ha. The resolution of the digital elevation model (DEM) is 30 × 30 m or ~0.09 ha, while the effective resolution of the terrain variables derived from it is even larger, given the local window for which the variable is calculated, and the smoothing of the terrain that was necessary to correct DEM errors (see below). It is possible that the resolution of a DEM could be too coarse to estimate landform shape at the appropriate scale, and hence key topographic variables related to the distribution of potential soil moisture and solar insolation (discussed by Hutchinson, Chapter 2, and Wilson et al., Chapter 5).

14.1.3 Error in Digital Elevation Models

In addition to these issues of spatial resolution or scale, errors in the terrain grid also need to be considered. Nonsystematic and systematic errors are commonly found in DEMs, and are related to the method used to derive the elevation grid. These errors are amplified when first- and second-order difference operations are applied to derive slope, aspect, and curvature (discussed extensively elsewhere; see Zevenbergen and Thorne 1987, Skidmore 1990, Weibel and Heller 1991, Brown and Bara 1994, and Chapters 2–5, this volume). This can confound any expected relationships between vegetation pattern and terrain-mediated site conditions (Davis and Goetz 1990, for example). Errors in the form of vertical or horizontal striping are frequently minimized by applying a smoothing or averaging filter. This could potentially introduce error into the elevation values, although this has not been found to be a serious problem (Brown and Bara 1994, Giles and Franklin 1996).

14.1.4 Modeling Methods

A final issue to consider when modeling the relationship between ecological patterns and terrain attributes is the form of the model. There are a number of modeling approaches that have been used which have different strengths in terms of their theoretical basis, predictive power, and simplicity. These have been reviewed elsewhere (Franklin 1995) but will be briefly outlined here. Rule-based models based on ranges of physiological tolerances (limiting factors) have been used at a range of spatial scales. Empirically derived rule-based methods have also been used to predictively map vegetation classes from terrain variables, especially in conjunction with remotely sensed data (Franklin et al. 1986, Bolstad and Lillesand 1992). Generalized linear models (GLMs) using a Gaussian response function for species abundance data have a basis in niche theory, although the Gaussian assumption has been questioned, and different forms of a GLM have been fit (Austin et al. 1994). GLMs with a logit link function are frequently used with species presence/absence data or categorical data (vegetation type). Classification (and regression) tree (CRT) is a binary, divisive monothetic classification method (Breiman et al. 1984, Venables and Ripley 1994) that has proved useful for exploratory analysis of ecological data (Michaelsen et al. 1987, 1994, Davis and Dozier 1990), and while it has no formal basis for inference (Clark and Pregibon 1992), it has been used predictively (Lees and Ritman 1991, D. M. Moore et al. 1991, Franklin 1998). Other machine learning methods, such as artificial neural networks (ANNs) and genetic algorithms, have also been used for predictive mapping of biotic distributions as well as classification of remotely sensed data (discussed in Franklin 1995).

14.2   CASE STUDIES

We used CRT, an exploratory, machine-learning classification technique, to develop models relating terrain variables and spectral reflectance variables derived from satellite imagery to (1) chaparral species associations and (2) riparian vegetation types for a study area in the Laguna Mountains of the Peninsular Ranges in San Diego County, California, USA. We derived the terrain variables from a digital elevation model and inverted the classification tree models to produce predictive maps of vegetation distribution.

Our objective was to use the relationship between environmental site characteristics and species distributions to develop methods for mapping vegetation for large areas with greater floristic detail and comparable or improved accuracy when compared to standard large-area mapping techniques (based on other methods of combining satellite imagery and DEMs). The resulting maps are needed for ecosystem inventory and for simulation modeling of vegetation dynamics. These particular plant formations, chaparral and riparian vegetation, were found to be difficult to map using remote sensing and geographic information system (GIS)-based techniques (Franklin and Woodcock 1997, Franklin and Stephenson 1996). This is due to complex landscape patterns of species composition, in the case of chaparral, and the small areal extent of riparian stands, which appeared as fine-grained linear features on the landscape, relative to the gridded data and image processing methods being employed to delineate map units.

14.2.1 Study Area

The study was conducted in the Descanso and Palomar Ranger Districts of the Cleveland National Forest (Figure 14.1). The Forest Service lands encompass the foothill and mountain peaks of the Peninsular Ranges, including Palomar, Cuyamaca, and Laguna mountains. The terrain is steep and dissected with elevations ranging from less than 200 to more than 1900 m. The Peninsular Ranges are a westerly-tilted fault-block, composed mainly of Cretaceous granitic, dioritic, and gabbroic rocks of the Southern California batholith, greatly fragmented by internal faulting (McArthur 1992). The climate is Mediterranean, with cool wet winters and warm to hot, dry summers. Annual precipitation is strongly correlated with elevation, and increases toward the northwest in the study area; it ranges from roughly 400 mm at the lowest elevations to over 1200 mm at Mt. Palomar. Coastal sage scrub and chaparral vegetation dominate lower and middle elevations, with oak woodland and grassland in valley bottoms, riparian forest along stream courses, and introduced annual grassland where land is managed for grazing. Conifer forests and montane meadows occur above ~1700 m (Beauchamp 1986, Gordon and White 1994).

Chaparral is the sclerophyllous, mainly evergreen shrub vegetation found throughout the Mediterranean-climate region of southwestern North America (Hanes 1977, Keeley and Keeley 1988). Summer–deciduous coastal sage scrub is found at lower elevations in coastal southern California (Mooney 1977). Chaparral occurs from sea level to 2000 m, especially on rocky slopes, in regions where the annual rainfall is 200–1000 mm, and typically forms a closed canopy 1–4 m tall. Coastal sage scrub occurs mainly below 700 m and has been displaced by urbanization in the coastal zone (Westman 1981a). Empirical studies have shown the distribution of chaparral and coastal sage scrub species to be related to potential soil moisture and potential evapotranspiration, as well as substrate (Wells 1962; Kirkpatrick and Hutchinson 1980, Poole and Miller 1981, Westman 1981b, Miller et al. 1983, Keeley and Keeley 1988).

Riparian ecosystems are among the most biologically diverse and threatened in western North America (Hubbard 1977, Katibah 1989, Franzreb 1990, Ferren and Fiedler 1993). Riparian plant species distributions have been shown in many regions to be related to hydrotopographic factors, such as channel width, stream gradient, and drainage basin position. These, in turn, are related to characteristics of the hydrologic regime—timing and magnitude of flow and duration of saturated conditions—and substrate texture, which affect seedling establishment and survival (Graf 1982, Harris 1987, Baker 1989, Stromberg and Patten 1990, Smith et al. 1991, Johnson 1994 and references therein).

14.2.2 Materials and Methods

14.2.2.1 Digital Terrain Data
Topographic variables were derived from a mosaic of 28 USGS 7.5′ 30-m resolution DEMs. DEM errors in the form of horizontal and vertical striping usually result from the profiling method used to generate the

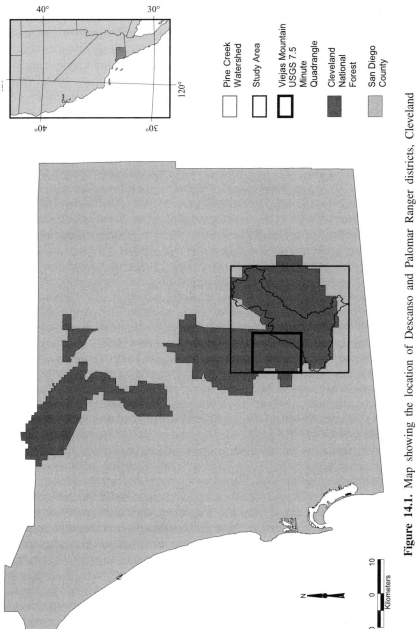

Figure 14.1. Map showing the location of Descanso and Palomar Ranger districts, Cleveland National Forest, within San Diego County. Also shown are the boundaries of the Pine Creek Watershed, a rectangular subarea of the Descanso District illustrated in Figure 14.8, and the boundary of the Viejas Mountain 7.5′ quadrangle shown in Figures 14.3–14.5 and 14.10.

Pine Creek Watershed

Study Area

Viejas Mountain USGS 7.5 Minute Quadrangle

Cleveland National Forest

San Diego County

DEM, with manual profiling from photogrammetric sources typically causing the greatest error. The range in quality among the quadrangles used in this study was substantial, with some having horizontal striping apparent when the elevation values were viewed (as in Figure 14.2), some having striping that became apparent in a curvature image, and some having no striping. We smoothed the entire DEM mosaic twice with a 3 × 5 filter to remove the striping, with obvious implications for the resulting (altered) elevation values and all derived topographic variables. We reasoned that for the purpose of vegetation modeling, the relative, rather than absolute values of elevation were important because most of the topographic variables derived were based on local (neighborhood) operations. However, an edge-preserving filter (Giles and Franklin 1996) would have preserved ridges and other topographic features related to vegetation patterns.

The following topographic variables were then derived from the elevation grid (Table 14.1): slope, aspect, potential solar insolation (Figure 14.3), upslope catchment area, steady-state topographic wetness index (Figure 14.4; TWI of I. D. Moore et al. 1991), three curvature measures, and distances to the nearest stream and ridge as a measure of hillslope position. Note that hillslope position can also be expressed as a relative distance from ridge or stream, related to the ordinal classes typically used in soils and geomorphology studies: ridge, midslope, toeslope, etc. (Skidmore 1990).

The first nine primary and secondary topographic variables (Table 14.1) and their derivations were described in Chapters 3 and 4. Slope, aspect, and curvature were each calculated using a 3 × 3 moving window. Slope was calculated using a second-order finite difference algorithm. Surface curvature was calculated by fitting a fourth-order polynomial to the 3 × 3 window and taking the second derivative of that surface (Zevenbergen and Thorne 1987); profile (downslope) and plan (across-slope) curvature were calculated using the methods specified by I. D. Moore et al. (1991, 10). The smoothed DEM was processed to remove spurious pits. The FD8 algorithm in TAPES-G was used to generate upslope catchment area (see Chapter 3). Upslope catchment area was then used to derive the distance to stream or ridge; threshold values were used to define ridges (<3 upstream pixels) and streams (>50) where those values were selected by interactively viewing the upslope catchment area map with a digital line map of streams (a USGS digital line graph or DLG). Then the distance to the nearest stream or ridge was calculated for each grid cell using a cost-distance function (in ARC/INFO's GRID module).

The potential solar insolation (PSI) variable used in the modeling was calculated for the winter solstice (representing the winter quarter, November–January) using the Solarflux AML (Hetrick et al. 1993a, b, Dubayah and Rich 1995). This variable showed a stronger empirical correlation with vegetation patterns than the potential insolation for any other season, or annual PSI, in our study, but probably because illumination differences between north- and south-facing slopes are most pronounced in winter due to low sun angle. In fact, the difference in energy balance and potential evapotranspiration between north- and south-facing slopes is most pronounced during the spring and summer (Miller et al. 1983).

There was multicolinearity among the terrain variables, and while our modeling

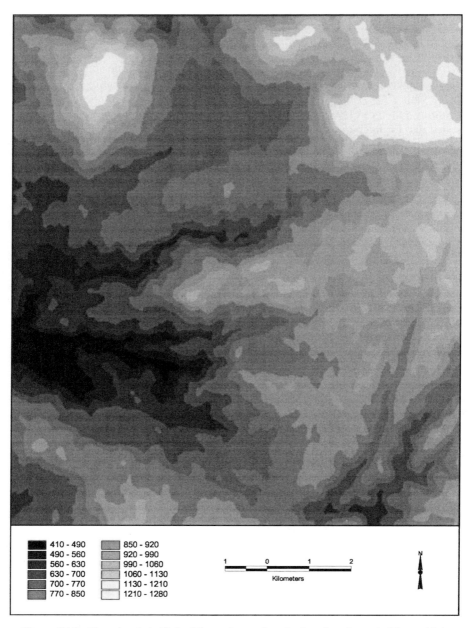

Figure 14.2. Elevation (m), Viejas Mountain quadrangle, location shown in Figure 14.1.

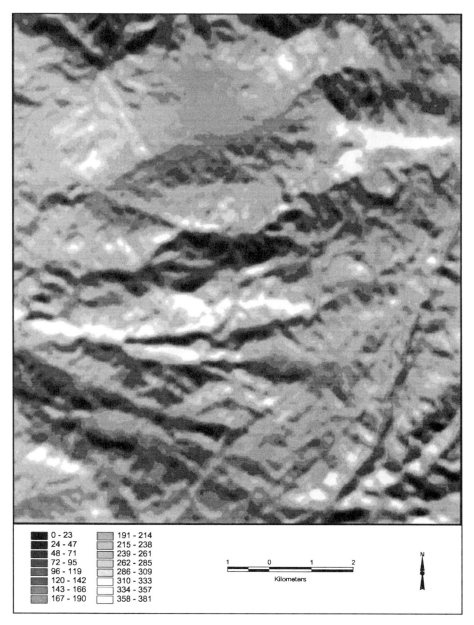

0 - 23		191 - 214	
24 - 47		215 - 238	
48 - 71		239 - 261	
72 - 95		262 - 285	
96 - 119		286 - 309	
120 - 142		310 - 333	
143 - 166		334 - 357	
167 - 190		358 - 381	

Figure 14.3. Potential solar insolation for 21 Dec (J/m²), Viejas Mountain quadrangle, location shown in Figure 14.1.

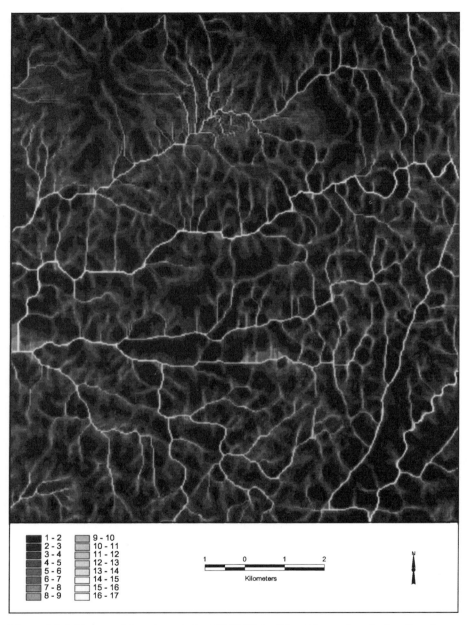

Figure 14.4. Topographic wetness index (TWI) Viejas Mountain quadrangle, location shown in Figure 14.1.

TABLE 14.1   Terrain and Spectral Explanatory Variables Used in This Study

Variables	Source/Software	Chaparral[a]	Riparian[b]
	Terrain		
Elevation	USGS 7.5′ DEM	1, 2, 3	v, l
Slope	Tapes-G	1, 2	v, l
Aspect	Tapes-G		
Potential solar insolation	Solarflux-Grid[c]	1, 2, 3	
Upslope catchment area	Tapes-G		v, l
Topographic wetness index	Grid	1, 2	v, l
Surface curvature	Grid		v, l
Profile curvature	Grid	1, 2	
Planform curvature	Grid	1, 2	
Distance to stream	Grid	1, 2	
Distance to ridge	Grid	1, 2	
	Spectral		
NDVI	Image processing	1	v, l
TM brightness[d]	Image processing	1	
TM greenness[d]	Image processing	1	
TM wetness[d]	Image processing	1	
Red (TM 3)	Landsat TM		l
Near-IR (TM 4)	Landsat TM		v, l
Mid-IR (TM 5)	Landsat TM		v

[a] 1, 2, and 3 refer to models 1–3 (see text).
[b] v, vegetation; l, life-form model.
[c] Solarflux was written in Arc Macro Language by Hetrich et al. (1993a, b). Grid, ARC/INFO's GRID module.
[d] TM brightness, greenness, and wetness were derived using the coefficients for the TM tassled cap (Crist and Cicone 1984).

method is nonparametric, it makes little sense to use explanatory variables that are highly redundant. The variable pairs plan curvature and profile curvature ($r = -0.83$) and surface curvature and plan curvature ($r = 0.80$) were (not surprisingly) strongly and linearly correlated in our training dataset, and therefore only one from each pair was used in the various models discussed below.

14.2.2.2 *Satellite Imagery*   *Landsat* Thematic Mapper (TM) images acquired for the study area on 22 September 1988 and 27 August 1990 were used for the chaparral and riparian models, respectively. Both images had been geometrically rectified (using *Landsat* ephemeris data), terrain corrected and georeferenced to UTM projection by a third-party vendor. The USGS DEMs are also georeferenced and in UTM projection so no rubber sheeting was performed. This may have introduced additional noise into the models; a previous study had noted misregistration between the 1988 image and the DEM on the order of one pixel (Shandley 1993). Multispectral radiance variables derived from the TM images were used as predictors because

reflectance patterns are frequently related to vegetation structure or physiognomy (Curran 1980, Jensen 1983). For the chaparral models, orthogonal multispectral indices related to overall landscape brightness (reflectivity), greenness (green leaf vegetation amount), and wetness (vegetation/soil-moisture content) were calculated from the six reflective TM wavebands using the TM Tassled Cap transformation (Crist and Cicone 1984). The normalized difference vegetation index (NDVI; Figure 14.5) was derived from the red and infrared radiances (Jensen 1996, 182). The structure and appearance of riparian vegetation in southern California contrasts strongly with the surrounding upland vegetation. In the riparian models, the uncalibrated radiances of the TM bands 3, 4, and 5 were also tested (see C. A. Gray (1995) and McCullough (1995) for details).

14.2.2.3 Chaparral Vegetation Field Data

The primary source of data on chaparral vegetation was a sample of field observations collected by the U. S. Forest Service for the purpose of ecological type classification in southern California (Gordon and White 1994). Percent cover by species was visually estimated for all woody plants in a 10-m-radius plot (~0.031 ha) whose location was subjectively chosen, based on ease of access and lack of recent disturbance (only mature stands were sampled). Plots were assigned to one of the five chaparral classes used in this study (Table 14.2) guided by a TWINSPAN-based cluster analysis (Hill 1979). Over 1000 locations were sampled in Southern California, but only 285 fell within the Descanso and Palomar Ranger Districts. Consequently, we also included 105 photointerpreted observations of chaparral class membership derived from a stratified random sample of 0.81-ha (90 × 90-m) plots used to quantify the accuracy of an earlier, remote-sensing-based vegetation map (Shandley 1993). Another 70 subjectively located observations of chaparral vegetation type were available from a windshield survey of variable-sized stands (averaging roughly 2 ha), where vegetation class was assigned on the basis of visual estimates of cover for 1–4 dominant species per stratum. The total sample size, 461, was not evenly distributed among classes but was roughly proportional to their area (Table 14.2).

14.2.2.4 Riparian Vegetation Field Data

Field observations of riparian vegetation composition were again derived from an independent effort by the U. S. Forest Service (Cleveland National Forest, unpublished data). Every public stream reach in the Pine Creek watershed (within the Descanso Ranger District) was surveyed and the percent cover of all species was visually estimated (T. White, personal communication). Thus, the plot size was variable, and plot extent was set by the surveyor (a new plot was started when the vegetation composition changed). The plots were transcribed manually as a line (the length of the stream reach) on a topographic map. This sample consisted of 168 observations, which were assigned to five life-form categories using a set of decision rules (shown in Table 14.3; detailed in C. A. Gray 1995). Each observation was then assigned to a riparian vegetation class within its life-form based on species dominance (Table 14.3).

To train a classification tree, we required a sample of nonriparian locations on the landscape and, following the logic of Aspinall and Veitch (1993) for obtaining a sam-

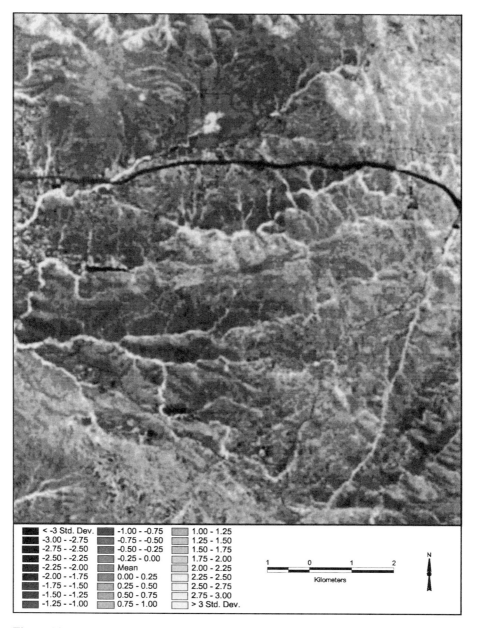

< -3 Std. Dev.	-1.00 - -0.75	1.00 - 1.25
-3.00 - -2.75	-0.75 - -0.50	1.25 - 1.50
-2.75 - -2.50	-0.50 - -0.25	1.50 - 1.75
-2.50 - -2.25	-0.25 - 0.00	1.75 - 2.00
-2.25 - -2.00	Mean	2.00 - 2.25
-2.00 - -1.75	0.00 - 0.25	2.25 - 2.50
-1.75 - -1.50	0.25 - 0.50	2.50 - 2.75
-1.50 - -1.25	0.50 - 0.75	2.75 - 3.00
-1.25 - -1.00	0.75 - 1.00	> 3 Std. Dev.

Kilometers

Figure 14.5. TM Tassled Cap greenness, derived from *Landsat* TM imagery, Viejas Mountain quadrangle, location shown in Figure 14.1.

TABLE 14.2   Chaparral Vegetation Class Definitions Used in This Study[a]

Label	Class Name	n_obs[b]	Class Description
SS	Coastal sage scrub Holland 32510	62	Soft-wooded sub-shrubs, dominated by *Artemesia californica, Eriogonum fasiculatum, Salvia* sp., *Malosma laurina;* on steep xeric low-elevation mainly coastal sites.
CD	Southern mixed chaparral Holland 37120	40	Mission manzanita (*Xylococcus bicolor*), *Ceanothus tomentosus,* and/or *Cneoridium dumosum* present as indicators; similar to northern mixed chaparral but on sites with lower precipitation and more moderate temperatures; coastal foothills of San Diego County and northern Baja California, Mexico.
CA	Chamise chaparral Holland 37200	62	1–3 m tall dominated (>80% relative cover) by *Adenostoma fasiculatum;* on xeric slopes and ridges, and somewhat drier sites or lower elevations than northern mixed chaparral.
CQ	Northern mixed chaparral Holland 37130	170	In southern California typically *Adenostoma fasiculatum* is present and codominant (average 30% relative cover) with various *Ceanothus* and *Arctostaphylos* species, *Cercocarpus betuloides,* scrub oak, etc.
CS	Scrub oak chaparral Holland 37900	127	Dense, tall chaparral co-dominated by scrub oak (*Quercus berberidifolia,* formerly *Q. dumosa*) occurring on mesic sites.

[a] Definitions are modified from Holland (1986) and labels follow Forest Service (1994), updated from Forest Service (1981). Note that Holland has been superseded by Sawyer and Keeler-Wolf (1995), and a translation between the two classifications can be found in the latter.

[b] Numbers of observations in the training sample.

ple of "absence," we buffered the USGS digital line graph of streams to one 30-m-grid cell on either side and randomly sampled 200 nonriparian cells (outside the buffer). This could have resulted in some commission errors, especially given the errors in the DLG stream coverage (noted below), but this did not seem to significantly affect the model. The classification tree models for riparian prediction were developed (trained) using this Pine Creek dataset.

A less-detailed riparian survey had been undertaken by the Forest Service (unpublished data) in the Descanso Ranger District outside of the Pine Creek watershed. Only points (107) were sampled (not reaches), cover for dominant species was estimated, and hydrotopographic variables were not recorded in the field. We again supplemented this with a random sample of 107 nonriparian locations, and this second sample was used to test the model. Owing to the different dimensions of the plots and survey methods used in the two samples, their average characteristics were quite different. For example, the average tree, shrub, and herb cover for the Pine Creek versus Descanso sample was 9, 38, and 30% versus 33, 22, and 31%, respectively. We

TABLE 14.3   Riparian Vegetation Class Definitions Used in This Study[a]

Life Form	Class Name/Label	n_obs[b]	Class Description
Tree (>10% cover by tree species)	Coast live oak riparian forest (CLO)	16/20	Greater than 40% cover of *Quercus agrifolia;* can be associated with understory of chaparral shrubs. (Holland 61310)
	Arroyo willow forest (AWF)	12/6	Greater than 60% cover of *Salix lasiolepis* that averages >6m high. (Holland 61320)
	Sycamore/ cottonwood–willow riparian forest (SyW)	9/3	Sycamore woodland (Holland 62400: dominated by *Platenus racemosa* with overstory cover >25%, scattered *Salix* spp. and *Populus freemontii* may be present, understory of *Baccharis* spp.); AND (aggregated with) Southern cottonwood–willow riparian forest (Holland 61330: tall, open forest dominated by *Populus freemontii, P. trichocharpa, Salix* spp., *Platenus racemosa* can also be present)
	White alder riparian forest (WAF)	3/1	Dominated by *Alnus rhombifolia* with shrubby deciduous understory (Holland 61510)
Shrub (Not as above [NAA] and >20% shrub cover)	Southern willow scrub (WiS)	40/32	Dense (>60% cover) broadleaf winter–deciduous riparian thicket dominated by several species of *Salix* that average less than 6 m tall (Holland 63320)
	Mulefat scrub (MuS)	23/14	Strongly dominated by *Baccharis salicifolia* and *B. viminea;* early-seral, maintained by frequent flooding (Holland 63310)
	Montane scrub (MoS)	8/0	Winter–deciduous shrubby riparian thicket: *Salix* spp., *Alnus* spp. (Holland 45200)
Herb (NAA and >50%herb cover)	Herbaceous vegetation (Herb)	16/9	Grass and forb cover >50% and shrub cover <20%, tree cover <10%
NFC (NAA and >50% barren)	Natural flood channel (NFC)	32/19	Unvegetated or sparsely vegetated drainages (>50% barren)—streambed
NoR	Nonriparian (NoR)	200/104	Other land cover (see text)

[a] Definitions are modified from Holland (1986) based on decision rules in O'Leary et al. (1994) noted in the first column. Holland has been superseded by Sawyer and Keeler-Wolf (1995), and a translation between the two classifications can be found in the latter.

[b] Numbers of observations in the training/test sample.

made the assumption that the Pine Creek sample is representative of the entire Ranger District and that these differences are entirely due to the survey methods and data enumerators. Therefore, we adjusted some of the decision rules for life-form assignment to >45% tree cover and >12% shrub cover (compare with Table 14.3). This additional uncertainty about the test data limited our ability to validate these models.

14.2.2.5 Data Analysis The approach used in classification tree analysis is to grow a large "tree" that is overfit to the training data, and then to "prune" it to a more parsimonious tree whose size can be determined by cross-validation (Breiman et al. 1984, Venables and Ripley 1994). The tree model is developed by iteratively partitioning the dataset into two subsets using the threshold value of the one explanatory variable that yields the greatest increase in class homogeneity of the subsets evaluated by an information statistic. For the chaparral sample, three models were explored based on (1) selected topographic and spectral variables, (2) topographic variables only, and (3) only the two topographic variables used in most ecological gradient analyses: elevation and potential solar insolation (used as a surrogate for topographic moisture) (Table 14.1).

Chaparral models were both pruned and evaluated on the basis of a cross-validation estimate of classification accuracy (because there was no independent dataset available for validation). Cross-validation was accomplished by iteratively dividing the data set into 10 parts, using 90% of the data to develop a model and the remaining 10% to test that model as it is sequentially pruned from its maximum size to its single "root" node (Clark and Pregibon 1992). The mean classification error for each number of nodes (based on all tree models in the cross-validation) indicates the size of a tree that yields robust predictions (for which the estimated error is minimized) and an estimate of the model error (the average error at the selected number of nodes).

For the Pine Creek riparian vegetation sample two models were created using the explanatory variables listed in Table 14.1: one using the riparian class as the dependent variable, and one where the observations were aggregated into life-form (Table 14.3). These two riparian models were assessed by applying them to the test data set from the Descanso Ranger District and calculating classification accuracy (proportion of the observations correctly classified).

14.2.3 Results

14.2.3.1 Chaparral Vegetation A classification tree based on topographic and spectral variables (chaparral model 1; Table 14.1) with 55 terminal nodes classified 81% of training data correctly, but cross-validation indicated that this model would classify 62% of an independent data set correctly. The number of observations per terminal node ranged from 5 (the minimum number allowed) to 47. Although individual vegetation class accuracies could only be calculated for the training data, they were comparable to the overall accuracy (69–87%). Model 1 was pruned to a more parsimonious 18-node model (accuracy 61%; not shown for brevity) that selected the following variables: greenness, elevation, NDVI, distance to ridge, TWI, PSI, slope, and plan curvature. A classification tree with 49 terminal nodes based only on topographic variables (Table 14.1; model 2) classified 78% of the training data accurately and had an accuracy of 52% predicted by cross-validation. Model 3, based solely on elevation and December potential solar insolation, pruned to only 13 terminal nodes (Figure 14.6), classified 64% of training data correctly, and cross-validation indicated 56% accuracy, or little loss of predictive power when compared with models 1 and 2.

Further, the distribution of classes in two-dimensional environmental gradient space (model 3) reflected patterns noted in the ecological literature (Figure 14.7), while models 1 and 2 (not shown; see McCullough 1995) produced some counterintuitive decision rules (assigning a more xeric vegetation type to a more mesic part of environmental gradient space). However, the 17 decision rules in the pruned model 1 did make ecological sense (scrub oak on lower insolation slopes, coastal sage scrub on steeper, higher insolation slopes) with one exception (chamise chaparral on slopes with concave plan curvature). When the models are inverted to produce a vegetation map, it is interesting to note the differences in the spatial patterns of the predictions between models 1 and 3 (Figure 14.8, see color insert). The more complex model 1 predicts a pattern with more high-frequency spatial variation, although cross-validation indicates that it is only slightly more accurate (62 vs. 56%) than model 3.

14.2.3.2 Riparian Vegetation

A classification tree (not shown) with 23 terminal nodes and 8 independent variables (upslope catchment area, TWI, elevation, band 5 radiance, curvature, band 4 radiance, NDVI, and slope) correctly classified the riparian vegetation type of 81% of the training data but only 53% of the independent data set used for model validation. Most of that accuracy resulted from correct prediction of nonriparian versus riparian vegetation on the basis of a threshold value of upslope catchment area (individual class accuracies for vegetation classes worse than expected by chance).

Aggregating the dependent variable to life-form (increasing the number of observations in each class, Table 14.3) produced a tree model (pruned to 12 terminal nodes, Figure 14.9) with 82% accuracy, based on five independent variables (upslope catchment area, slope, NDVI, TWI, band 3). This model yielded 71% accuracy when applied to new data. While the individual class accuracies are low (Table 14.4), the decision rules are supported by the ecological literature. The topographic–hydrological variables, and spectral variables related to herbaceous and woody vegetation amount, were most important in predicting riparian vegetation patterns. The predicted map (Figure 14.10, see color insert) illustrates the strong relationship between predicted life-form and position in the drainage basin in this model (riparian vegetation is highlighted in the background image due to its strong infrared reflectance).

14.2.4 Discussion

Was this method of combining terrain data and satellite imagery for vegetation mapping an improvement over existing methods? Are the results of these models, the predicted vegetation maps, of adequate accuracy for ecological inventory, land management, and ecosystem modeling?

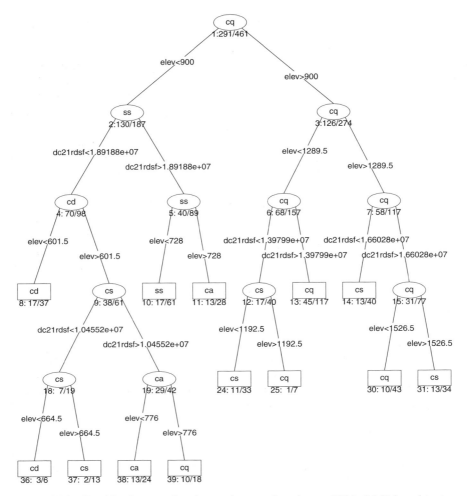

Figure 14.6. Classification tree for chaparral vegetation classes (Table 14.2) based on two explanatory variables: elevation and potential solar insolation (model 3). Ellipses are interior nodes (splits that do not assign class labels), and rectangles are terminal nodes (classes are assigned). Class labels of the predominant (interior) or predicted (terminal) class appear inside the ellipses/boxes (see Table 14.2 for definition of class labels). Numbers below the boxes are the node number and fraction of observations incorrectly classified at that node.

14.2.4.1 *Chaparral Models*   As noted earlier, our goal was to map shrub vegetation types that are difficult to distinguish owing to their physiognomic, and thus spectral, similarity, and the tremendous niche overlap among their dominant species (Westman 1991, Franklin 1998). These vegetation types corresponded to Anderson level IV land cover (Anderson et al. 1976) or vegetation types at the Alliance level (Federal Geographic Data Committee Vegetation Subcommittee 1996). While ~60% mapping accuracy sounds abysmal, 40–80% class accuracy, with confusion between

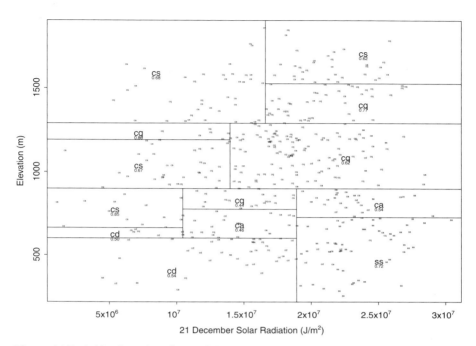

Figure 14.7. A bivariate plot of potential solar insolation versus elevation for the chaparral data where each observation is labeled by its vegetation class. The rectangular boxes show the locations of the threshold values of the variables used as decision rules in the classification tree (Figure 14.6). Large labels are the predicted class within that portion of data space, and the small number is the proportion of the observations correctly classified at that node.

similar classes, is typical in the vegetation mapping literature (Franklin 1991, Bauer et al. 1994, White et al. 1995, and many other examples). Higher accuracies (~80–95%) are usually reported for land cover at Anderson level II or III (e.g., Dicks and Lo 1990, Congalton 1991, Bolstad and Lillesand 1992) or vegetation classes that are physiognomically or phenologically distinct (in different Formations or Groups, Federal Geographic Data Committee Vegetation Subcommittee 1996), or dominated by species with well-separated niches. An unpublished vegetation map of the study area (Shandley 1993), generated by stratifying the region into broad elevation classes and conducting unsupervised classification of TM data within each elevation zone, resulted in class accuracies comparable to or slightly higher than our results for 3/5 classes, substantially lower than ours for one class (scrub oak), and an inability to identify one class that we were able to model (southern mixed chaparral).

Thus, we conclude that tree-based predictive mapping did no worse, but perhaps only marginally better in this example, than other approaches to combining topographic and satellite data for identifying vegetation types (e.g., Franklin et al. 1986, Woodcock et al. 1994). This is probably due not only to the broad overlap in spectral reflectance patterns and site preferences among these vegetation types, but also to the

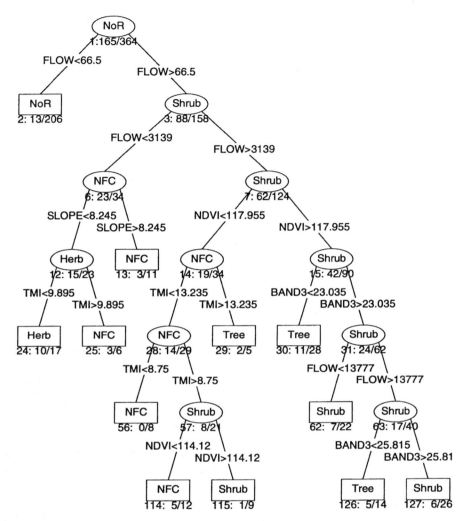

Figure 14.9. Classification tree model for riparian vegetation life-form classes (Table 14.3). See Figure 14.6 for description of tree diagram.

small data set that was available. D. M. Moore et al. (1991) found that they needed >1250 observations to develop a classification tree of 80% accuracy for 30 forest community types within a 100-km² area. CART holds promise, especially when used with improved DEMs (Chapter 2) and a vegetation sample based on randomly located plots from a large regional survey. The hierarchical interactions among environmental variables are easily expressed in the tree model, and variable selection can be examined and guided by known ecological principles (discussed below).

TABLE 14.4 Error Matrix Showing the Life-form Class Assignment of Descanso Riparian Plots (test sample) Measured in the Field Versus That Predicted by the Classification Tree Model

Field	Predicted Tree	Shrub	Herb	NFC	NoR	Total Plots	Producer's % Accuracy
Tree	**15**	5	4	4	0	28	54
Shrub	10	**25**	8	5	2	50	50
Herb	2	2	**2**	1	0	7	29
NFC	3	4	4	**6**	2	19	32
NoR	0	0	0	4	**100**	104	96
Total plots	30	36	18	20	104	208	
User's % accuracy	50	69	11	30	96		71

Note. Classes are defined in Table 14.3.

14.2.4.2 Riparian Models

Low accuracy for the riparian model of the 10 vegetation types probably resulted from the following factors: (1) ambiguities and errors in assigning the field plot data to vegetation classes, (2) differences between the training and test field data sets, described above, (3) inadequate numbers of observations per class for the classification tree method (<10 in 5 out of 10 classes), (4) coarse resolution of the remotely sensed and digital terrain data relative to the phenomena being mapped, and (5) (the "real" error) lack of a correlation between the topographic and vegetation variables. This case study serves as a cautionary tale, illustrating how important a large, well-designed field survey is and how dangerous it can be to use field data collected for some other purpose in this type of modeling. We strongly recommend that in the future, regional field surveys of riparian or other vegetation (Austin and Heyligers 1989, Bourgeron et al. 1995) be designed with explicit consideration for subsequent vegetation classification, modeling, and mapping. Both sample size and plot area relative to the resolution of terrain models and other mapped environmental variables are important considerations.

For example, the predictive map of riparian life-form (Figure 14.10) could be used as a basis for stratification for further field sampling (Strahler 1981, Michaelsen et al. 1994). There is some locational error in the modeled location of riparian vegetation caused by subpixel misregistration between the imagery and DEM (Figure 14.10). In addition, the resolution of the DEM results in the location of maximum upslope catchment area, for example, being offset by one pixel. In spite of these problems related to resolution, registration, and the errors in the DEM that was used, the resulting map is useful for defining the riparian vegetation zone when compared with other currently available data sources. For comparison, the streams represented on the DLG for the study area are shown. They match the location of the riparian zone more precisely in some cases, but many higher order streams are not mapped, and occasionally streams occur in the DLG where none exist on the landscape. Thus, predicting the location of riparian vegetation by buffering this stream coverage, the only other digital map data source currently available, would result in many errors of omission and some errors of commission.

14.3 CONCLUSIONS

Topographic variables are related to vegetation patterns in the southern California mountains. In spite of the resolution of and errors in data, the models for chaparral vegetation types and riparian life-forms resulted in maps with 52–71% overall accuracy and explained 29–96% of the variance in vegetation type or life-form composition. Substrate and disturbance (fire history) would be expected to explain additional variance in the distribution of vegetation classes, if accurate digital maps of these variables were available.

Elevation and potential solar insolation were the most important variables for explaining patterns of species composition in chaparral vegetation, although multispectral reflectance, slope, curvature, TWI, and topographic position explained additional geographic variance. Drainage basin position (upslope catchment area), slope, and spectral variables related to herbaceous (NDVI) and woody (band 3 red reflectance) vegetation amount were the most important variables in explaining riparian vegetation patterns.

Topographic variables derived from DEMs are used in almost every published example of predictive mapping of terrestrial vegetation at the regional scale, while at the continental to global scale biotic distributions are frequently predicted on the basis of climate variables (reviewed in Franklin 1995). Complex topographic variables, and terrain-based predictions of soil water and energy balance (outlined in Chapters 3 and 4), used in conjunction with interpolated climate variables, should have a stronger relationship to biotic distribution than the indirect gradients, elevation, slope, and aspect (Franklin 1998), although in either case there are limits to the predictive power of these habitat models.

This modeling approach is suitable for data exploration and predictive vegetation mapping and could also be used for testing inferences about biogeographical distributions. CRT has some interesting properties that can be useful for ecological studies and predictive mapping. Because the tree structure is hierarchical, it can capture hierarchical interactions that can occur in ecological data (explained in Michaelsen et al. 1987, 1994). Also, the proportional class membership of the observations at the terminal nodes can be used to assign a probability of that class occurring in each grid cell falling into that node. Therefore, a probabilistic map, or series of map layers, can be produced instead of a traditional thematic map of discrete class membership (see Goodchild 1994). The probabilistic maps may better characterize our uncertainty about static realizations of biotic distributions, and in our empirical modeling methods.

In the case studies, CRT did not produce predicted maps with accuracies substantially higher than those reported in the literature for other methods of combining satellite and terrain data. However, it provided an interpretable method for layering topographic and spectral variables in order to identify vegetation types. It took some of the guesswork out of a priori stratification based on elevation classes or geographically discrete "natural regions" (areas within which terrain–vegetation relationships are expected to remain constant, subjectively delineated by an "expert" based on reconnaissance field work (Franklin et al. 1986, Woodcock et al. 1994)). Subsequent research suggests that, in either case, achieving higher accuracy for large-area, fine-scale, categorically detailed vegetation maps requires some degree of photointerpretative or field work-based editing of the map labels generated by classification or modeling techniques (Franklin and Woodcock 1997).

CHAPTER 15

Automated Land Cover Mapping Using *Landsat* Thematic Mapper Images and Topographic Attributes

Jonathan M. Wheatley, John P. Wilson, Roland L. Redmond, Zhenkui Ma, and Jeff DiBenedetto

15.1 INTRODUCTION

Land managers must know the geographical distribution of landscape components such as vegetation, soils, and terrain to manage natural resources effectively. The need for natural resource mapping has long been acknowledged:

> A vegetation map not only serves as a record of what exists when it is made, but also as a starting point for the study of changes, whether natural or brought about by human activity. It serves to arouse public interest of a country in its wild vegetation, which ought to be recognized as a national possession not to be lightly destroyed or wasted, and it indicates the localities which are most suitable for the nature reserves which every country should have. The making of such maps should be part of the national stock-taking which is the duty of every modern community (Tansley and Chipp 1926).

The production of detailed resource maps was costly, time-consuming, and not easily standardized before the advent of digital remote sensing and computerized image analysis. Such maps were compiled from decades of fieldwork that could not be updated quickly. Geographic information systems (GIS) and remote sensing now provide tools for mapping large areas at spatial resolutions as fine as 30 m, and potentially allow these maps to be updated efficiently (Tueller 1989, Franklin 1995, Gerrard et al. 1997).

Landsat multispectral remote sensing attempts to identify vegetation based on its spectral reflectance, measured in the satellite's seven wavelength bands, ranging

Terrain Analysis: Principles and Applications, Edited by John P. Wilson and John C. Gallant.
ISBN 0-471-32188-5   © 2000 John Wiley & Sons, Inc.

from blue visible light to far-infrared thermal wavelengths (Niemann 1993). Generally, green leaves absorb light in the 0.4- to 0.7-µm wavelength range, but make an abrupt transition to reflectance at 0.7 µm, scattering light effectively between 0.7 and 1.3 µm (Gates et al. 1965, Ripple 1985). In the semiarid grassland and badland cover types that are common in the Little Missouri National Grassland (LMNG) of western North Dakota, the observed reflectance may result from a combination of live vegetation, dead vegetation, and bare soil. In addition, the spectral signature may vary due to differing angles of illumination, the health of vegetation (affected by precipitation and grazing), and the seasonal life cycle of vegetation (Tueller 1989). The ability of *Landsat* Thematic Mapper (TM) to distinguish vegetation types is often limited by the spectral and spatial resolution of the detector (Niemann 1993).

Two general types of classification are used: unsupervised and supervised. In the former, the data are clustered into arbitrary categories, but in the latter, ground-truth is used as training data to divide the multidimensional data space into categories based on the observed properties of the training data set (Jensen 1996). Provided that adequate, unbiased training data are available, supervised methods are far superior to the arbitrary classes generated with unsupervised methods.

Numerous statistical clustering methods have been used to group multispectral data into land cover classes, but ultimately, the quantity and quality of ground-truth data used for the supervised classification of vegetation classes may be more important than the classification method used (Congalton 1991). In one such study, Zhuang et al. (1995) compared minimum distance, maximum likelihood, and neural network classification accuracy for six land cover classes in a mixed cropland, rangeland, and broadleaf forest site in Indiana. The three methods produced similar accuracies ($\pm5\%$) for each of the land cover classes, with the exception of the bare soil class, where the neural network classifier was 10% more accurate than the other two methods. As with many remote sensing applications, accuracies of 85–95% could be achieved only by lumping land cover types into very broad categories: water, bare soil, one forest class, one grassland class, and two crop classes (Zhuang et al. 1995).

Most automated remote sensing classification methods are not able to distinguish ecologically distinct vegetation types of interest to the land manager. Anderson et al. (1976) defined a hierarchy of land cover types: Land managers are typically interested in Anderson level III vegetation categories (the species level), but automated classification accuracies for these classes are generally quite low, ≤65–75% (Skidmore and Turner 1988). This result occurs because there is no simple correlation between vegetation type and spectral signature. Considerable confusion arises because dissimilar vegetation types may have similar spectral signatures. Furthermore, a given vegetation type may have a seasonally variable signature. Grasslands, in particular, show rapid responses to varying precipitation amounts (Pickup et al. 1994). Regardless of future improvements in remote sensing technology, such as imaging spectrometers, which produce a reflectance spectrum for each pixel in the image, there will always be some inherent confusion between spectral reflectance and land cover (Wilkinson 1996).

An alternative approach to identifying vegetation by species is to characterize it by ecological parameters, such as leaf area index (LAI) or above-ground biomass, that are less arbitrarily related to the plant's spectral reflectance than genus and species (Anderson et al. 1993, Paruelo and Lauenroth 1995). However, establishing a universal relationship between remote sensing spectral indices, such as the Normalized Difference Vegetation Index (NDVI), and ecological parameters has sometimes proved difficult. In Great Plains grasslands at Mandan, ND, Aase et al. (1987) found that the relationship between NDVI and LAI varied according to the intensity of cattle grazing, making it impossible to translate NDVI into LAI without knowing the grazing intensity. Friedl et al. (1995) reported similar problems in the Konza Prairie, KS. The rapid temporal variation of the spectral signature and interplay of forces, such as drought and grazing, would seem to provide special challenges for those interested in automated mapping of rangeland cover types over large areas.

Topographic attributes are increasingly utilized in remote sensing classifications. In one such study, Niemann (1993) used three topographic attributes (elevation, slope gradient, aspect) and location (an easting to measure distance along a gradient between the interior and coastal biogeoclimatological zones) with *Landsat* TM information to identify eight conifer classes in the old-growth forest of southwestern British Columbia. Only three of the eight classes (which varied in terms of species composition, age class, and crown closure) yielded accuracy results > 50%. The low overall accuracy (~45%) was attributed to two factors: (1) Different conifer stands displayed similar spectral reflectances, and (2) many of the age differences were related to past land use practices or disturbance history that was not well correlated with site characteristics. In another study, Joria and Jorgenson (1996) used ancillary data layers of elevation, slope, landform type, solar radiation, and riparian zones in a postclassification rule-based model to classify 14 arctic tundra cover types. Classification accuracy remained at ~50% when these topographic attributes were used. One potential advantage of the statistical clustering method used in the first application is that it does not involve assumptions about the relationship between vegetation and terrain, whereas rule-based classifiers may involve subjective rules and broad assumptions about terrain–vegetation correlations. Often, rule-based classifiers are specific to a particular area, whereas general statistical clustering is less site-dependent.

We explored whether topographic attributes could be used to improve the accuracy of remote sensing land cover maps in the LMNG. We compared the accuracy of vegetation maps prepared with an existing remote sensing classification method and varying numbers of topographic attributes to determine whether application of these methods is worthwhile over the entire Little Missouri region. Topographic attributes were examined because topographic data were available (unlike soil and climate data) at an appropriate scale in this study area. Modern soil surveys are available for less than 50% of the study area, and where 1:24,000 scale soil maps are available, they do not resolve individual wooded valleys and small drainages, but lump together much of the highly dissected badlands as "mixed badlands complex" (Aziz 1989). The climate of western North Dakota does not exhibit the dramatic elevation-induced variations seen in more mountainous terrain: The climate within the study area is quite homogenous (Owenby and Ezell 1992) and is therefore of little use in differentiating vegetation types. We were particularly interested in two vegetation-related parameters that may be improvements over the basic parameters of elevation, slope, and aspect: (1) the spatial pattern of soil moisture as affected by slope and upslope contributing area, and (2) solar radiation as affected by topographic shading from nearby features. Both indices mimic physical processes not addressed by slope and aspect alone. Thus, one might expect these secondary topographic attributes to correlate with vegetation better than slope and aspect. We also examined the sensitivity of predicted land cover maps to the level of spatial aggregation and types of source data that were utilized.

15.2 DESCRIPTION OF STUDY AREA

Our study area was a 1700-km^2 portion of the LMNG (Figure 15.1) delineated by five 7.5′ USGS quadrangles (i.e., the Tracy Mountain, Cliffs Plateau, Deep Creek North, Spring Creek, and Juniper Spur quads). This area was selected for four reasons. First, it has relatively high vertical relief (farther east the terrain is very flat, the USGS digital data errors become more troublesome, and the effects of terrain on vegetation are less dramatic). Second, these five quadrangles contained the highest density of ground-truth data. Third, the area included all the major landscape types of the Little Missouri (i.e., rolling uplands, wooded draws, badlands, river floodplains, and terraces). Fourth, they contained the Third Creek catchment that drains into the Little Missouri River and we needed to be able to delineate upslope contributing areas to calculate the topographic wetness index.

The erosional processes that created the North Dakota badlands began more than 600,000 years ago, when the continental ice sheet displaced the Little Missouri River southward into a shorter, steeper channel. The Little Missouri and its tributaries continue to cut into the surrounding plain. The eastern boundary of this erosion bisects the study area, dividing it into two dramatically different physiographic regions: the gently rolling Missouri Slope Upland to the east and the Little Missouri Badlands to the west (Bluemle 1975).

The bedrock is entirely sedimentary, consisting of Upper Cretaceous and Tertiary shale strata. The Bullion Creek Formation, which underlies most of the study area, consists of alternating layers of sandstone, shale, and lignite. Flat-topped buttes (monadnocks) of the more resistant sandstone rise 150 m above the surrounding peneplain. Total vertical relief in the study area is less than 250 m. The Little Missouri has incised a channel 100–200 m below the surrounding terrain. Most of the wooded draws have V-shaped cross sections, but the larger channels have floodplains that are several hundred meters wide, and the Little Missouri River itself has a floodplain up to one kilometer wide. Pleistocene river terraces, containing material eroded from the Rocky Mountains, cover areas up to 10 km^2 near the main channel of the Little Missouri River. A distinctive feature of this area is the presence of naturally fired clay (clinker, locally named "scoria"), which is more erosion resistant than unbaked materials and forms steep-sided knobs 20–40 m high (Bluemle 1975). In places these knobs, together with depressions caused by the collapse of burned lignite beds, form large areas of hummocky terrain. This highly complex terrain is

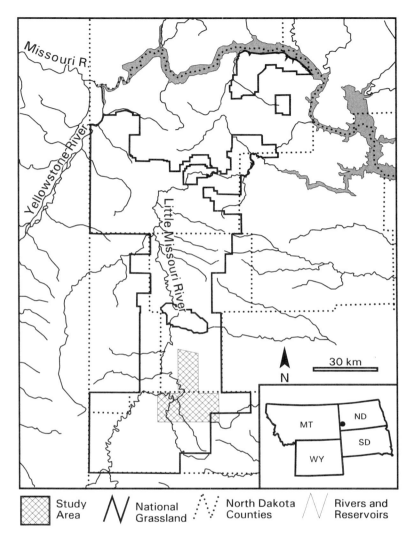

Figure 15.1. Study area map.

barely resolved at the 30-m cell size of the *Landsat* TM images and USGS digital elevation models (DEMs).

Soil is largely absent in the badlands because the steep slopes are continually eroding. Where the parent materials are sufficiently weathered to be considered soil, Inceptisols and Entisols are common, such as the Cherry (fine-silty, mixed, frigid Typic Ustochrepts) and Cabbart (loamy, mixed (calcareous), frigid, shallow Ustic Torriorthents) series, two of the more common soils in the wooded draws (Thompson 1978). The floodplains of the larger creeks are also classified as Entisols, though they may be deep soils, such as the Korchea series (fine-loamy, mixed (calcareous), frigid Mollic Ustifluvents). The rolling uplands, which are mostly grassland and cropland, contain moderately deep soils such as Entic and Typic Haploborolls.

The southern half of the study area has been mapped at a scale of 1:24,000 as part of the Slope County soil survey (Thompson 1978). Most wooded draws are not distinguished from their surroundings on these soil survey maps, and many of these features are mapped as the Badlands–Cabbart complex. Flatter upland areas and wide floodplains within the badlands are delineated, and cropland on the rolling uplands is mapped at greater detail. The computerized North Dakota State Soil Geographic Database (STATSGO, 1:250,000 scale) soil maps cover the remainder of the study area and show even less detail, but they do distinguish broad areas of badlands from the surrounding rolling uplands (United States Department of Agriculture–Soil Conservation Service 1993).

North Dakota is located near the geographic center of North America and has a semiarid midlatitude steppe climate (BSk in the Köppen classification). Cold, dry winters (−15°C mean monthly average temperature) alternate with warm summers (21°C mean monthly average temperature) that average 120 frost-free days (Owenby and Ezell 1992). Precipitation is greatest in the early part of the growing season: 50% of the 400 mm annual precipitation occurs in April, May, and June, compared to <30 mm in the three winter months (Owenby and Ezell 1992). The temporal variation of precipitation from year to year at a given station is far greater than the spatial variation across North Dakota. At Dickinson Experimental Farm, just east of the LMNG, annual precipitation has ranged from 170 to 800 mm over the period 1904–1993 (Hydrosphere 1993). The rainfall distribution is skewed, with drier than average years occurring more frequently than wetter years. The moisture balance in the growing season shifts dramatically from year to year. At Williston Experimental Farm, the nearest station for which data are complete, a dry year such as 1988 yielded 9 mm precipitation and 290 mm observed pan evaporation in the month of July. This moisture balance is reversed in wet years such as 1993 when 208 mm of precipitation and 136 mm observed evaporation were reported (NOAA 1993a). The wet weather was also accompanied by much cooler temperatures: In July 1993 the mean daily maximum was 9.3°C cooler than in July 1988.

The LMNG is also a transition zone between western and eastern plant species (Rudd 1951). The region is at the western limit of the range of eastern hardwoods, such as green ash (*Fraxinus pennsylvanica*) and bur oak (*Quercus macrocarpa*), and at the eastern limit of western species, such as Rocky Mountain juniper (*Juniperus scopulorum*) and ponderosa Pine (*Pinus ponderosa*) (Little 1971). Furthermore, the area is the southernmost extent of some boreal forest trees, such as balsam poplar (*Populus balsamifera*) and paper birch (*Betula papyfira*). Desert plants with ranges centered on the Great Basin, such as cactus (*Opuntia fragilis*), yucca (*Yucca glauca*), saltbush (*Atriplex* sp.), skunkbrush (*Sacrobatus vermiculatus*), saltgrass (*Distichlis stricta*), and sagebrush (*Artemesia tridentata*) are also widespread in the Little Missouri region (Rudd 1951). However, they may not grow in patches large enough to be detected by remote sensing.

The climate is sufficiently arid and hot in the growing season that most tree species are unable to grow on south-facing slopes, with the possible exception of ponderosa pine (*Pinus ponderosa*). Broadleaf trees such as green ash (*Fraxinus pennsylvanica*) and shrubs such as snowberry (*Symphoricarpos occidentalis*) are largely confined to woody draws near watercourses (Table 15.1). Brown (1993) mapped species composition of grasses in the Great Plains and found western wheatgrass (*Agropyron smithii*) and needle-and-thread (*Stipa comata*) to be the most common "cool-season" grasses (C3 photosynthesis path), and blue grama (*Bouteloua gracilis*) is the only "warm-season" (C4 photosynthesis path) grass in western North Dakota. Crested wheatgrass (*Agropyron desertorum*), an introduced bunchgrass from the Russian steppe, was abundantly planted to prevent erosion on former croplands abandoned during the Dust Bowl.

TABLE 15.1 Land Cover Type Codes and Descriptions (as delineated by DiBenedetto)

2100	**Cropland, cultivated pasture, and hay**
3110	**Low grass/low cover:** Blue grama (*Bouteloua gracilis*), needle-and-thread (*Stipa comata*), and threadleaf sedge (*Carex filifolia*) with low biomass productivity (less than 800 kg/ha), low ground cover, and extensive bare ground. Occurs on ridgetops and clay outwash areas below buttes.
3111	**Low grass/moderate cover:** Needle-and-thread, western wheatgrass (*Agropyron smithii*), and blue grama with medium biomass productivity, moderate ground cover, and less bare ground than cover class 3110. Typically occurs on hillslopes.
3201	**Mesic upland shrub:** Snowberry (*Symphoricarpos occidentalis*), buffaloberry (*Sheperdia argentea*), chokecherry (*Prunus virginiana*), and creeping juniper (*Juniporus horizontalis*) with occasional young green ash (*Fraxinus pennsylvanica*) trees. Occurs on north-facing slopes and in depressions and draws.
3309	**Sagebrush:** Silver sagebrush (*Artemisia americana*) with snowberry as a co-dominant shrub in some locations and a western wheatgrass/blue grama understory. Occurs mainly in valley bottoms and on terraces adjacent to stream channels.
4102	**Upland broadleaf forest:** Dominated by green ash and found mostly on steep, north-facing slopes associated with badlands or deeply incised channels and draws.
4206	**Ponderosa pine forest:** Dominated by ponderosa pine (*Pinus ponderosa*) with occasional Rocky Mountain juniper (*Juniperus scopulorum*), green ash, snowberry, and chokecherry. Occurs on shallow soils associated with ridges and hillslopes.
4214	**Rocky Mountain juniper forest:** Dominated by Rocky Mountain juniper with occasional green ash. Occurs on very steep, north-facing badland slopes.
5000	**Water**
6102	**Riparian broadleaf forest:** Dominated by green ash. Occurs along river bottoms and narrow draws.
6201	**Riparian grass and forb:** Western wheatgrass, baltic rush (*Juncus balticus*), and a mixture of forbs. Occurs along channel bottoms and draws.
6202	**Riparian shrub:** Snowberry, buffaloberry, and chokecherry with occasional young ash trees. Occurs in channel bottoms, valley bottoms, and narrow draws.
7601	**Badland:** Scattered patches of grass and big sagebrush (*Artemisia tridentata wyomingensis*) shadscale (*Atriplex confertifolia*), green rabbitbrush (*Chrysothamnus visidiflorus*), and greasewood (*Sarcobatus vermiculatus*). Occurs on steep, south-facing badland slopes and in outwash areas at the base of buttes and slopes.

15.3   METHODS AND DATA SOURCES

15.3.1   *Landsat* TM Image Classification

A cloud-free *Landsat* Thematic Mapper image (path 34, row 28) dated 5 July 1989 was used because no more recent cloud-free images were available for the growing season at the start of this study. This image was terrain-corrected and georeferenced to an Albers conical equal area projection by Hughes STX Corporation. We did not evaluate the spatial accuracy of this terrain correction, but it appeared to register very well with the other digital data for the area, including roads and streams from U. S. Forest Service 1:24,000 scale cartographic feature files.

Land cover was classified using a multistep method developed for mapping large (10,000–100,000 km^2) areas (Ma and Redmond 1993, Ma et al. 2000). The first step used bands 3, 4, and 5 for each 30-m *Landsat* TM pixel and applied an unsupervised clustering algorithm, which is closely related to the widely used color-quantization scheme for approximating three-color composite images in a single 8-bit color image (Heckert 1982, Ma et al. 2000). The resulting vegetation classes were color-coded to resemble their appearance in a red–green–blue (RGB) color composite of the three bands, allowing easy visual inspection of the classification, and comparisons with the original image.

Image pixels were spatially grouped into three ecologically relevant patch sizes to eliminate the visual confusion of "salt-and-pepper" pixels during the second step. Three minimum map unit thresholds of 0.36 ha (5 cells), 0.81 ha (9 cells), and 2.0 ha (23 cells) were used to evaluate the trade-offs between map clarity and detail. Care must be taken to avoid removing ecologically important features such as woody draws and riparian corridors, which are prevalent in the LMNG. Vegetation classes that did not occur in patches as large as these minimum map units were eliminated, resulting in fewer classes in the aggregated image. An additional advantage of aggregation is that it greatly speeds subsequent computation and reduces data storage requirements because later classifications are performed on the raster polygons rather than individual image pixels (Ma and Redmond 1996). The remaining four *Landsat* TM bands (1, 2, 6, and 7) and selected topographic attributes were then averaged within each raster polygon to obtain mean attribute values for each patch.

The third step consists of a supervised nonparametric classification that assigns a cover type label to each patch based on the nearest distance in multidimensional parameter space to one of the ground reference "training" plots. In this study, 9, 10, or 11 data layers were used, depending on how many topographic attributes were included in the classification. The advantage of this "nearest member of group" classification is that it does not depend on any assumptions about the statistical distribution of parameters. Several cumulative topographic attributes, such as upslope contributing area and the two topographic wetness indices, were not normally distributed.

A fourth, optional step provides a postclassification sorting of riparian and upland land cover classes with similar spectral signatures using a stream buffer defined by an elevation threshold. The stream channels can be input either from USGS digital line graph (DLG) hydrography data or from drainage channels computed by terrain analysis software such as TAPES-G, and the buffer is calculated with the same DEM data used to calculate other topographic attributes (see below). The supervised, nonparametric classification (third step) is repeated using a smaller minimum map unit inside the riparian buffer in some applications. In this study, we used an elevation difference of 5 m to define the width of the riparian zone and we examined whether the reclassification of land cover types to the appropriate riparian or upland designation (based on their occurrence inside or outside this zone) improved overall performance.

15.3.2 Digital Elevation Models

The five USGS 7.5' 30-m DEMs (Figure 15.1) were joined and reprojected in ARC/INFO from the original Universal Transverse Mercator (UTM) projection to the same Albers projection as the *Landsat* TM image. The digital elevation data conformed to USGS map accuracy level one that allows a mean elevation error of 7 m in a 7.5' quadrangle (United States Geological Survey 1993). However, the systematic and spatially patterned nature of the error can sometimes be confused with real terrain features in an area such as the LMNG where the total vertical relief is as little as 20 m in some quadrangles. Several examples of the types of error that can occur in USGS elevation data are illustrated in Figure 15.2, which shows a 16×22-km area located at the southern boundary of the study area. Parts of six USGS 7.5' map quadrangles are included. Three types of error are visible. First, there is a quasi-periodic "ripple" of 250- to 300-m wavelength and up to 7-m amplitude, oriented roughly, but not exactly, east–west. Second, there is a quasi-random lumpiness of approximately 20-ha patch size, which varied in amplitude from quadrangle to quadrangle. Third, there are linear discontinuities of up to 15 m in elevation, running north–south and east–west, which often, but not always, agree with quadrangle boundaries. The amplitude of all three types of error varied substantially from quadrangle to quadrangle; the orientation and spatial frequency varies less dramatically. Over water bodies, such as reservoirs, the errors have apparently been removed by the USGS, as they were not present.

The ripple originated from USGS manual profiling of photogrammetric stereomodels (U. S. Geological Survey 1993) and is actually modulated by terrain. Human operators made negative errors (resulting in lower elevations) when scanning in an uphill direction and positive errors when scanning downhill (Band 1993b), so that the phase of the ripple is altered by terrain. This error pattern is therefore difficult to remove using standard de-striping algorithms such as Fourier filtering (Simpson et al. 1995). If the ripple were a perfect sinusoid superimposed on terrain, it could simply be subtracted from the DEM, but the variable amplitude, wavelength, and phase make this impossible. A disadvantage of the commonly used rectangular box-average method of stripe removal (Brown and Bara 1994) is that it removes real terrain features, such as some east–west trending woody draws in the LMNG (see Chapters 2 and 5 for more detailed discussions of DEM sources and errors). The 7.5' USGS 30-m DEMs are by far the best elevation data available for our study area, so precise quantification and correction of errors, or the side effects of ripple removal methods, cannot be undertaken, because there are no reference data.

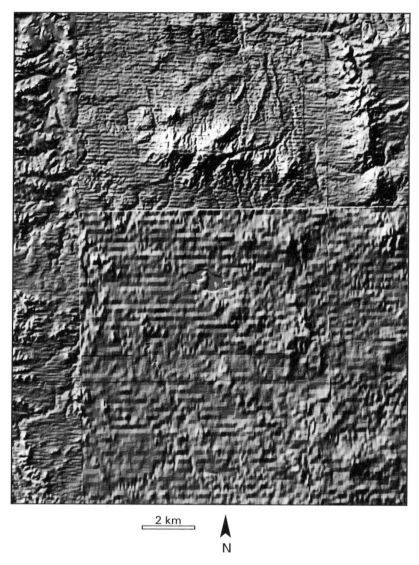

2 km

N

Figure 15.2. Shaded elevation map showing DEM errors in part of study area.

15.3.3 Terrain Analysis

The standard suite of primary topographic attributes described in Chapter 3 were calculated with TAPES-G using the following options. First, depressions present in the
raw DEM caused by elevation errors such as the ripple pattern and/or the terrain
being insufficiently resolved at the 30-m cell size of the DEM were removed. The
narrow wooded draws in the LMNG are only a few cells wide, so the bottom of the

channel is not always sampled, producing a chain of "spurious pits" along the drainage path. Second, the FRho8 multiple-flow direction method was used to simulate flow across upland areas with upslope contributing areas of less than 20,000 m^2. Third, the Rho8 single-flow direction algorithm was used to simulate flow across "channel" cells with upslope contributing areas of 20,000 m^2 or more.

This maximum cross-grading area threshold of 20,000 m^2 was chosen because it provided the best match with the blue stream lines recorded on 7.5' USGS quadrangles. These stream lines were not totally consistent with catchment area since they were derived from photo interpretation. Another approach is to adjust the TAPES-G threshold to give channels that correspond to the pattern of woody draw vegetation seen in the *Landsat* TM image. This yielded a lower catchment area threshold, 10,000 m^2, because wooded draw vegetation occurred in drainages smaller than those delineated with blue stream lines on topographic maps. One disadvantage of the lower threshold would have been the creation of some spurious channels caused by DEM errors. Typically, these spurious channels trend east–west in the orientation of the "ripple" error, but sometimes north–south quadrangle boundaries produce errors as well (Figure 15.2).

A quasi-dynamic topographic wetness index was calculated with DYNWET (Barling et al. 1994; Chapter 4) and the slope and upslope contributing area outputs from TAPES-G. DYNWET also requires average soil depth, saturated hydraulic conductivity, effective (drainable) porosity, and average drainage time between rainfall events as inputs. A mean soil depth of 0.5 m and a modal soil texture of silt loam were estimated from the Slope County soil survey (Thompson 1978). A saturated hydraulic conductivity of 15 mm/h (Rawls and Brakensiek 1989) and drainable porosity equal to 25% of soil volume (Ratliff et al. 1983) are representative for this soil texture. In reality, soil properties vary across the landscape, but the version of DYNWET implemented for this study assumed that they could be approximated by these spatial averages. A mean drainage time of 10 days was chosen to reflect the typical interval between major precipitation events in the LMNG.

SRAD (Wilson and Gallant 2001; Chapter 4) calculates short-wave (visible) and long-wave (infrared) radiation budgets at each cell in a DEM. For this study, only the incoming short-wave solar radiation was used, because the other components of the radiation budget depend on vegetative cover and to use them in the determination of remote sensing vegetation type would invoke circular reasoning. SRAD uses a three-component approximation of the sky-brightness function. It computes the sum of direct-beam solar radiation, a circumsolar diffuse component scattered from within a few degrees of the solar direction (both affected by topographic shading), and the wider-angle diffuse radiation reflected from blue sky, clouds, and surrounding terrain. To compute this diffuse component, SRAD requires an input parameter, β, that represents the mean transmission coefficient for sunlight passing through clouds. The β parameter was calculated from 1961–1990 mean monthly solar radiation and percent-of-possible-sunshine measured at the nearest station to the LMNG in Bismarck, ND (NOAA 1993b). Incident radiation was computed at four 90-day intervals throughout the year, then added to obtain the mean annual short-wave solar radiation incident on the land surface.

15.3.4 Ground-Truth Data

Measurements of dominant overstory vegetation were available from plot data collected by the U. S. Forest Service between 1987 and 1994 at 97 field sites in the five-quadrangle study area. Each site was associated with a vegetation patch visible on the *Landsat* TM image, and each plot was assigned a cover type label from a hierarchical system based on Anderson et al. (1976) (Table 15.1). To increase the sample size for land cover types, such as riparian broadleaf forest, that were poorly represented in the field data, an additional 76 ground-truth sites were obtained from 1:24,000-scale air photos dated 17 August 1981. Some land cover types, such as riparian shrub, sagebrush, and juniper forest, were nevertheless still poorly represented (five plots or less) in the combined data set because they could not be identified on aerial photographs.

15.3.5 Performance Evaluation

We took the training data set (173 observations) and removed one plot at a time, using the remaining data to classify each one. If the plot held back was correctly classified by the others, then it was counted along the diagonal of the error matrix; if not, it was counted in the row or column where it was "confused." Steele et al. (1998) have since proposed a new evaluation method based on a true bootstrap process; however, there is very little difference between the error matrices and percent agreements reported here and those that would be obtained with this new method. The accuracy of the overall classification and individual land cover categories were both of interest. In addition, a pixel-by-pixel comparison of the land cover maps can be used to estimate a percent agreement between any pair of maps, indicating how much the addition of topographic attributes or a change in the size of the minimum map unit affects the maps. The percent agreement was computed with the CON function

TABLE 15.2   Summary Statistics for Topographic Attributes

Topographic Attribute	Study Area				Ground-Truth Sites			
	Min	Max	Mean	SD	Min	Max	Mean	SD
Elevation (m)	695	1024	825	46	731	899	821	40
Slope (%)	0.0	98.9	9.2	8.2	0.0	47.7	10.3	9.2
Upslope contributing area (000 m^2)	0.9	300	23.7	64.7	0.9	300	57.5	98.8
Steady-state topographic wetness index	4.03	24.50	8.00	2.28	5.02	16.96	8.34	2.13
Quasi-dynamic topographic wetness index	1.64	2.67	2.54	0.17	2.17	2.67	2.53	0.14
Incident short-wave solar radiation (W/m^2)	77	196	153	10.6	125	185	152	10.5

in ARC/INFO, which can be applied to either the vegetation maps or stream buffers in raster (image) format.

15.4 RESULTS

15.4.1 Terrain Attribute Maps

Terrain parameters were computed for an area somewhat larger than the five-quadrangle remote sensing area (Figure 15.1) to insure that cumulative hydrologic parameters were complete. The only parts of the TAPES-G upslope contributing area map that do not have complete catchment coverage are the one-cell-wide drainage channels of the Little Missouri River and two of its tributaries. USGS 30-m DEMs were not available for the headwaters of these catchments. However, the channels had very large values of upslope contributing area by the time they entered the study area, and the channels were not resolved in the remote sensing image, so this had a negligible effect on final classification accuracy. The upslope contributing areas computed at the locations of each of the ground-truth sites were based on complete elevation data.

Output grids from TAPES-G, DYNWET, and SRAD are shown in Figures 15.3–15.8. The differences between the gently rolling upland areas at the eastern edge of the study area and the rugged wooded draws and badlands in the western two-thirds of the study area are evident in most of these maps. The abrupt transition between the rolling uplands in the east and the highly dissected Little Missouri drainage to the west is particularly prominent in the slope map (Figure 15.3). Flat areas, shown in dark tones, such as floodplains, river terraces, and depressions in rolling upland, are clearly visible in the slope map (Figure 15.3). They are also visible as areas of large upslope contributing area (Figure 15.4) and as wet areas (light tones) in both the steady-state and quasi-dynamic topographic wetness index maps (Figures 15.5 and 15.6). Ridge lines appear as dark areas (low upslope contributing area) and valleys, floodplains, and depressions appear as light tones (high upslope contributing area) in the upslope contributing area map (Figure 15.4). The distribution of both topographic wetness indices follows a pattern of high wetness in flatter terrain and lower wetness on steeper slopes (Figures 15.3, 15.5, and 15.6).

The upslope contributing area and topographic wetness maps (Figures 15.4–15.6) show numerous spurious linear drainage features trending east–west and confirm that elevation data quality was visibly lower in the northern and eastern USGS quadrangles. Other discontinuities at quadrangle boundaries are visible in the left center part of Figure 15.4, where the flat river terraces and floodplains are truncated by an elevation jump across a quadrangle boundary.

The aspect and short-wave solar radiation maps (Figures 15.7, 15.8) reveal similar patterns; however, topographic shading on north-facing slopes produced a more complex spatial pattern in the short-wave solar radiation map. The white areas in the aspect map show areas with zero slopes for which aspects could not be calculated. In the solar radiation map (Figure 15.8), north-facing slopes receiving 80–100 W/m^2

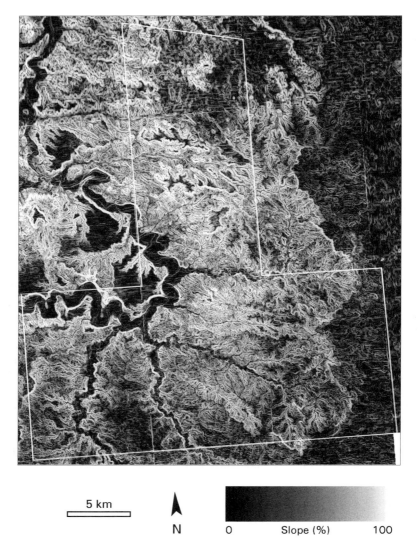

Figure 15.3. Slope map calculated with TAPES-G.

average annual solar radiation are shown in dark tones and southern exposures receiving 150–190 W/m² are shown in lighter tones. More variation in radiation intensity is seen in the badlands (center of map) compared with the gentler rolling terrain to the east. Figure 15.8 also shows a high-frequency ripple in elevation over part of the study area, plus a more pronounced low-frequency ripple in the northern and eastern quadrangles outside the study area.

The magnitude and range of values calculated for each of the topographic attributes mapped in Figures 15.3–15.8 are summarized in Table 15.2. These statistics por-

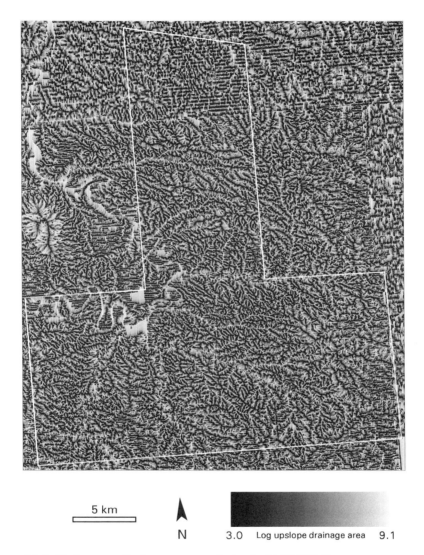

5 km

N

3.0   Log upslope drainage area   9.1

Figure 15.4. Upslope contributing area map calculated with TAPES-G.

tray a relatively flat study area (9.2% average slope, 329 m total relief). Topographic wetness (8.0 mean, 2.3 standard deviation and 2.5 mean, 0.2 standard deviation for steady-state and quasi-dynamic topographic wetness index, respectively) and incident short-wave solar radiation (153 and 11 W/m^2 mean and standard deviation, respectively) varied little in upland areas (i.e., in those cells not traversed by channels). The corresponding statistics for the 170 ground-truth plots showed that these locations omitted those cells containing the minimum and maximum values reported for most topographic attributes. These sites also exhibited slightly steeper slopes,

5 km

N   4.0   Wetness Index   24.5

Figure 15.5. Steady-state topographic wetness map calculated with DYNWET.

larger upslope contributing areas, and higher steady-state topographic wetness indices compared to the five quadrangle remote sensing study area as a whole.

15.4.2 Land Cover Classification Without Topographic Attributes

The accuracy of the initial land cover classification increased slightly (from 54 to 57%) when the minimum map unit was increased from 0.4 to 0.8 ha and was unaltered when the minimum map unit was increased again to 2.0 ha (Table 15.3). Six

5 km

N   1.64   Wetness Index   2.67

Figure 15.6. Quasi-dynamic topographic wetness map calculated with DYNWET.

additional ground-truth sites were correctly classified in each instance. Correct identifications fall along the diagonal in Table 15.3 and incorrect classifications fall elsewhere. Increasing the minimum map unit produced minor changes in the accuracy of
individual land cover classifications as follows. First, the number of mesic upland
shrub (3201), Rocky Mountain juniper (4214), and riparian grass/forb (6201) sites
correctly identified increased as map unit size increased. Second, the number of
upland broadleaf forest (4102), water (5000), and badland (7601) sites correctly
identified decreased as minimum map unit size increased. Third, the number of crop-

Figure 15.7. Aspect map calculated with TAPES-G.

land (2100), low grass (3110, 3111), and riparian broadleaf forest (6102) sites correctly identified varied as minimum map unit size increased.

However, the number of sites affected was small (± 3 sites) for all but two of the classes (mesic upland shrub and riparian broadleaf forest). The errors reveal numerous trends as follows. Several cropland sites were classified as riparian broadleaf forest, shrub, or grass/forb. Numerous low grass/low productivity sites were assigned to the low grass/medium productivity category and vice versa. Several mesic upland shrub sites were classified as riparian broadleaf forest. Upland

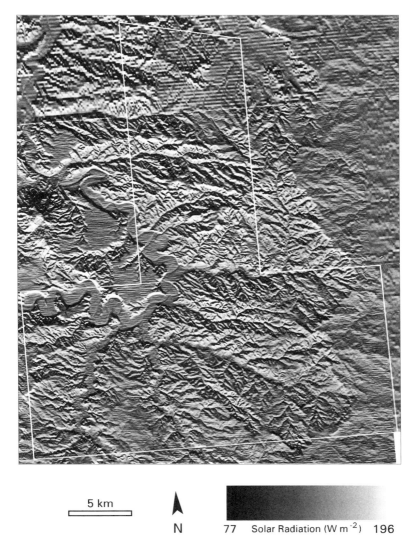

5 km

N

77   Solar Radiation (W m^{-2})   196

Figure 15.8. Mean annual short-wave solar radiation map calculated with SRAD.

broadleaf forest was often classified as riparian broadleaf forest. Riparian broadleaf forest was often assigned to the cropland, upland broadleaf forest, or one of the other riparian land cover classes (Table 15.3). Several riparian grass/forb sites were assigned to cropland or one of the other riparian land cover classes. Overall, these results suggest that there may be no disadvantage to using the larger mapping unit in the study area, and this level of aggregation was used to evaluate the results of adding topographic attributes to the final two steps in the classification method as reported below.

TABLE 15.3 Error Matrices for Land Cover Classifications Prepared With Initial Classification Method and Three Minimum Map Units

A. 0.4-ha Minimum Map Unit[a]

Field Classification	Remotely Sensed Classification													Total
	2100	3100	3111	3201	3309	4102	4206	4214	5000	6102	6201	6202	7601	
2100	12 (67%)					1					2	2	1	18
3110		16 (67%)	8											24
3111		13	7 (33%)											21
3201			1	0 (0%)	2	1	2			3	1			10
3309			1	1	3 (60%)									5
4102						3 (25%)	1			7		1		12
4206				1			7 (88%)							8
4214			1			1		3 (60%)						5
5000									3 (100%)					3
6102						8				24 (60%)	4	4		40
6201	1			2						3	2 (22%)	1		9
6202	1					1				3		0 (0%)		5
7601													13 (100%)	13
Total	14	29	18	5	5	15	10	3	3	40	9	8	14	173

B. 0.8-ha Minimum Map Unit[b]

Field Classification	Remotely Sensed Classification													Total
	2100	3100	3111	3201	3309	4102	4206	4214	5000	6102	6201	6202	7601	
2100	13 (72%)		1			1					1	1	1	18
3110		17 (71%)	6								1			24
3111		11	9 (43%)								1			21
3201				2 (20%)	1		1			5	1			10
3309				1	4 (80%)									5
4102	1					3 (25%)	1			6	1			12
4206					1		7 (88%)							8
4214			1					4 (80%)						5
5000									3 (100%)					3
6102			1	3	1	10				19 (48%)	4	2		40
6201			1							3	4 (44%)	1		9
6202	1									1	1	2 (40%)		5
7601	1												12 (92%)	13
Total	16	28	19	6	7	14	9	4	3	34	14	6	13	173

(continued)

375

Table 15.3 (*Continued*)
C. 2.0-ha Minimum Map Unit [c]

Field Classification	Remotely Sensed Classification													Total
	2100	3100	3111	3201	3309	4102	4206	4214	5000	6102	6201	6202	7601	
2100	11 (61%)		1			1				3	1		1	18
3110		15 (63%)	4						1	2			2	24
3111		11	6 (29%)	1						1			2	21
3201		1	1	5 (50%)		1				2				10
3309			1		3 (60%)						1			5
4102				3		2 (17%)			1	5		1		12
4206				1			7 (88%)							8
4214								5 (100%)						5
5000						1	1		1 (33%)					3
6102	3	1	1	1		4				26 (65%)	2	2		40
6201	2									2	5 (56%)			9
6202	1									1	1	2 (40%)		5
7601	1		1										11 (85%)	13
Total	18	28	15	11	3	9	8	5	3	42	10	5	16	173

[a] Overall agreement = 54% (numbers in parentheses indicate percent correct in each category).
[b] Overall agreement = 57%.
[c] Overall agreement = 57%.
Note. See Table 15.1 for descriptions of individual land cover classes.

15.4.3 Adding Topographic Attributes

Table 15.4 shows the effects of adding quasi-dynamic topographic wetness index and incident short-wave solar radiation to the land cover classification based on a 2.0-ha minimum map unit. Overall and individual class accuracies changed very little (± 1–3 sites in each instance), and there was little change in the errors of omission and comission. The explanation for this result is evident in Table 15.5 which shows how all of the topographic attributes except slope exhibited a large amount of overlap between land cover classes. Both secondary topographic attributes, in particular, showed a very large overlap between vegetation classes (Table 15.5). Small ranges in small samples ($n \leq 5$) may simply reflect a lack of data, but a large range in a small sample size does provide evidence of a large variance. Classes 3309 (sagebrush), 4214 (Rocky Mountain juniper forest), and 6202 (riparian shrub) have sample sizes of five or less. The quasi-dynamic topographic wetness indices reached similar maximum values of 2.64–2.67 in all classes except Rocky Mountain juniper forest, and three classes (3110, low grass/low cover; 4102, upland broadleaf forest; 7601, badland) span the entire range of 2.20–2.67 (Table 15.2). Thus, land cover classes were not well separated by this parameter, and this was reflected in the slight reduction in classification accuracy when it was added to the *Landsat* TM data.

In addition, some of the most seriously confused classes, such as 2100 (cropland) and 6102 (riparian broadleaf forest), showed largely overlapping distributions in both quasi-dynamic topographic wetness index and incident short-wave solar radiation (Table 15.5). If steady-state topographic wetness (Figure 15.5) is substituted for quasi-dynamic topographic wetness (Figure 15.6), a similar overlap exists between cropland and riparian broadleaf forest. In the case of incident short-wave solar radiation (Figure 15.8), the riparian forest class has the largest range of values. In contrast, the slope attribute used in the supervised classification process did exhibit interesting variations between land cover classes (Figure 15.3). None of the 18 cropland sites occurred on slopes greater than 10%, whereas 6 of the remaining 11 classes had average slopes > 10% and one (4214, Rocky Mountain juniper forest) occurred on slopes exceeding 30%. Thus, slope did show less overlap between some land cover classes than the secondary topographic attributes used in this study (Table 15.5).

15.4.4 Use of Stream Buffers to Delineate Riparian and Upland Cover Classes

The topographic attributes calculated with TAPES-G can be used to delineate the channel system, since the channel initiation threshold (drainage density) can be adjusted to delineate additional channels that are not identified in published USGS hydrography DLGs. The maximum cross-grading area specified in TAPES-G to switch from multiple to single flow directions with the *FD8/F8* and *FRho8/Rho8* options was treated as the channel initiation threshold in this study. A maximum cross-grading area of 20,000 m^2 was chosen because it provided the best match with the stream lines recorded on 7.5′ USGS quadrangles. These "channel" cells were

TABLE 15.4 Error Matrices for Land Cover Classifications Prepared With Modified Classification Method, 2.0-ha Minimum Map Unit, and One or Two Selected Topographic Inputs

A. *Landsat* TM Bands Plus Elevation, Slope, and Quasi-dynamic Topographic Wetness Index[a]

Field Classification	Remotely Sensed Classification													Total
	2100	3100	3111	3201	3309	4102	4206	4214	5000	6102	6201	6202	7601	
2100	10 (56%)		1							4	1	1	1	18
3110		16 (67%)	4		1				1	1			1	24
3111		12	5 (24%)	1						1			2	21
3201			2	5 (50%)		1				2				10
3309			1		3 (60%)						1			5
4102				2		4 (33%)			1	5				12
4206							7 (88%)		1	1				8
4214								5 (100%)						5
5000						1	1		1 (33%)					3
6102	4	1	1	2		5				25 (63%)		2		40
6201	2									2	5 (56%)			9
6202	1									1	1	2 (40%)		5
7601	1		1										11 (85%)	13
Total	18	29	15	10	4	11	8	5	3	42	8	5	15	173

B. *Landsat* TM Bands Plus Elevation, Slope, Quasi-dynamic Topographic Wetness Index, and Incident Short-Wave Solar Radiation[b]

Field Classification	*Remotely Sensed Classification*													Total
	2100	3100	3111	3201	3309	4102	4206	4214	5000	6102	6201	6202	7601	
2100	10 (56%)		1			1				3	2		1	18
3110		17 (71%)	4		1				1	1				24
3111		11	6 (29%)	1						1			2	21
3201		2	1	4 (40%)		1				1	1			10
3309				1	3 (60%)					1				5
4102	1			2		2 (17%)			1	5	1	1		12
4206							7 (88%)				1			8
4214						2	1	2 (40%)						5
5000				1	1				1 (33%)					3
6102	3	1	1	2		4				26 (26%)	1	2		40
6201	1						1			2	5 (56%)			9
6202	1										1	3 (60%)		5
7601	1		1										11 (85%)	13
Total	17	31	14	11	5	10	9	2	3	40	11	6	14	173

[a]Overall agreement = 57%.

[b]Overall agreement = 56%.

Note. See Table 15.1 for definitions of individual land cover classes.

TABLE 15.5 Topographic Attribute Summary Statistics by Land Cover Type

Variable	Min	Max	Mean	SD	Min	Max	Mean	SD	Min	Max	Mean	SD
	Cropland (n = 18)				**Low grass/low cover (n = 24)**				**Low grass/moderate cover (n = 21)**			
Elev (m)	734	882	807	48	778	899	846	35	778	889	834	41
Slope (%)	1.7	9.1	4.6	2.6	0.7	35.5	7.1	7.0	0.1	20.8	7.0	5.6
Aspect (deg.)	0	315	122	109	0	315	144	104	0	346	138	108
Upslope (000 m²)	0.90	204.2	32.4	57.2	0.90	190.5	15.7	39.4	0.90	300.0	20.3	64.6
SS TWI	6.92	11.46	8.48	1.43	5.45	11.22	7.70	1.31	5.93	13.12	8.09	1.70
QD TWI	2.34	2.67	2.59	0.11	2.21	2.67	2.46	0.14	2.32	2.67	2.54	0.11
Solar (W/m²)	128	165	153	8.4	139	180	155	10.3	135	172	150	9.1
	Riparian grass/forb (n = 9)				**Mesic upland shrub (n = 10)**				**Riparian shrub (n = 5)**			
Elev (m)	806	860	840	20	773	829	798	20	750	807	775	21
Slope (%)	1.3	19.4	5.5	5.7	2.1	29.5	11.3	8.2	0	19.2	5.6	7.8
Aspect (deg.)	0	342	131	122	0	281	138	112	0	270	176	113
Upslope (000 m²)	1.7	300.0	167.9	156.7	0.9	300.0	79.8	118.0	11.8	300.0	166.8	131.3
SS TWI	5.83	14.46	10.39	3.01	5.78	12.32	8.70	2.08	6.51	14.37	9.52	2.94
QD TWI	2.48	2.67	2.60	0.07	2.37	2.66	2.57	0.10	2.63	2.67	2.66	0.02
Solar (W/m²)	140	168	154	8.2	136	160	149	7.9	153	165	157	5.5
	Sagebrush (n = 5)				**Badlands (n = 13)**				**Upland broadleaf forest (n = 12)**			
Elev (m)	741	780	768	16	787	886	833	31	785	872	835	29
Slope (%)	0.3	7.1	3.6	2.9	2.1	42.7	19.7	14.6	4.8	35.1	18.5	7.7
Aspect (deg.)	0	236	158	91	0	227	119	86	11	315	149	121
Upslope (000 m²)	0.9	300.0	141.0	147.0	0.9	7.8	2.0	1.8	6.6	300.0	69.5	100.0
SS TWI	7.29	14.02	11.53	2.81	5.08	8.82	6.25	1.19	6.61	14.11	8.48	2.11
QD TWI	2.64	2.67	2.66	0.01	2.21	2.67	2.41	0.18	2.20	2.64	2.45	0.14
Solar (W/m²)	132	153	141	8.3	142	163	155	6.2	134	164	152	9.7
	Riparian broadleaf forest (n = 40)				**Ponderosa pine forest (n = 8)**				**Rocky Mountain Juniper (n = 5)**			
Elev (m)	731	878	813	41	778	810	797	10	833	872	846	16
Slope (%)	1.1	25.8	10.3	5.3	5.6	16.5	11.0	3.8	20.1	47.7	34.2	11.5
Aspect (deg.)	0	333	191	91	7	349	156	151	274	350	309	32
Upslope (000 m²)	1.8	300.0	90.4	110.3	0.9	4.9	2.6	1.5	1.7	17.4	8.9	7.0
SS TWI	5.53	16.96	8.89	2.07	6.1	7.5	6.7	0.48	5.02	7.69	6.47	1.28
QD TWI	2.34	2.67	2.56	0.11	2.33	2.65	2.54	0.12	2.17	2.54	2.34	0.16
Solar (W/m²)	130	185	154	11.9	129	168	148	15.1	125	164	144	18.4

used with the 5-m elevation threshold to delineate riparian and upland areas. This method designated 16.0% (272 km^2) of the study area as riparian (Figure 15.9) compared to only 11.4% (194 km^2) using the 0.5° by 1° USGS hydrography DLGs (Figure 15.10). The DEM errors reported earlier produced some spurious drainages trending east–west in low-relief areas (see east side of Figure 15.9). About 117 km^2 was designated as riparian by both methods notwithstanding these problems. Agreement was generally good in the middle sections of well-defined, wooded draws, but worse in source areas and in flatter rolling upland terrain (cf. Figures 15.3, 15.9, and 15.10). However, the choice of 5 m as the elevation threshold may have overestimated riparian areas in both instances because this threshold produced very wide buffers (≤0.5 km) in areas of low relief, such as river terraces and rolling upland (cf. Figures 15.9 and 15.10).

The computed riparian zones did not match the satellite-based land cover classification very well. For example, only 39% of the 365 km^2 assigned to riparian land cover classes with the initial classification method (2.0-ha minimum map unit) was located in the riparian areas delineated with the TAPES-G stream network and the 5-m elevation threshold. Similarly, 47.5% of the TAPES-G riparian area was assigned to nonriparian land cover classes with the initial classification method, and these areas should be removed from the modeled riparian zone. We had hoped to be able to use the riparian zones to separate out riparian and upland land cover classes with similar spectral signatures. However, the five upland broadleaf forest training sites classified as riparian broadleaf forest and four riparian broadleaf forest training sites classified as upland broadleaf forest with the initial land cover classification (2.0-ha minimum map unit; Table 15.3C) were not resolved when the USGS and TAPES-G stream buffers were applied. Overall, these results indicate that different methods and data sources will delineate very different riparian zones and these may not help with the correct identification of cover classes that have similar spectral signatures but grow in different landscape positions. Hence, the riparian zones computed with this step were not used to revise the land cover maps discussed in the next section.

15.4.5 Evaluation of Land Cover Maps

Land cover maps for a representative area in the southeast part of the study area are reproduced in Figure 15.11 (see color insert). Some confusion between cropland and riparian forest remains in all cases: For example, the rectangular cultivated field in the lower left corner of the maps is erroneously classified as ponderosa pine or riparian forest, even when topographic attributes are added to the classification. Changing the spatial aggregation from 0.4 to 0.8 ha and then 2.0 ha altered some map unit identifications, but many misclassifications still remained.

Although overall classification accuracy varied little between different levels of spatial aggregation and topographic attribute combinations, a pixel-by-pixel comparison of the vegetation maps shows substantial differences between them. Percent agreement between the 36 unique combinations of minimum map unit and classification method is summarized in Table 15.6. The addition of topographic attributes holding minimum map unit size constant generated 67–89% agreement, which

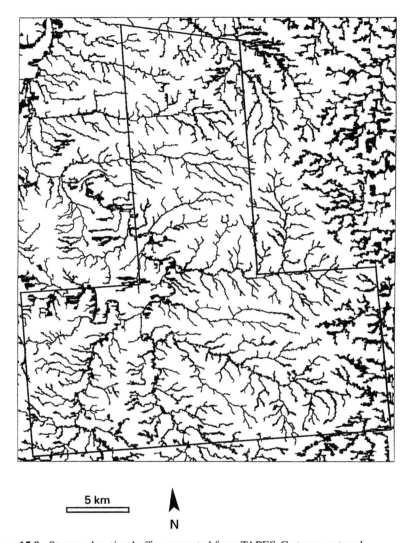

Figure 15.9. Stream elevation buffer computed from TAPES-G stream network.

means that up to 33% of the pixels in the map were reclassified with the addition of one or two secondary topographic attributes. Figures 15.11A–C show how the classification changes with the addition of topographic attributes for the 0.4-ha mapping unit. Varying the minimum map unit produced even lower agreement, only 34–44%; in other words, between 56 and 66% of the pixels were reclassified when the level of spatial aggregation was varied from 0.4 to 2.0 ha. This is evident in Figures 15.11A, D, and E, where the spatial pattern of vegetation patches varies greatly, depending on the level of spatial aggregation. Ultimately, the choice of minimum map unit is a mat-

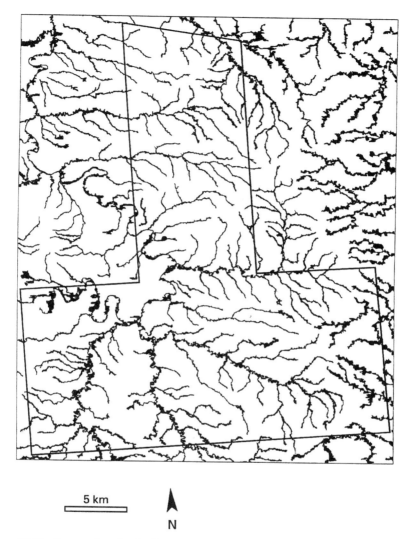

5 km

N

Figure 15.10. Stream elevation buffer computed from USGS hydrography DLGs.

ter of user preference, but since the aggregation process cannot be reversed, it may be advisable to underestimate the size of mapping unit, to preserve ecologically valuable information in small vegetation patches.

Although there is no loss of overall accuracy in a 2.0-ha aggregation, the spatial pattern of wooded draw vegetation in the LMNG is more difficult to discern in the aggregated image. For example, the narrow strips of riparian broadleaf forest (class 6102), shown in magenta in the bottom part of Figure 15.11C, are visible with a 0.4-ha minimum map unit, but are replaced with three large patches when the 2.0-ha min-

TABLE 15.6   Percent Agreement Based on Pixel-by-Pixel Comparison of Land Cover Maps

Classification	L0.4	L0.8	L2.0	W0.4	W0.8	W2.0	S0.4	S0.8	S2.0
Landsat[a] (0.4 ha)	—								
Landsat[a] (0.8 ha)	41.7	—							
Landsat[a] (2.0 ha)	43.2	36.8	—						
Wetness[b] (0.4 ha)	84.8	41.2	43.1	—					
Wetness[b] (0.8 ha)	41.8	87.5	36.8	42.8	—				
Wetness[b] (2.0 ha)	43.2	36.4	89.4	43.8	37.5	—			
Solar[c] (0.4 ha)	67.1	39.9	40.9	69.8	40.5	41.7	—		
Solar[c] (0.8 ha)	39.9	70.3	35.6	39.2	73.8	35.9	41.3	—	
Solar[c] (2.0 ha)	40.9	34.4	69.6	41.7	35.9	71.6	42.0	39.0	—

[a]Initial (unmodified) classification method.
[b]Modified classification method plus quasi-dynamic topographic wetness index and specified minimum map unit.
[c]Modified classification method plus quasi-dynamic topographic wetness index, short-wave solar radiation, and specified minimum map unit.

imum map unit is used (Figure 15.11E). These large riparian broadleaf forest patches were classified as either upland broadleaf forest (class 4102), Rocky Mountain juniper forest (class 4214), riparian grass and forb (class 6201), or riparian shrub (class 6202) in the three classifications produced using a 0.4-ha minimum map unit (cf. Figures 15.11A–C, E).

15.5   DISCUSSION

The overall results produced in this study (~55% classification accuracy using 12 land cover classes plus water) matches that achieved in other remote sensing classifications of land cover. Higher levels of accuracy can be achieved by merging vegetation classes with similar spectral signatures. Table 15.7 shows how a reduction of the number of land cover classes from 12 to 7 increased overall classification accuracy to approximately 70% in this instance. However, this strategy is helpful only where land managers do not want or need to delineate large numbers of land cover classes.

We added two additional topographic attributes to our classification method in hopes of improving accuracy without reducing the number of land cover classes. We were clearly unsuccessful and the negligible improvement in classification accuracy gained by adding quasi-dynamic topographic wetness index, incident solar radiation, and riparian buffers to distinguish upland and riparian vegetation classes would not seem to justify the use of topographic attributes in remote sensing vegetation mapping. Six sets of issues should be investigated further before applying terrain analysis to a large geographic area such as the LMNG:

- Utilization of additional ground-truth data and/or satellite imagery
- Inclusion of topographic attributes in the first stage of the classification method used here

TABLE 15.7 **Error Matrices for Generalized Land Cover Classifications Prepared With Modified Classification Method and One or More Selected Topographic Attributes**
A. *Landsat* TM Bands Plus Elevation and Slope (2.0-ha minimum map unit)[a]

Field Classification	Remotely Sensed Classification							Total
	2100	3110 3111 6201	3201 6202	3309 7601	4102 6102	4206 4214	5000	
2100	11 (61%)	2		1	4			18
3110, 3111, 6201	2	41 (76%)	1	4	5		1	54
3201, 6202	1	3	7 (47%)		4			15
3309, 7601	1	3		14 (78%)				18
4102, 6102	3	4	7		37 (71%)		1	52
4206, 4214			1			12 (92%)		13
5000					1	1	1 (33%)	3
Total	18	53	16	19	51	13	3	173

B. *Landsat* TM Bands Plus Elevation, Slope, and Quasi-dynamic Topographic Wetness Index (2.0-ha minimum map unit)[b]

Field Classification	Remotely Sensed Classification							Total
	2100	3110 3111 6201	3201 6202	3309 7601	4102 6102	4206 4214	5000	
2100	10 (56%)	2	1	1	4			18
3110, 3111, 6201	2	42 (78%)	1	4	4		1	54
3201, 6202	1	3	7 (47%)		4			15
3309, 7601	1	3		14 (78%)				18
4102, 6102	4	2	6		39 (75%)		1	52
4206, 4214					1	12 (92%)		13
5000					1	1	1 (33%)	3
Total	18	52	15	19	53	13	3	173

(*continued*)

Table 15.7 (*Continued*)
C. *Landsat* TM Bands Plus Elevation, Slope, Quasi-dynamic Topographic Wetness Index, and Incident Short-Wave Solar Radiation (2.0-ha minimum map unit)[c]

Field Classification	Remotely Sensed Classification							Total
	2100	3110 3111 6201	3201 6202	3309 7601	4102 6102	4206 4214	5000	
2100	10 (56%)	3		1	4			18
3110, 3111, 6201	1	43 (80%)	1	3	4	1	1	54
3201, 6202	1	5	7 (47%)		2			15
3309, 7601	1	1	1	14 (78%)	1			18
4102, 6102	4	3	7		37 (74%)		1	52
4206, 4214		1			2	10 (77%)		13
5000			1	1			1	3 (33%)
Total	17	56	17	19	50	11	3	173

[a]Overall agreement = 71%.
[b]Overall agreement = 72%.
[c]Overall agreement = 71%.
Note. See Table 15.1 for descriptions of individual land cover classes.

- Quantification of the DEM error over the entire Little Missouri region and how it is propagated with computed topographic attributes
- Exploration of the viability and impact of using a more sophisticated topographic wetness index that incorporates the effects of spatially variable precipitation and evapotranspiration
- Inclusion of a more sophisticated buffer algorithm in which the elevation threshold is a function of local vertical relief and/or upslope drainage area, and validation of the stream elevation buffer results with ground-truth data
- Inclusion of additional ancilliary data layers in the classification

The bootstrap results used to assess classification accuracy in this chapter were originally generated to identify weaknesses in the training data set. The large omission and commission errors in Tables 15.3 and 15.4 suggest that a larger training set may have helped. Similarly, multitemporal analysis of *Landsat* TM or AVHRR imagery throughout the growing season might have helped to resolve some of the confusion between the vegetation classes. Cropland, for example, is likely to exhibit rapid changes in spectral signature due to plowing and harvest, compared with the more constant signature of broadleaf forest in the summer months. Time-series remote sensing has improved accuracy in other large-scale mapping projects (e.g., Pickup et al. 1993, Samson 1993, Fuller et al. 1994). However, in the Konza Prairie, KS, a grassland site similar to the LMNG, Oleson et al. (1995) found that cropland and forest had considerable overlap in spectral signature at all times during the May to October growing season. Both of these options—collecting more ground-truth data and using multitemporal satellite imagery—were prohibited due to limited project funds.

Some improvement might have been gained by modifying the classification method to accommodate multidimensional combinations of the *Landsat* TM bands and selected band transforms as well as the original TM bands. In one such study, Lauver and Whistler (1993) utilized *Landsat* TM2, TM4, TM5, TM7, and NDVI to identify 77 previously unknown natural grassland areas in Anderson County, KS. The six 30-m TM bands plus NDVI and the TM tasseled cap features of brightness, greenness, and wetness (Crist and Cicone 1984) were examined in this study, and the results demonstrated that their final methodology was faster and more accurate than aerial photographs and aerial surveys. Similarly, Anderson et al. (1993) utilized a single *Landsat* TM scene and showed that it is possible to relate vegetation indices to green biomass measurements when data are combined in greenness strata for a semi-arid rangeland study area in northeastern Colorado.

The methodology of Lauver and Whistler (1993) is instructive for the current study because they performed a multistage classification and filtered the output map to produce a final map of potential high-quality grasslands of 2 ha or larger. We averaged topographic attributes before the supervised classification was performed. Ma and Redmond (1996, personal communication) observed a loss of continuity in riparian corridors at the 2.0-ha aggregation in other parts of the LMNG. The incorporation of topographic attributes in the initial classification step may have improved classification accuracy in this study. At large minimum map units, the computed topographic attributes are averaged over a large area and this may diminish their impact on the classification process. Dividing the initial classification into classes based on topographic attributes will identify landscape units identified in traditional maps (see descriptions in Table 15.1 for examples) and may therefore improve classification accuracy. Dikau (1989) has proposed a landscape classification scheme that divides catchments (hillslopes) into 16 landform classes based on plan and profile curvature, two of the attributes computed with TAPES-G. Burrough et al. (2000a, b), in contrast, have proposed an automated method of landscape classification that incorporates spatial sampling methods, statistical modeling of the derived stream topology, and fuzzy *k*-means using the diagonal metric. Either approach might have generated better predictions, especially if steps were taken to measure and/or eliminate DEM errors.

USGS DEM error varies greatly from quadrangle to quadrangle, and may produce unacceptable errors in topographic attributes in the rolling upland areas of the Little Missouri Grassland. The root mean square error (RMSE) of 7 m that is often reported for 7.5′ USGS quadrangles provides a global measure of vertical accuracy that is of little value in assessing the question of topographic attribute uncertainty (Kumler 1994, Hunter et al. 1995, Hunter and Goodchild 1996). Small errors in the horizontal positions and/or elevations of landforms represented in DEMs may lead to large errors in computed primary and secondary topographic attributes (e.g., Wilson et al. 1998). In the LMNG, the 7.5′ USGS elevation data are by far the best available: Quantification of errors is hampered by the lack of a reference data set of accurate elevations with which to compare USGS data. Florinsky (1998) has proposed a series of measures of root mean square errors for slope, aspect, plan, and profile curvature that can be mapped and used to depict the spatial distribution of errors within study sites. Burrough et al. (2000a, b) have suggested adding random errors to DEMs, calculating selected attributes multiple times, and calculating cell averages to minimize these errors. Visual inspection of topographic attribute maps can, of course, provide an excellent qualitative impression of whether the linear patterns of DEM error obliterate terrain features.

A more sophisticated topographic wetness index that incorporates the combined effects of spatially variable precipitation and evapotranspiration might be more successful in discerning vegetation types as well (e.g., Moore et al. 1993e). The level 3 analysis incorporated in WET uses spatially variable net radiation to compute potential evapotranspiration at each grid cell and then determines soil-water content using a set of functional relationships based on soil-water content, evapotranspiration, and deep drainage (see Chapter 4 for additional details). Both evapotranspiration and deep drainage are dependent on soil-water content using this approach. Similarly, the TAPES-G derived buffer should be compared and calibrated, if possible, with ground-truth data such as field maps or air photos showing riparian areas before proceeding with a stream buffer calculation over a larger area.

Bendix (1994) examined the role of scale-specific environmental factors in shaping the pattern of riparian vegetation for two mountainous streams in southern California. Three transverse-scale (distance above water table, flood severity, substrate texture) and five longitudinal-scale variables (elevation, valley orientation, valley width, fire history, lithology) were examined. The regression results showed that the transverse and longitudinal variables jointly influenced the vegetation patterns and observed patterns along different reaches reflect a subtle combination of overlapping gradients. These results may be site-specific and the contribution of longitudinal variables might be reduced in flatter landscapes such as the LMNG.

This study suggests that local (i.e., site-specific) scale thresholds may be required to guide the choice of the channel initiation threshold in TAPES-G and the height that riparian zones extend above stream level. The channel initiation threshold determines the density of drainage channels and the height threshold determines the width of the buffers. Adjusting these values to match the observed riparian vegetation may result in a more useful stream buffer. The calculated stream buffers reproduced in Figures 15.9 and 15.10 suggest that a single height above stream level for the entire area may not be able to generate suitable buffers in both rolling upland and wooded draw areas. The buffer height might be estimated as a function of local vertical relief, so that the height is increased in areas of large vertical relief and reduced in areas of low relief. Alternatively, the roughness (vertical relief divided by area) of each catchment could be computed and used to adjust the buffer threshold used with individual streams. Finally, the threshold could also be scaled by upslope drainage area within each of the catchments if longitudinal-scale variables were thought to exert a large impact on riparian vegetation patterns.

A combination of one or more of these improvements to the terrain analysis methods may improve results. Additional data layers, such as soil properties and property ownership, might be of some use in improving performance as well. Moore et al. (1993b) have demonstrated a strong correlation between soil properties and computed topographic attributes, so soil mapping may not add much new information. In addition, soils have not been mapped in county soil surveys at sufficient spatial resolution to distinguish the small wooded draws that affect vegetation patterns in the Little Missouri Grassland (Thompson 1978). In other areas, where geologic parent material is more diverse than in the LMNG, soil mapping might improve classification accuracy when incorporated into the classification. Property ownership maps might be of use in identifying cropland, which occurs almost exclusively on private land, although riparian areas also occur on private land. These riparian areas meander through the complex checkerboard pattern of land ownership in the LMNG, making identification by property ownership difficult. This state of affairs illustrates how disturbance and succession are intertwined in the processes that produce spatial patterns of vegetation in these types of landscapes.

15.6 CONCLUSIONS

The negligible improvement in classification accuracy gained by the use of a quasi-dynamic topographic wetness index and average incoming short-wave solar radiation does not appear to justify the considerable effort involved in calculating these topographic attributes. From this analysis, it appears that the addition of secondary topographic attributes to remote sensing classifications of semiarid grasslands provides no significant increase in accuracy. It was hoped that distinctions between similar spectral types, such as riparian and upland broadleaf forest in the LMNG, could be made with terrain analysis, but this hypothesis is not supported by this study. East of the Little Missouri, in areas with less vertical relief, such as the Grand River, Cedar River, and Sheyenne National Grasslands, topographic attributes may be of even less use, since the DEM errors are large compared to the vertical relief. The low relief and sparse vegetation cover of the LMNG suggest that additional ground-truth data and multitemporal satellite imagery were required. Both of these options were prohibited due to limited project funds. The inclusion of topographic attributes in the supervised classification, quantification of DEM error over the entire Little Missouri region, and inclusion of additional ancillary data layers may have helped to offset these shortcomings. The results of the current study do demonstrate (1) the benefits of using modern terrain analysis tools to identify channel systems, and (2) that the final land cover maps must be used with care because different source data and levels of spatial aggregation will predict different patterns of existing vegetation.

CHAPTER 16

Towards a Spatial Model of Boreal Forest Ecosystems: The Role of Digital Terrain Analysis

Brendan G. Mackey, Ian C. Mullen, Kenneth A. Baldwin, John C. Gallant, Richard A. Sims, and Daniel W. McKenney

16.1 INTRODUCTION

Boreal forest ecosystems comprise approximately 80% of Canada's 650 million ha of forested land (S. L. Gray 1995). Plant species diversity in the boreal forest is low relative to forest regions at lower latitudes. The full expanse of the Canadian boreal forest contains fewer than 2000 vascular and nonvascular plant species (Sims and Addison 1994) and there are only 15 species of trees—the dominant life-form of forest vegetation communities (Sims and Addison 1994, Farrar 1995). Despite their relative floristic simplicity, these forests exhibit surprisingly complex spatial patterns of community organization when examined on a landscape-wide basis. In fact, perhaps the most distinctive feature of the boreal forest is its landscape mosaic, with the scale of patches ranging from tens of meters to tens of kilometers.

In this chapter we examine the extent to which the spatial distribution of canopy tree species in a boreal mixed-wood landscape in Ontario, Canada, is correlated with environmental conditions related to the local topography. Using data from a field study site located at Rinker Lake, Northwestern Ontario, we empirically explore topographic control of species' distributions and boreal forest patterns, and the extent to which these effects can be adequately captured by digital terrain analysis.

The distribution of forest attributes is conventionally mapped used aerial-photo interpretation (API) and/or remote sensing analysis (RS), validated by field survey observations. Both API and RS have the admirable advantage of providing a landscape-wide inventory of contemporary land cover characteristics. However, neither technique provides assessments of total species composition or vegetation structure, beyond those

Terrain Analysis: Principles and Applications, Edited by John P. Wilson and John C. Gallant.
ISBN 0-471-32188-5   © 2000 John Wiley & Sons, Inc.

that can be observed in the canopy. A full account of forest ecosystems therefore requires the application of additional analytical and modeling approaches.

Bonan and Shugart (1989) presented a conceptual model of boreal forest ecosystems that stressed the roles of climate, soil moisture, substrate, fire, and plant life histories. Most would agree that the spatial patterns of canopy tree species are the result of this general set of determining processes. However, the exact species mix and linkages of cause and effect have proven difficult to establish in any given landscape situation. Hence, predicting the spatial distribution of boreal forest tree species has proven an elusive goal. At least two hypotheses can be invoked to explain the spatial distribution of boreal forest tree species: the niche hypothesis and the disturbance hypothesis.

16.1.1   The Niche Hypothesis

This hypothesis proposes that the spatial distributions of tree species are strongly partitioned according to their response to critical environmental resources. Strong spatial gradients in these resources, together with markedly different niche responses between species, will lead to strong and readily predictable spatial patterning.

The niche of a species (Hutchinson 1957, see also Austin 1985) is defined by the set of environmental conditions to which it is adapted. The fundamental niche identifies environmental conditions that meet the physiological requirements of the species (hence it is also called the physiological niche). The realized niche (or ecological niche) represents those conditions where the species actually occurs, given competition and other factors such as the species' biogeographic history. A species' niche response to a single environmental gradient is commonly defined by minimum and maximum environmental values outside of which the species cannot survive, and a narrower optimal range contained within these tolerance thresholds that corresponds with conditions that are optimal for its growth, reproduction, and competitiveness.

The niches of different species can relate in various ways; for example, tolerance values may overlap but optimum values are distinct. All other factors being equal, if two species' niches match, then they should both be equally likely to occur in those locations where the corresponding environmental conditions occur. The realized (ecological) niche can be equal to but is usually a subset of the fundamental (physiological) niche.

As noted by Nix (1982), complete niche specification is probably impossible to calculate for most species. However, it is possible to quantify a species' response to the critical portion of its total niche that is defined by the primary environmental regimes. This defines its potential environmental domain. The primary environmental regimes refer to the set of biophysical processes that determine the distribution and availability of heat, light, water, and mineral nutrients. These regimes define critical axes in a species' niche, since they constitute the fundamental resources required by all photosynthesizing plants.

The distribution and availability of these resources is a function of various factors related to climate, substrate, and topography. Mackey (1996) discussed the set of hierarchically scaled processes that determine resource delivery to wild plants, and noted the role of toposcaled processes. The notion of toposcale was introduced by Linacre (1992) and falls between the mesoscale (or regional scale) and the microscale. In the forest ecology context, toposcaled processes are associated with the influences of local topography, such as the redistribution of water in the landscape and the effects of slope and aspect on solar radiation. At certain scales, the biota also play a role in the primary environmental regimes; for example, vegetation cover influences the water balance through its effects on surface radiation and evaporation. Digital elevation models play a critical role in the generation, particularly at the mesoscale and toposcale, of spatially reliable estimates of these primary environmental resources.

Determining the niche of a long-lived tree species is a difficult task. Controlled growth experiments in an environmental chamber (e.g., Forman 1964) are impractical for species that live for many decades. An alternative, empirically based approach attempts to calculate a species' niche based on field observations of its natural occurrence, and estimates of environmental attributes from in situ observations (e.g., soil pit data) or ex situ environmental estimates (e.g., using climate models or digital elevation models). If presence-only data are available for the occurrence of a species, then its niche can be defined by the maximum and minimum environmental values associated with the limits of its distribution, with optimal conditions being defined by the range of environmental values associated with the highest density of field survey sites (for example, the Bioclim method of Nix (1986)). If presence–absence data are available to describe a species occurrence, then statistical models can be calculated using generalized linear models (e.g., Austin et al. 1990).

16.1.2 The Disturbance Hypothesis

This hypothesis proposes that a tree species' distribution is dominated by processes related to disturbance. A species' postdisturbance response is determined by the type and severity of the disturbance event, initial conditions of establishment, and the species' life-history attributes (see discussion in Bonan and Shugart 1989). Disturbances can kill canopy trees, creating opportunities for the germination and growth of new plant stock. Most boreal ecosystems are subject to cyclic regimes of catastrophic disturbance—primarily by fire, but insect infestations and wind are also important. Stand-replacing fires kill much of the above-ground component of the existing vegetation, creating the physical and biological conditions within which a succeeding plant community will develop. According to the disturbance hypothesis the postdisturbance species' composition at a site will be a function of the fire regime (fire intensity, frequency, seasonality, and fire type, see Gill (1975)), the extent to which the existing vegetation has been damaged or killed by the fire, the availability of seed and regenerating stock, and their life-history attributes (Noble and Slatyer 1980).

While not denying niche theory, the disturbance hypothesis suggests that the niches of canopy tree species in boreal ecosystems overlap to such an extent that they have no significant sorting effect on the species' spatial distributions. It can also be argued that the local range in primary environmental regime values is too narrow to provide any leverage for niche separation. Rather, the determining factors are key life-history attributes that determine postfire response. The variation in light response of species is of particular interest. Boreal tree species germinants vary in their tolerance to light levels. Those species that initially establish after a stand-replacing fire require high light levels, while species that tolerate low light levels have the potential to regenerate under mature forest canopies. Given this, the species composition at a site should be related to the postfire successional stage and hence time since the last fire (see Bergeron and Dansereau 1993), with shade-intolerant early successional species dominating immediately postfire, and shade tolerant species dominating after a shade-producing canopy has developed.

If fire does dominate tree distributions in the boreal forest, the question arises as to whether the fire regime itself is subject to environmental constraints, for example, acting through the effect of slope on fire intensity and the effects of soil moisture and radiation on fuel moisture and fuel loads. The spatial distribution of fire in the landscape can be modeled as an entirely stochastic process, with every location having an equal probability of fire, subject only to the constraint of the availability of fuel as a function of time since last fire (see McCarthy et al., in press). At mesoscales, the relationship between various atmospheric conditions and the risk of fire has been well documented, in particular, wind speed, humidity, and temperature (e.g., the Canadian Fire Weather Index as used in Flannigan and Van Wagner (1991)). Such mesoscaled atmospheric conditions are highly variable over short time periods, and hence are difficult to model spatially. In addition, climatic processes are not correlated to a significant degree with local topography in the relatively low-relief boreal landscapes. These mesoscaled processes, while important, are not considered further here. Our interest is in the effect of processes influenced by the local topography.

At the landscape or toposcale, few studies have examined fire behavior within the perimeter of a fire. Eberhardt and Woodward (1987) studied boreal forest fires in Alberta and found that up to 18% of the landscape could remain unburnt within the fire boundary. Shafi and Yarranton (1973) noted that spatial variability in the intensity of fire can affect the understorey composition of postfire boreal stands that have similar prefire floristic compositions. Within-fire heterogeneity may be due to various topographic-related factors, such as slope angle and slope length, that affect fire intensity and rate of spread. Local topography exerts a degree of control on water flow through a catchment and is correlated with areas of run-on and runoff. The soil-moisture and radiation regimes, in turn, affect fuel production and flammability.

Local topography, therefore, can influence tree species distributions in a number of ways. It has the potential to influence the spatial patterning of niche gradients and to create spatial heterogeneity in the fire regime.

16.2 THE STUDY AREA

The Rinker Lake research area is a 900-km² landscape located about 100 km north of Thunder Bay, Canada (Figure 16.1). The approximate coordinates at the center of the

Research Area are 49° 10′N, 89° 20′W. Details of its environmental characteristics can be found in Sims et al. (1997) and Mackey et al. (1996). It lies wholly within the Lake Nipigon Ecoregion of the Boreal Shield Ecozone (Ecological Stratification Working Group 1995). The humid to perhumid, moderately cold, boreal climate of this Ecozone is generally characterized by warm, rainy summers and cold, snowy winters. The research area receives about 75 cm of total precipitation annually, with about 50 cm occurring as rainfall (Environment Canada 1982b). Annual snow fall averages 250 cm (Environment Canada 1982b), but average snow depth rarely exceeds 75 cm. Mean annual daily temperature is about 0°C, with mean daily minimums of around −6°C and mean daily maximums of 6.5°C (Environment Canada 1982a).

Figure 16.1. Location of Rinker Lake Research Area (RLRA) within the Province of Ontario, Canada.

Landscape features of the research area reflect the effects of major glaciations, the last of which ended approximately 10,000 years ago (Zoltai 1965). Hundreds of water bodies and rivers cover some 103.4 sq km (11.5%) of the research area. The physiography of the Research Area is primarily bedrock controlled, characterized by undulating hills of low to moderate relief and by low parallel ridges orientated in a northeast to southwest direction. Elevation ranges between 360 and 560 m above sea level. The underlying bedrock includes Achean greenstones and granites, as well as diabase and unmetamorphosed shales and limestones from the Proterozoic era.

Surficial materials in the Research Area are predominantly of glacial origin. These deposits range in thickness from a thin veneer less than 1-m to more than 30-m depth, but on average are probably less than 5 m thick. Glacial till is the most common surface sediment, covering 68% of the terrestrial portion of the Research Area. Approximately 17% of the land area is covered by glaciofluvial deposits, comprising flat outwash plains and positive-relief, ice-contact features, such as eskers and kames. There are also limited expanses of fine-grained, mostly silt-sized, glaciolacustrine sediments, which were deposited in local glacial lakes. Recent materials include organic deposits, covering about 12% of the land area, sandy alluvium, and coarse angular talus.

The research area is located within the boreal forest region of Canada (Rowe 1972). This region is characterized by extensive conifer forests of black spruce (*Picea mariana* [Mill.] BSP.) and white spruce (*Picea glauca* [Moench] A. Voss), jack pine (*Pinus banskiana* Lamb.), and balsam fir (*Abies balsamea* [L.] Mill.). It also supports mixed-wood stands, composed of a mixture of conifer and northern hardwood species such as trembling aspen (*Populus tremuloides* Michx.) and white birch (*Betula papyrifera* Marsh.). The most common species composition within the mature forests of the research area consists of a mixture of black and white spruce, balsam fir, white birch, and trembling aspen. The second most frequent cover type comprises a mixture of black spruce and jack pine. In June 1992, about 48% of the land portion of the research area was covered by conifer-dominated mature forest, while hardwood and hardwood-dominated mixed woods covered about 34%, and approximately 7.5% was dominated by wetlands and other minor vegetation classes.

The Research Area lies within the bounds of the Abitibi-Consolidated Inc. Spruce River Sustainable Forest Licence. Although timber harvesting has been conducted within the research area over the last 40 years, the majority of industrial activity has occurred since 1985, resulting in rapid changes in the relative proportions of mature forest to young age classes.

16.3   WHY DIGITAL TERRAIN ANALYSIS?

The general nature of the landscape patterns in the research area is apparent in Figure 16.2 (see color insert), an infrared aerial photograph of a portion of the study area. This spatial complexity reflects, to a large degree, the heterogeneity in the bedrock and surficial geology. These diverse geological formations generate a variety of surface landforms and soil parent materials. The resultant soils of varying chemistries, textures, and water-retention capacities affect many aspects of forest ecology.

Sims et al. (1989) identified 38 forest community types for Northwestern Ontario, and recognized that the spatial distribution of these vegetation communities is strongly related to landform, topographic position, and substrate conditions that influence the moisture and nutrient regimes. Given the demonstrable influence of bedrock and surficial geology on landform, why not simply develop appropriately scaled spatial data sets of these attributes as the basis for predictive modeling of boreal forest ecosystems, rather than rely on digital terrain analysis?

Data about the spatial distribution of bedrock and surficial geology in complex boreal landscapes is a difficult, time-consuming, and expensive exercise, involving extensive field survey and sampling, along with laboratory and remote sensing analyses (Sims et al. 1997). Indeed, many key substrate attributes are simply not amenable to remote sensing techniques. By comparison, digital elevation data can now be readily derived for most of boreal Canada from 1:20 000 scaled topographic mapping. Even if adequate geological data could be readily obtained for a landscape, the causal linkages between the geology, plant response, and forest pattern are complex and difficult to establish.

In summary, the geology of the Canadian boreal zone is spatially heterogeneous and includes features that are not reflected in the surface topography. Some attributes are difficult to quantify, especially on a landscape-wide basis, and uncertainty remains about which attributes are relevant to specific ecological questions. Given these constraints, there is considerable value in exploring the role that digital terrain analyses can play in capturing the influence of topography on boreal forest processes.

16.4   METHODS

The aim of these analyses was to examine whether the spatial distributions of selected boreal tree species corresponded with differing environmental conditions. The niche hypothesis and the potential for within-fire heterogeneity are underpinned by the spatial distribution of environmental attributes that are related to the primary environmental regimes.

Here we focused on environmental resource distribution at the toposcale, as defined by terrain attributes, and microscaled attributes obtained from ground field survey. The set of DEM-derived and ground survey environmental attributes used in these analyses is given in Table 16.1. These attributes are related in various ways to the distribution and availability of heat, light, water, and mineral nutrients. In particular, all are direct or indirect measures in some degree of the water regime. Radiation is used here in terms of its role in the process of evaporation rather than as an index of photosynthetically active radiation. Note that no attempt is made here to integrate these variables into a process simulation of water availability as discussed in Chapter 4.

**TABLE 16.1   List of DEM-Derived and Ground Survey Environmental Attributes
Used in the AMF models and Domain Analyses**

Attribute	Label	Scale	Units
DEM			
Catchment area	—	Topo	m^2
Short-wave incoming radiation	SWR	Topo	$MJ/m^2/day$
Topographic wetness index	TWI_dem	Topo	Dimensionless index
Elevation percentile	E%	Topo	Dimensionless index
Ground Survey			
Topographic wetness index	TWI_gs	Topo	Dimensionless index
Topographic position	TOPOS	Topo	Ordinal scale
Slope angle	—	Micro	Degrees
Soil depth	—	Micro	cm
Dominant soil texture of C horizon	CTEXT	Micro	Ordinal scale
Soil-moisture regime	SMR	Micro	Ordinal scale

The spatial distributions of three selected tree species were investigated in relation to environmental attributes related to topography and substrate. Observations of the current distribution of these species were obtained from ground-based survey and remotely sensed data (air-photo interpretation and classified *Landsat* Thematic Mapper (TM)). Values for the environmental attributes were obtained both in situ, from ground survey, and ex situ, using digital terrain analysis. Correlations were analyzed using two methods: algorithms for monotonic functions and a two-dimensional domain analysis technique (Mullen 1995). These methods are detailed below.

16.4.1   Ecology of the Target Tree Species

Three tree species were selected for analysis: jack pine (*Pinus banksiana*), trembling aspen (*Populus tremuloides*), and black spruce (*Picea mariana*). A brief description of some aspects of the autecology of the three target species follow. Further details can be found in Sims et al. (1990).

16.4.1.1 Jack Pine (Pinus banksiana)   Jack pine is an early successional, evergreen conifer tree species, reaching heights of up to 30 m at maturity. It is a shade-

intolerant species, requiring full sunlight at all stages of its life cycle to achieve opti-mum growth. Where a seed source is present, early stand growth is rapid after a major disturbance such as fire. Jack pine is well adapted to periodic fires, typically develop-ing even aged stands after a stand-replacing fire. The heat of a forest fire causes the serotinous cones of jack pine to open and release seed that germinates readily on burned ground surfaces. Although an early successional species, jack pine is relatively long-lived and will persist into later successional stages where, in the absence of fire, it is eventually replaced by black and white spruce as well as balsam fir.

*16.4.1.2 Trembling Aspen (***Populus tremuloides***)* Trembling aspen is a relatively short-lived, broad-leaved hardwood tree species that can reach heights up to 34 m at maturity. It is a shade-intolerant species that requires full sunlight for growth and survival. It is an early successional species that is typically replaced, in the absence of fire, by balsam fir, white spruce, and/or black spruce. Reproduction occurs both vegetatively and by seed production, but root suckering is the predomi-nant form of reproduction and can occur at all ages. Both young and mature trees are susceptible to fire damage, but clonal colonization of burned areas through suckering occurs immediately after a fire.

*16.4.1.3 Black Spruce (***Picea mariana***)* Black spruce is a slow-growing, long-lived evergreen conifer tree species, averaging between 15 and 20 m tall at matu-rity. It has an intermediate tolerance to shade but grows best in full sunlight. It is fre-quently classified as an early successional species but also fulfils the role of a midsuccessional species in mixed boreal forests. After a stand-replacing disturbance, such as fire, black spruce readily develops into even-aged stands, often in association with jack pine. In boreal mixed woods, it will occupy a secondary canopy under trem-bling aspen. In older black spruce stands growing on more fertile sites, balsam fir tends to increase in relative abundance, whereas black spruce maintains its dominance on nutrient poor sites. In upland conifer stands, in the absence of fire, black spruce may remain over time as a major stand component, often accompanied by balsam fir and jack pine. Black spruce reproduces both vegetatively and by seed production.

16.4.2 Data Sources

This project used a variety of data sources, including ground survey data, aerial-photo interpretation data, and remotely sensed data. A digital elevation model (DEM) and terrain analysis were used to provide additional landscape-wide toposcaled envi-ronmental data.

16.4.2.1 Ground Survey Data At the Rinker Lake study area, 99 plots were surveyed between 1992 and 1995, each with an area of 10×10 m. Survey method-ologies followed those developed for the Northwestern Ontario Forest Ecosystem Classification program (Sims et al. 1989). Detailed quantitative descriptions were compiled on vegetation composition and abundance, tree growth, forest floor cover, and humus characteristics. Measurements were also made of soil physical and chemical attributes, local drainage patterns, and site moisture status. Detailed soil attributes were described from a 1 × 1-m pit. Slope angle (percent), aspect, and topographic position were also recorded for each plot. An attempt was also made to measure in situ the upslope contributing area by walking upslope from each plot and flagging the visually perceived immediate drainage boundary.

The standard approach to analyzing plant–environment relations is to sample the spatial distribution of plants in a landscape using fixed-sized quadrats. In the boreal landscape, the convention has been to use 10 × 10-m plots. This is argued on various grounds, but, in particular, that it represents the scale at which canopy gaps appear, promoting stand regeneration (Shugart 1984). This scale also reflects the small crown size of boreal trees, and fits within the fine-scaled patchiness of boreal forest ecosystems.

16.4.2.2 Digital Elevation Data   A digital elevation model for the area was developed at a 20-m grid resolution using digital input data from the Ontario Base Map topographic series at a scale of 1:20,000 (Sims et al. 1997). The input data were elevation contours at intervals of 10 m, stream vectors, spot heights, and lake polygons. The DEM was produced using the ANUDEM procedure (Hutchinson 1988, 1989b), following computational procedures described by Mackey et al. (1994). Figure 16.3 shows an image and histogram of the DEM. The spikes in the histogram stem from biases in the interpolated DEM associated with using elevation contour data, since there is a concentration of data at contour intervals and a lack of data at intervening elevations (see Chapter 2, Hutchinson and Gallant).

16.4.2.3 API and Remotely Sensed Data   Vegetation cover data were obtained from API maps and land cover data was obtained from analysis of *Landsat* TM data. The source of the vegetation data was the provincial Forest Resource Inventory (FRI) dataset, an API-derived product that provides mapped coverage of most of the commercially forested areas in Ontario. FRI map coverages for the Rinker Lake research area at 1:20,000 provide detailed information on forest cover types, canopy species composition, and potential wood volumes. The FRI canopy species composition data was used to derive coverages of the spatial extent of the three target tree species. Areas where the target species composed 50% or greater of the canopy were classified as presence of that species. The resultant spatial distribution maps of jack pine, trembling aspen, and black spruce are shown in Figures 16.16a, 16.17a, and 16.18a.

Land cover information was derived from two classified *Landsat* TM satellite images dated 20 June 1992 and 11 September 1993, using unsupervised clustering techniques supplemented by ground survey data and photo-interpretation of infrared aerial photographs (Sims et al. 1997). The land cover data were used to identify lakes and cutovers (areas that had been subject to clearfell logging). These areas were excluded from the analysis. Figure 16.4 shows the mask of lakes, cutover areas, and the portion of the research area where no FRI data were available.

16.4.2.4 Derived Terrain Attributes   The TAPES-G suite of terrain modeling programs (Moore et al. 1993e, Gallant and Wilson 1996) was used to calculate an

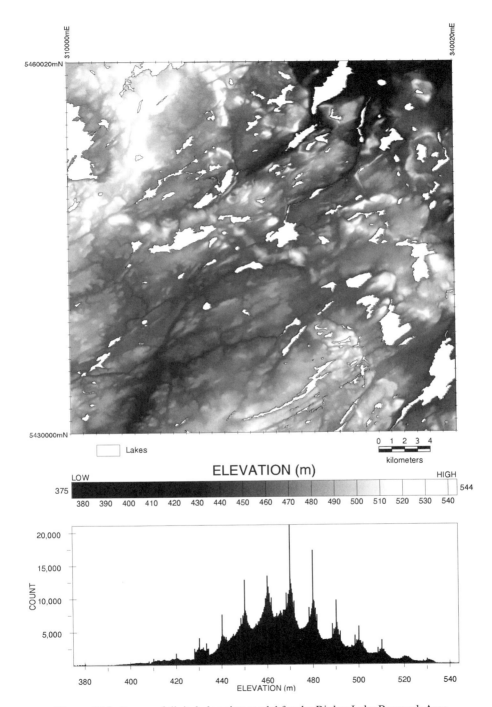

Figure 16.3. Image of digital elevation model for the Rinker Lake Research Area.

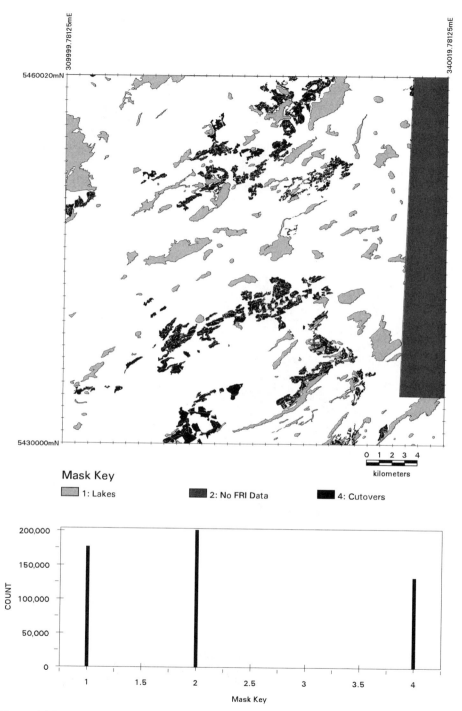

Figure 16.4. Image of data set masks for the Rinker Lake Research Area, showing the distribution of lakes, cutover areas (clearfell logged), and the portion of the research area where no FRI data were available.

array of primary and compound terrain attributes (*sensu* I. D. Moore et al. 1991). Primary attributes used in this study include slope, aspect, and catchment area. The compound terrain attributes used were (1) the topographic wetness index and (2) an estimate of incoming short-wave radiation. These were supplemented by the Elevation Percentile index of Gallant (1996), as an estimate of local topographic position. The development of this index was influenced by the terrain shape index discussed by McNab (1989).

The topographic wetness index (TWI) (Beven and Kirkby 1979, I. D. Moore et al. 1991), as used here, is defined in Equation 4.40. The index provides a measure of potential discharge per unit area modified by slope angle as a surrogate for the hydraulic gradient. TWI has been used to model the potential spatial distribution of water in a landscape for the purposes of predicting vegetation response (e.g., Mackey 1994a). However, the model assumes steady-state conditions and spatially invariant conditions for both infiltration and transmissivity. It also assumes that there is no complex or deep subsurface drainage, so that subsurface flows track surface morphology.

The SRAD program (Moore et al. 1993e, Wilson and Gallant 2001) was used to generate gridded estimates of incident solar radiation. Cloudiness and other atmospheric conditions are factored into the short-wave irradiance estimates. The terrain effects of slope angle, aspect, and topographic shading are calculated from the DEM and used to modify the estimates of short-wave irradiance. The use and calibration of SRAD for the research area are given in McKenney et al. (1999).

The Elevation Residuals program of Gallant (1996) was used to calculate the elevation percentile index ($E\%$) as a local measure of topographic position. This method analyzes the properties of each cell in a DEM using a circle of a user-defined radius around the center of each grid cell. The elevation values inside this circle are the context for the analysis of that point. In this study, a radius of 100 m was used to capture the average distance from tops of hills and ridges to the bottoms of adjacent valleys. As described by Gallant (1996), the elevation percentile index measures the ranking of the elevation of the central point compared to all points in the context circle, and ranges in value from 0 to 1. If the point is the lowest in the circle it will be given the value 0, and if it is the highest it will be given the value 1. If all points are exactly the same height a value of 0.5 is assigned. The elevation percentile index ($E\%$) has no units.

Figures 16.5–16.8 show images of the gridded coverages developed for short-wave radiation, the topographic wetness index, catchment area, and the elevation percentile index. Histograms of each attribute are given at the bottom of the figures.

16.4.3 Analytical Techniques

A useful distinction was made by Austin and McKenzie (1988) between exploratory and confirmatory data analysis. Exploratory data analysis provides a quantitative analysis of trends and patterns in a data set. Confirmatory data analy-

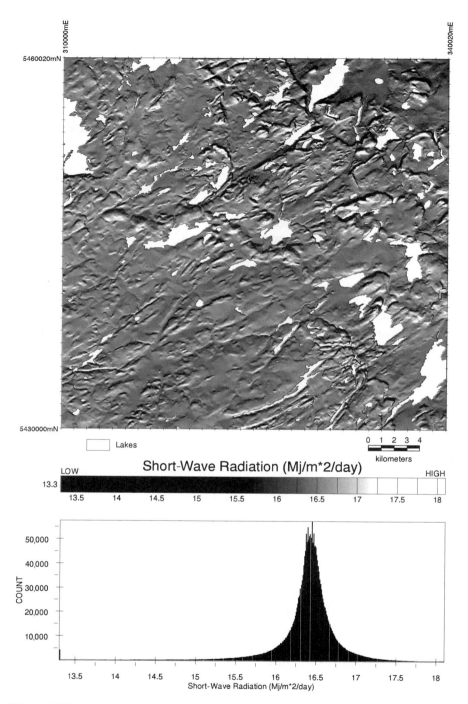

Figure 16.5. Image of gridded short-wave radiation values for the Rinker Lake Research Area.

404

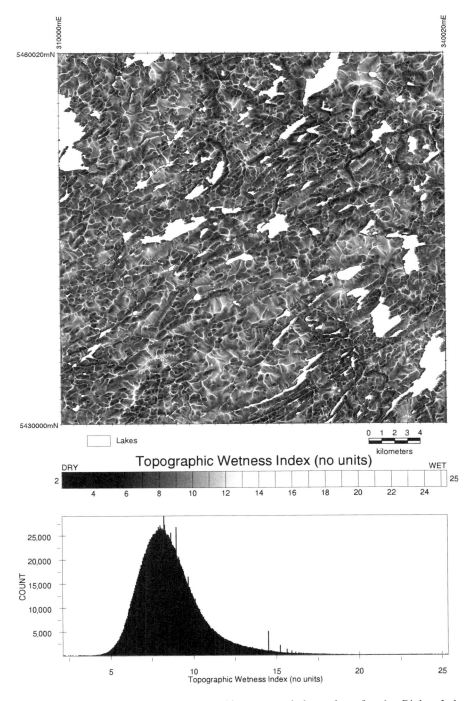

Figure 16.6. Image of gridded topographic wetness index values for the Rinker Lake Research Area.

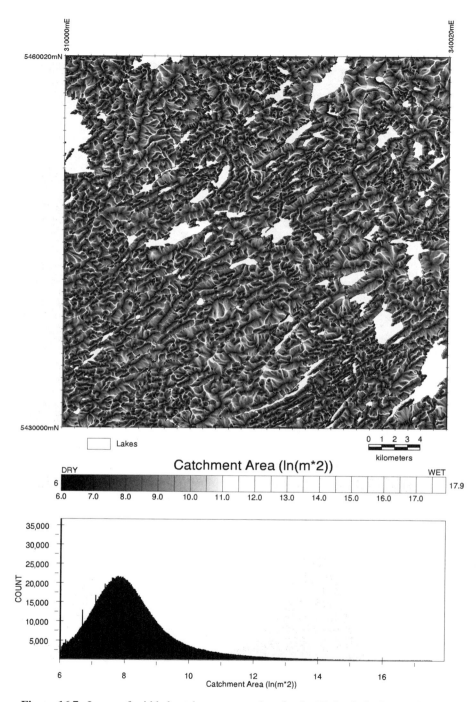

Figure 16.7. Image of gridded catchment area values for the Rinker Lake Research Area.

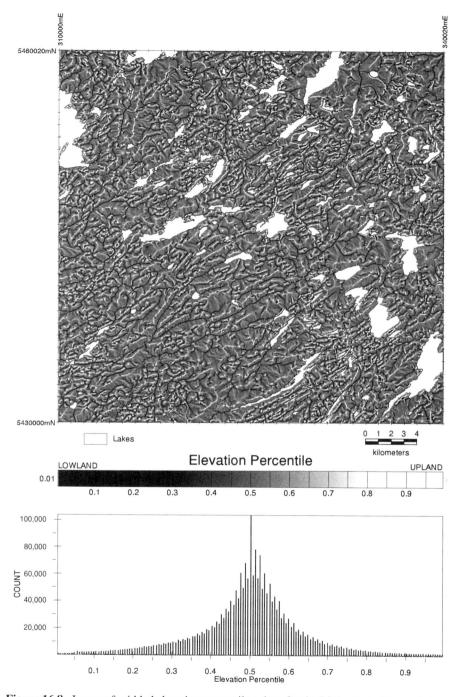

Figure 16.8. Image of gridded elevation percentile values for the Rinker Lake Research Area.

sis allows for formal testing of null hypotheses, and in the context of vegetation science, includes the application of statistical modeling procedures, such as generalized linear modeling (McCullagh and Nelder 1989). There are significant problems for confirmatory data analysis associated with the application of standard statistical modeling procedures to landscape-based analyses (e.g., the lack of independence in gridded data).

This study focused on the use of exploratory data analysis to examine vegetation–environmental relations. Two exploratory data analysis techniques were used: Algorithms for Monotonic Functions (Bayes and Mackey 1991) and the domain analysis procedure developed by Mullen (1995). These are discussed in the following sections.

16.4.3.1 Algorithms for Monotonic Functions

Algorithms for Monotonic Functions were presented by Bayes and Mackey (1991). The technique is a statistical method and was used to analyze the presence–absence of jack pine at the 99 ground survey points. The set of ground survey and DEM-derived environmental attributes used as explanatory variables in these analyses are given in Table 16.1. The AMF algorithms provide maximum likelihood estimates of response variables, subject only to the assumption that the function relating the response and explanatory variables is monotonic. Pairs of explanatory variables were modeled using the two-dimensional AMF modeling procedure. A diagnostic E statistic is also calculated and can be used as a statistic for comparing different models, in a similar manner to the r^2 statistic in linear regression. The larger the E statistic, the better the model fits the data.

16.4.3.2 Domain Analysis

The other exploratory data analysis technique used here was the domain analysis procedure developed by Mullen (1995). This method was used to explore the spatial distribution patterns of the three target species: jack pine, trembling aspen, and black spruce. The procedure uses patch-based data on the distribution of the target species to analyze the environmental domain within which the species occurs. It quantifies the observed probability of a tree species occurring within a two-dimensional domain, which is defined by the combination of two selected environmental variables.

The domain analysis requires as input three gridded data sets for the research area: one showing the spatial occurrence of the tree species (in this case, tree species distributions obtained from the FRI data set), and one each for the two selected environmental variables. A gridded data mask was used to exclude cells that were water bodies, cutovers, or lacked FRI data (Figure 16.4).

The domain analysis procedure involves a number of steps:

Step 1. A gridded data set is required showing the spatial distribution of the target species.

Step 2. A pair of environmental variables is selected for analysis, and gridded estimates of each variable are generated for the study area.

Step 3. A two-way table is defined by the selected environmental gradients, and each gradient is divided into 100 equally spaced class across its range (this is an arbitrary number). This results in 10,000 environmental domains.

Step 4. The frequency of occurrence of all grid cells in the study area is then calculated for each domain.

Step 5. The frequency of occurrence of all grid cells that are labeled as the target tree species is then calculated for each domain.

Step 6. An observed probability of occurrence of the target tree species is then calculated for each environmental domain by dividing the frequency of occurrence for the target tree species by the frequency of occurrence of all grid cells. That is, within each of the 10,000 two-dimensional environmental domains, the total number of grid cells where a species is present is divided by the total number of grid cells in the landscape that occupy that domain.

The result is an observed probability of occurrence (0–100%) for each environmental domain in the two-way table. The observed probabilities can be interpreted as indicating those environmental domains where the target species is more or less likely to dominate.

16.5 RESULTS

16.5.1 AMF Analyses

These analyses were based on the observed presence of jack pine at the 99 ground survey plots. Table 16.2 shows the E statistic derived from AMF analyses of various combinations of environmental attributes for predicting the presence of jack pine. This set of combinations is not exhaustive and was intended to be only indicative. The first three models are based on explanatory variables derived from ground sur-

TABLE 16.2 *E* **Statistic for Indicative Combinations of Environmental Variables for Predicting Jack Pine Presence**

Model	Explanatory Variables	E Statistic
1	SMR and CTEXT	0.244
2	TOPOS and soil depth	0.259
3	SMR and slope angle	0.280
4	SMR and $E\%$	0.316
5	TWI_gs and $E\%$	0.325
6	Catchment area and $E\%$	0.348

Note. Attribute abbreviations are defined in Table 1. The larger the E statistic, the better the model.

vey data, that is, observations taken by a field worker in a 10×10-m plot. Models 4 and 5 combine DEM and ground survey data, while model 6 uses DEM-derived attributes only. All six models can be interpreted as capturing the effect of different components of the water regime on jack pine distribution.

The results demonstrate that there are major correlations between the presence of jack pine in the landscape and terrain-related attributes. Furthermore, they show that the best result (i.e., the highest E value) is obtained from attribute values calculated ex situ from the DEM. The DEM-derived attributes performed better than the soil profile and plot-based observations. Given the immense effort required to measure soil profile and topographic attributes in situ, using ground survey, this is a very interesting result.

There are errors in the ground survey data that need to be considered. Generally, it is difficult in complex terrain to accurately observe toposcaled features from ground survey. The catchment area component of TWI_gs was very difficult to measure in practice in the field. A reasonable accuracy was obtained for plots on upper slope positions as their upslope area was relatively small and slope-based boundary conditions readily identified. However, the farther downslope the plot, the larger the catchment area and the more indistinct the watershed boundary; hence, less certainty can be placed on the measurement.

Similarly, the DEM-derived elevation percentile index ($E\%$) performed better than the ground-surveyed topographic positions (TOPOS). This may also be related to problems associated with observing toposcaled phenomena at microscales. Figure 16.9 shows a subsection of the study area where contour lines are mapped along with the location of a selection of ground survey plots. The observed topographic position of each plot is indicated. The sequence of sites 2508–8001–2510–2509 shows correspondence between the ground survey TOPOS and the elevation contours. However, site 7002 was recorded as a crest, whereas it is adjacent to site 7003, which is more correctly recorded as an upper slope. About 15% of the 99 sites showed a similar discrepancy between ground observed TOPOS and the topography as modeled by the DEM.

These discrepancies largely reflect scale-related observer differences. For example, a plot may be located on the top of a mini-esker and hence be in a local crest position, while the mini-esker itself is located midslope on a much larger outwash landform unit. On the ground, such features are very distinctive at the observer scale of 10×10 m and may exert considerable influence on the immediate vicinity. However, they are below the resolution of the mapped elevation contours on which the DEM and subsequent terrain analyses are based.

Nonetheless, the bottom line is that the best AMF model explained only about 35% of the variance in the data. Clearly, one reason for this may be that key forcing functions exist that were not considered. However, another possibility worth exploring is that the scale of observation for the presence of jack pine (i.e., a 10×10-m plot) is simply the wrong scale at which to obtain data for analyzing the effect of toposcaled processes. This suggests that more landscape-wide analytical procedures are needed.

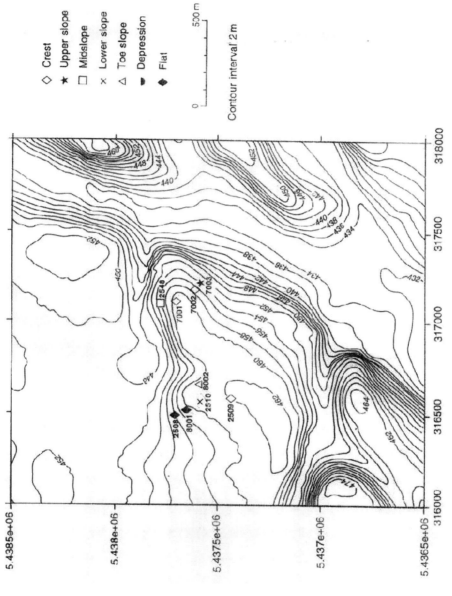

Figure 16.9. Subsection of the study area where elevation contour lines are mapped along with the location of a selection of ground survey plots. The observed topographic position of the plots is also indicated.

411

16.5.2 Domain Analyses

As noted above, this analytical procedure quantifies the observed probability of a tree species occurring within a defined two-dimensional environmental domain. Once again, only indicative analyses were undertaken rather than an exhaustive exploration of all possible terrain combinations. Results are presented here for jack pine, trembling aspen, and black spruce, and as a function of (1) the topographic wetness index (TWI) and elevation percentile ($E\%$), and (2) short-wave radiation (SWR) and elevation percentile ($E\%$).

Figures 16.10, 16.11, and 16.12 show, respectively, the observed probabilities for jack pine, trembling aspen, and black spruce as a function of TWI and elevation percentile ($E\%$). All three species have almost complete overlap in terms of where they occur. However, black spruce is clearly dominant in a domain that is quite distinct from the domains of jack pine and trembling aspen. For the latter two species there is considerable overlap in their dominant domains. Observed probabilities for jack pine are considerably lower than those for the other two species, reflecting its relative scarcity in the Rinker Lake landscape. Jack pine and trembling aspen are dominant at higher elevation percentile values and lower TWI values. This is equivalent to upper-slope positions with small catchment areas and/or high slope angles. Black spruce is dominant in the lower end of the landscape (note in Figure 16.6 that TWI values of greater than 15 are rare, and that most of the $E\%$ values are between 0.4 and 0.6). Aside from the core area of observed probability, a scatter of high probability values is also apparent on all these figures. These are an artifact of very low numbers of cells in these domains.

Figures 16.13, 16.14, and 16.15 show, respectively, the observed probability for jack pine, trembling aspen, and black spruce as a function of short-wave radiation (SWR) and elevation percentile ($E\%$). Black spruce dominates in a domain where radiation is around 16.5 MJ/m^2/yr. As can be seen in Figure 16.5, this value corresponds with the majority of the Rinker Lake landscape, which is largely flat, and at the lower-end of watersheds. Both jack pine and trembling aspen are more dominant at slightly higher and lower values. These radiation values are found at the upper-slope positions where these two species have been shown to dominate (i.e., upland sites have both exposed and shaded slopes).

The observed probabilities presented in Figures 16.10, 16.11, and 16.12 can be used to generate spatial predictions of the likelihood of occurrence of the three species. This is achieved by locating in turn each grid cell in the spatial data base within the corresponding domain of the two-way table and assigning the calculated probability. In this case, the mask was turned off, so that probabilities were mapped for all grid cells except those that fell within lakes. These spatial predictions are shown in Figures 16.16b, 16.17b, and 16.18b (see color insert) and are in effect a spatial representation of the observed probabilities. They indicate where in the landscape the species is most likely to dominate. Note that all species are predicted as occurring across the entire landscape, but only with a high probability of occurrence within particular locations. The images in Figures 16.16a, 16.17a, and 16.18a show the location of FRI polygons where the species is recorded as composing 50% or

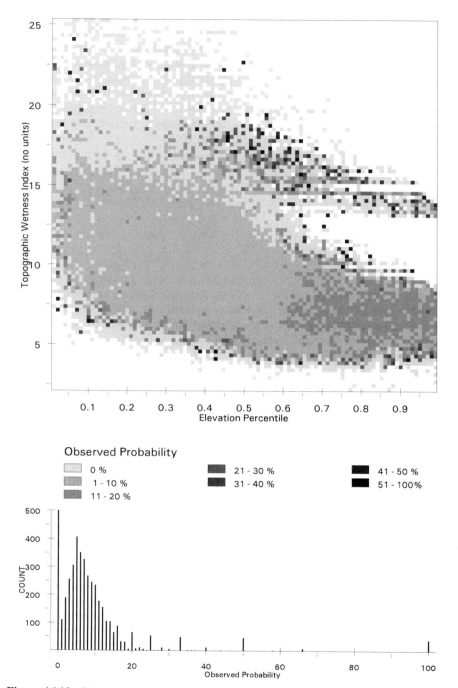

Figure 16.10. Observed probabilities for jack pine as a function of the topographic wetness index (TWI) and the elevation percentile index (*E%*).

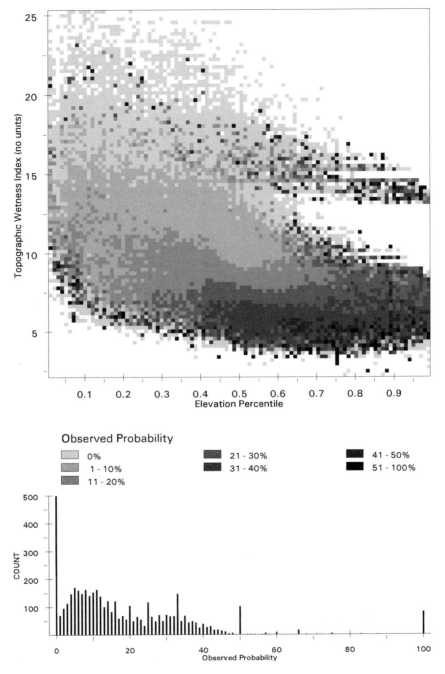

Figure 16.11. Observed probabilities for trembling aspen as a function of the topographic wetness index (TWI) and the elevation percentile index (*E%*).

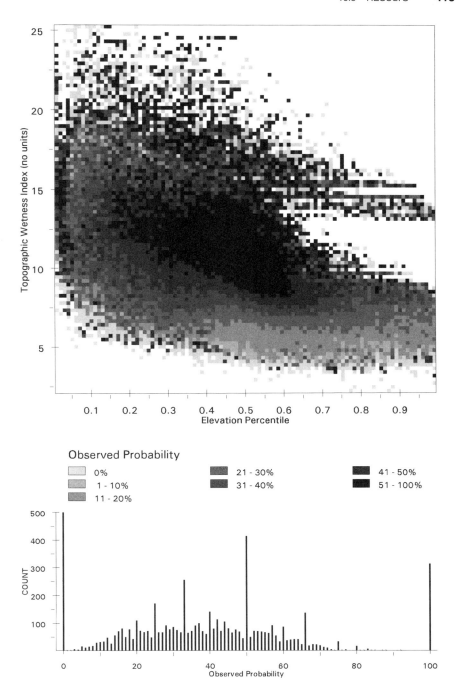

Figure 16.12. Observed probabilities for black spruce as a function of the topographic wetness index (TWI) and the elevation percentile index (*E%*).

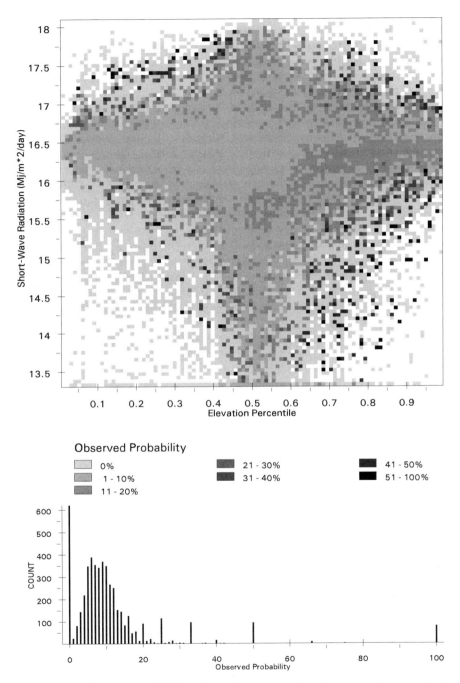

Figure 16.13. Observed probabilities for jack pine as a function of short-wave radiation (SWR) and the elevation percentile index (*E%*).

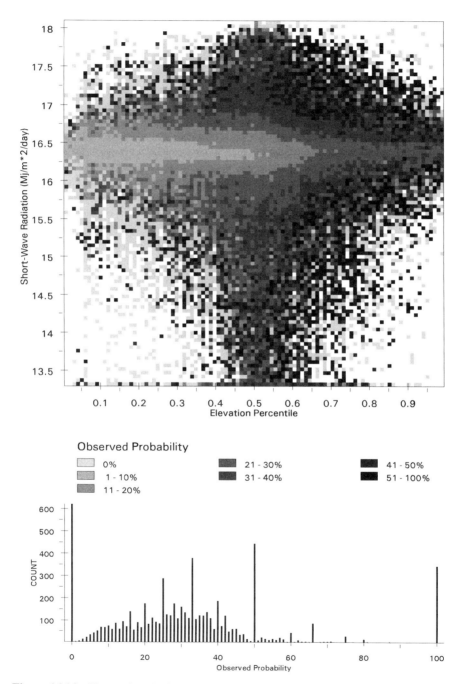

Figure 16.14. Observed probabilities for trembling aspen as a function of short-wave radiation (SWR) and the elevation percentile index (*E%*).

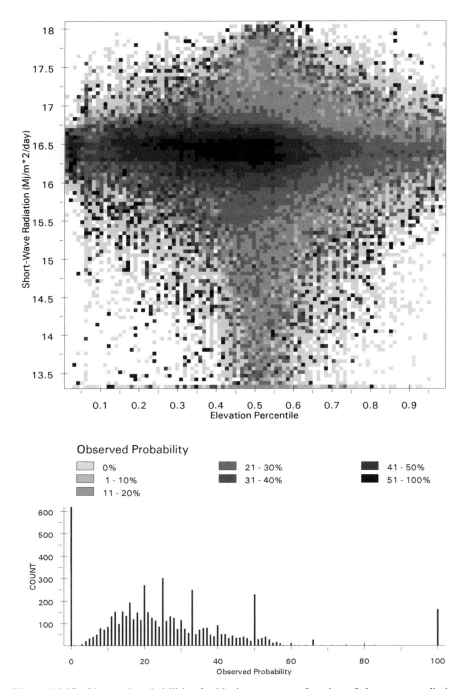

Figure 16.15. Observed probabilities for black spruce as a function of short-wave radiation (SWR) and the elevation percentile index (*E*%).

greater of the canopy. Note that this does not mean that the species necessarily occupies all the area of a polygon.

16.6 DISCUSSION

The analyses support elements of both the niche and disturbance (fire) hypotheses for the spatial distributions of these boreal tree species. One interpretation is that while the three tree species niches may overlap there is sufficient differentiation in the optimum portion of their realized niche to provide some controls on spatial patterning. Also, while gradients in the primary environmental resources across the boreal landscape are relatively shallow, water and nutrients are limiting in parts of the landscape for certain species such that they provide a moderate spatial sorting effect. From this perspective, the reason why trees species dominate in particular domains can be partly explained by invoking niche theory.

The correlations between the tree species and topography can also be explained by (1) spatial heterogeneity in the fire regime (in particular, intensity and frequency) and (2) different life histories that determine postfire response (e.g., vital attributes as defined by Noble and Slayter (1980)). As suggested above, the likelihood of fire across a landscape can vary spatially in relation to variables other than time since fire.

Gauthier et al. (1993) found in the southern part of the Canadian boreal forest in Quebec that jack pine stands can be subject to both lethal and nonlethal fires. Lethal fires produce even-aged stands, while nonlethal fires allow trees to survive fire episodes and create gaps for continuous recruitment from nonserotinous trees. The fire regime therefore comprised both long-return interval lethal fires and short-return interval nonlethal fires. This spatial heterogeneity was related to site conditions (with xeric sites being more multiaged) and topographic location with multiageness being more common on island rather than mainland stands.

Spatial variation in fire regimes therefore can be due to a variety of factors. For example, the presence of physical barriers or impediments to fire spread, such as rugged terrain or water bodies (see Hunter 1993). Also, topographic or substrate controlled variation in the moisture regime can have a range of effects, including modifying primary productivity, fuel loads, and flammability. Foster (1983), for example, found that a different fire regime was supported in lowland plains with many wetlands than in hillier areas.

Sites that support short fire return times should tend to be dominated by early succession species such as jack pine and trembling aspen that are shade intolerant and have vital attributes well suited to rapid postfire regeneration (e.g., serotinous cones and regrowth through root suckering). The digital terrain analysis presented here suggests that it may be possible to develop spatially explicit models of such site conditions.

There is also an interplay between niche response to the primary environmental regimes and the fire regime. Our analyses suggest that while the optimum component of the black spruce niche could be different from those of jack pine and trembling aspen, the three species' niches do overlap. However, this niche overlap does not necessarily translate to a complete spatial overlap, because the species may be temporally separated due to their moderately different fire responses and light tolerances. As the time since last fire (or other form of stand-replacing disturbance) lengthens, then, all other factors being equal, the likelihood of more shade-tolerant species, such as balsam fir and white spruce, will increase (for example, see the discussion of this topic in Bergeron and Dansereau (1993)).

Black spruce also exhibits diversity in growth form on different site locations that may be related to either a subspecies level difference in niche and/or fire response. Two growth forms of black spruce are commonly recognized by field workers: lowland and upland (e.g., as referred to in Kasischke et al. (1995)). The former occupy hydric sites, exhibiting very slow grow rates and low levels of total biomass production; the reverse is true for upland black spruce, which occupies better drained, more mesic sites. For example, our field data from Rinker Lake show that a mature lowland black spruce might be only 10 cm DBH (diameter at breast height) and 9 m in height, whereas a mature upland tree can grow to 80 cm DBH and 30 m in height. We would expect that both site conditions have differing fire regimes. Further research is needed to examine whether corresponding differences exist between the two forms in terms of vital attributes.

There are, of course, additional ecological processes that influence boreal forest succession and affect the spatial distribution of tree species that are not related to local topography. De Grandpre et al. (1993), for example, noted the presence of early successional understorey plants in old balsam fir stands as the result of openings in the canopy due to outbreaks of spruce budworm (although topography may influence the dispersal of budworm, at least at lower population levels).

16.6.1 The Domain Hypothesis

While the complex landscape mosaic of vegetation patches is evident in boreal forest ecosystems, attributing causal factors is difficult. Nonetheless, we argue that environmental domains can be identified across the landscape where the probability of a given species is more likely. A target species may not exclusively occupy a particular environmental domain, but will more likely dominate a given domain. These domains, while encompassing a potentially complex suite of ecological processes, can be defined in terms of terrain attributes calculated from a suitably scaled digital elevation model.

The idea that environmental domains exist in the boreal landscape within differing tree species' responses has important resource management implications. Where industrial timber harvesting occurs, silvicultural practices are designed to promote natural regeneration of commercially desirable tree species and to reduce the propagation of unwanted species. The analyses presented here suggest that it is possible to map domains where species with certain life-history traits relevant to silvicultural prescriptions are more likely to occur. Such analyses can provide a spatial context for harvest scheduling and postharvest silviculture plans.

A major limitation of both *Landsat* TM and aerial-photo-derived data is that these remotely sensed signals are dominated by canopy species and hence provide little if any measure of understorey and ground cover vegetation. From a biodiversity per-spective this is unfortunate, because it means there is no direct means of mapping these components of forest ecosystems. The domain hypothesis, however, provides a means for spatially extending ground survey observations about understorey and ground cover, since we believe that many of these plants are sensitive to toposcaled site conditions (see discussion in Mackey et al. 1996). This is supported by De Grand-pre et al. (1993), whose study based in southern boreal forests found that slope angle explained some of the variability in understorey plant composition and abundance. A combination of remotely sensed data and domain analysis therefore has the potential to provide a far more complete spatial inventory of boreal forest biodiversity.

It has been the convention, particularly since the adoption of the gap modeling approach, to assume that a 10×10-m plot is the standard spatial unit for analysis of boreal systems. There are certainly key ecological processes that occur at this scale. However, our analyses suggest that there are also a range of toposcaled processes that are critical in terms of the spatial structuring of boreal forests and that cannot be ade-quately accounted for by 100-m^2 sampling units. Such plots do not sample patch char-acteristics and boundary conditions. Hence, analyses based on these data are not necessarily able to identify important causal factors. Plot-based field survey, therefore, needs to be complemented by toposcaled methodologies such as domain analysis.

16.7 CONCLUSIONS

The development of relatively fine-scaled digital elevation models has made it possi-ble to undertake landscape-wide, quantitative analyses of toposcaled processes. The terrain analysis here is relevant to modeling the fire regime, the distribution of pri-mary environmental resources, and subsequent effects on biological productivity and species' distributions. Previously, analysis was restricted to the essentially subjective mapping of landscape patterns from aerial photographs and satellite imagery.

Digital terrain analysis provides the basis for identifying domains where the like-lihood of a species' occurrence is most likely. Many boreal tree species can occur throughout a landscape, but dominate within a more restricted range of conditions. The potential domain of a species may be the consequence of a set of related factors, including niche partitioning and differential response to disturbance regimes. Similar relations have been found for species' distributions in other terrestrial ecosystems (see Chapter 14, Franklin et al.).

The effects of terrain-controlled processes in contributing to spatial heterogeneity in the fire regime is only poorly studied. Few analyses of within-fire boundary behav-ior exist, and it remains a promising area for research. We note that our observations about terrain–fire relations are simply inferred from terrain–vegetation correlations and knowledge of plant life histories, and are not based on direct field observations of fire history.

Empirically derived plant–terrain relations defined at the Rinker Lake research area may not hold in other boreal landscapes with substantially different surficial geology. For example, jack pine may occur across a wider range of topographic positions in boreal landscapes with greater expanses of glacio-fluvial outwash. It is important to keep in mind the importance of soil parent material in defining both the moisture and nutrient regimes, and to recognize the geographic limits of inference that can be drawn from terrain-based domain analyses.

There is a limit to the ability of forest models to spatially predict the distribution of target species. A degree of stochasticity must be accepted and factored into simulation strategies. Viewing species distributions in a probabilistic sense and identifying environmental domains where certain ecological phenomena are likely to dominate provides the spatial modeling framework needed to implement more sophisticated landscape-modeling approaches. Models of boreal tree species distributions must ultimately be considered in terms of space and time dimensions. The temporal dimension of stand dynamics has been long recognized as reflected in forest succession or gap simulation models (see Botkin et al. 1972, Shugart 1984). The spatial dimension clearly requires more attention.

The contribution of the domain hypothesis is to provide a framework for the development of spatially distributed forest simulation models that can be inclusive of terrain-related resource distribution and disturbance. Our Rinker Lake case study illustrates how this can be implemented using digital terrain analysis. Su and Mackey (1997) described such a dynamic, spatially distributed forest simulation model for wet *Eucalyptus* forest in Tasmania, Australia. The model utilizes the domain theory suggested here to simulate mesoscaled and toposcaled environmental constraints on tree species' distribution, regeneration, and growth, and their interactions with the fire regime.

Boreal forests, like all of the world's forests, are subject to pressures from economic development and the many dimensions of global change. Digital terrain analysis is a critical tool for improving our resource and biodiversity spatial inventories and for simulating the likely ecological impacts of disturbance scenarios. Such information can assist in the difficult task of ensuring the wise use and sustainable management of these forest ecosystems.

CHAPTER 17

Future Directions for Terrain Analysis

John C. Gallant, Michael F. Hutchinson, and John P. Wilson

17.1 INTRODUCTION

This book describes the state of the art in DEM generation, terrain analysis methods, and applications. It provides a guide to the use of terrain analysis methods for the study of spatial patterns and processes in a range of landscapes and application domains. The first 5 chapters discussed digital elevation data sources, DEM construction, the calculation of selected primary and secondary topographic attributes, and the effects of data source, grid resolution, and flow-routing method on computed topographic attributes. The next 11 chapters described how these topographic attributes have been used in hydrological, geomorphological, and ecological modeling studies across a range of spatial scales. Topographic attributes were combined with a series of process and statistical models and were used to predict stormwater source areas, soil moisture and soil loss patterns, landscape elements, soil properties, landslide susceptibility, and vegetation cover in these chapters.

However, the individual chapters also exhibit varying rates of success in terms of their use of topographic attributes to characterize selected environmental processes and/or patterns. This state of affairs indicates that there remain a number of challenges to the routine application of terrain analysis for spatial prediction in many fields and that these problems can be divided into methodological issues, knowledge of relationships, and scaling.

17.2 METHODOLOGY AND DATA

While most of the primary terrain attributes are efficiently and accurately calculated by the existing methods, there are still some questions regarding the contributing area and specific catchment area attributes. Part of the difficulty is that they are inherently

Terrain Analysis: Principles and Applications, Edited by John P. Wilson and John C. Gallant.
ISBN 0-471-32188-5 © 2000 John Wiley & Sons, Inc.

difficult to calculate well, as the values can change by large amounts over small distances. The most efficient method, D8, is quite inaccurate, although useful for some purposes, while the more accurate methods, such as DEMON, are quite inefficient. There is considerable scope for development of better methods of calculation.

One area of terrain analysis that has been largely ignored is error propagation and the effect of variability of terrain within the bounds of a single-grid cell. DEMs are models of a real surface and are subject to error. These errors vary, depending on the landscape: For example, aspect and flow direction may be accurately represented on steep slopes but may be highly uncertain in flat areas. In current terrain analysis techniques, this error is ignored, and derived attributes are assumed to be accurate. Modifying the calculations and presentation of results to include estimates of error bounds may improve the usefulness of the results, or at least help reduce misinterpretation of the results.

Subgrid scale variability is a closely related issue. At fine scales, it may be safe to ignore subgrid scale variability for many purposes, but at the most useful resolutions, the variation in topography within the grid cell can be considerable. This is particularly problematic for compound attributes where the covariation of component attributes can distort the estimates at the grid scale. Erosion predictions based on stream power derived from whole-cell contributing area and slope are particularly awkward, since water generally converges to channels finer than the grid resolution and channel slope is often much lower than the slope of the adjacent hillslopes. Even at fine scales, calculation of stream power based on the assumption of uniform flow depth is problematic. Methods of accounting for this multiplicity of behaviors within a single cell are needed before these predictions can be readily applied.

Development of DEM generation methods to provide better terrain data is another important area of work. As pointed out in Chapter 2, the calculation of many topographic attributes depends on the representation of the shape of the terrain. This also relates to measures of accuracy, because elevation accuracy gives a poor assessment of shape accuracy. Conversely, and in this light not surprisingly, good control on shape can yield good control on elevation accuracy. This fact has guided the development of ANUDEM since its inception in 1988–89, and the primary assessment of accuracy has always been cast in terms of drainage as opposed to elevation accuracy. Recent changes to ANUDEM improve its ability to capture the shape represented by contours while reducing the bias to individual contour elevations, for example.

New DEM data sources are becoming more common, including global positioning systems, laser altimetry, and space-borne radar. At the time of this writing, the TOPSAR shuttle mission was orbiting the earth, acquiring 30 m resolution data for most of the earth's land surface. The quality of this data is yet to be established, particularly in relation to the representation of terrain shape and drainage structure, but high-resolution DEM data will continue to be more available in the next few years. This will enable much wider application of terrain analysis for a variety of purposes.

17.3 KNOWLEDGE OF RELATIONSHIPS

This book describes a number of applications of terrain analysis to spatial prediction of patterns in ecological systems, water in catchments, soil properties, and so on. In each case, a relationship is sought between the properties of interest and terrain attributes, and these relationships are built by statistical methods. The use of a soil-terrain model to estimate spatial patterns of soil organic carbon in Minnesota (Chapter 12) and the search for environmental conditions correlated with the spatial distribution of canopy tree species in the boreal forests of Canada (Chapter 16) illustrate just two of many possible approaches. There is no question as to the value of using topographic attributes for these types of applications, because there are numerous physical, chemical, and biological properties of landscapes that cannot be observed directly (Burrough and McDonnell 1998).

However, we have not yet built a consistent and interpretable picture of the link between various terrain attributes and the environmental parameters of interest. Each new application requires the collection of sample data and the construction of new relationships. The results summarized in Chapters 14–16, for example, show that our ability to transfer the predictions from one area to a different area with different characteristics is still very limited.

While the relationships derived by the statistical models appear to be useful, they are often not easy to interpret in terms of processes. Many of the environmental parameters have coevolved or, at least, vary together. For example, vegetation, soil properties, and water availability typically all vary down a hillslope in a systematic way that may be well described by terrain attributes such as contributing area and slope. The existence of such a relationship does not help to establish the causal links between the different components of the biogeochemical system.

To improve our ability to use terrain attributes to describe processes, we need to develop a better understanding of the links between terrain and spatio-temporally varying surface and atmospheric processes. We see the development of process-based terrain parameters and the use of statistics in calibrating process-based structures in hydrologically driven models as major ongoing activities. This is separate from the role of statistics in describing subgrid scale variability. The use of splines to calibrate process-based structures over empirically determined 2-D variation represents one possible approach (Moore and Hutchinson 1991). However, this may also require introducing some dynamics, such as seasonal variations, into the simple models underlying compound attributes, such as the topographic wetness index. The resulting models are likely to be more complex conceptually, but if they are calibrated with appropriate statistical methods, they should provide more accurate predictions of spatial patterns, such as topographic-dependent interpolation of surface climate (e.g., Hutchinson 1995, Running and Thornton 1996). These process-based models, which need not be more computationally complex, will provide a better understanding of the behavior of dynamic systems comprising linked hydrological, geomorphological, and ecological processes. The resulting models are more likely to be able to predict the response of such systems to changes induced by climate change or management.

17.4   SCALING

The sensitivity of terrain analysis results to the scale (or resolution) of the elevation data has been recognized for some time. Terrain attributes vary systematically as the resolution of the data changes (as illustrated in Chapter 5). The effect of this is that empirically derived relationships between terrain attributes and environmental parameters (presence of a tree species, thickness of soil, propensity for overland flow, etc.) are only valid at the scale for which they have been derived. Different resolution data, or even data at the same resolution but derived from source data at different scales, will require different models. Conceptually, there should be some underlying relationship between the real terrain and the parameters of interest that the terrain analysis is seeing through some scale-dependent filter. If we were able to describe that scale-dependent filter and know how it modifies the relationship, we would be able to know how the relationship appears at any scale. At this stage, little has been done beyond describing the change in terrain attributes with scale. The challenge is to model this scale-dependent filter, knowing that the filter itself will vary from one landscape to another. Ultimately, it is the "scales of interaction" of the various processes that are of real interest, because these affect our ability to predict the response of systems to environmental trends, like global warming and a variety of management interventions.

17.5   CONCLUDING REMARKS

Terrain analysis and related GIS and modeling techniques are being increasingly used for many applications in the environmental sciences, and the techniques are steadily advancing. Digital elevation data are rapidly becoming readily available at sufficiently fine resolution to be useful for a wide variety of purposes.

Future efforts that utilize terrain attributes to improve our understanding of the interactions between morphology and process across a variety of landscapes will probably incorporate concepts and tools drawn from dynamic modeling, geostatistics, and fuzzy classification. Procedures for creating generic, interactive models of key space-time processes, for example, may be combined with animated cartography and used to support the identification and visualization of new relationships linking morphology and process (e.g., Mitas et al. 1997, Burrough 1998). Similarly, the methods of geostatistics and spatial statistics provide rapidly expanding sets of tools for handling uncertainty and spatial correlation (Moore and Hutchinson 1991, Goovaerts 1997, Burrough and McDonnell 1998). Finally, the methods of fuzzy logic and continuous classification provide numerous concepts for dealing with inexact objects, as Irvin and Ventura have shown in Chapter 11.

Wilson and Burrough (1999) recently argued that the sustained advancement of knowledge in physical geography and related environmental fields is tied to the study of large quantities of data in a reproducible data-handling environment that extends from the field to the laboratory and the computer. The above-mentioned concepts, methods, and tools are required to help resolve differences, point out sources and consequences of uncertainty, and provide a range of alternatives for discussion and further evaluation. These concepts and tools may also advance the terrain analysis methods themselves to the extent that they will provide new opportunities for handling error and uncertainty and help with the identification of scale-dependent filters and handling of subgrid scale variability. Of one thing we can be certain, terrain attributes computed at regular intervals across large areas have featured prominently—and will continue to do so—in these types of applications in the future. Terrain analysis is likely to remain an important field of study, given this state of affairs.

REFERENCES

Aase, J. K., Frank, A. B., and Lorenz, R. J. 1987. Radiometric reflectance measurements of northern Great Plains rangeland and crested wheatgrass pastures. *Journal of Range Management* 40:299–302.

Ackermann, F. 1978. Experimental investigation into the accuracy of contouring from DTMs. *Photogrammetric Engineering and Remote Sensing* 44:1537–48.

Afyuni, M. M., Cassel, D. K., and Robarge, W. P. 1993. Effect of landscape position on soil water and corn silage yield. *Soil Science Society of America Journal* 57:1573–80.

Akima, H. 1978. A method of bivariate interpolation and smooth surface fitting for irregularly distributed data points. *ACM Transactions on Mathematical Software* 4:148–59.

Anderson, G. L., Hanson, J. D., and Hass, R. H. 1993. Evaluating *Landsat* Thematic Mapper derived vegetation indices for estimating above-ground biomass on semiarid rangelands. *Remote Sensing of Environment* 45:165–75.

Anderson, J. R., Hardy, R. E., Roach, J. T., and Witmer, R. E. 1976. *A Land Use and Land Cover Classification System for Use With Remote Sensor Data.* Arlington, VA: U.S. Geological Survey Professional Paper 964.

Anderson, K. E., and Furley, P. A. 1975. An assessment of the relationship between the surface properties of chalk soils and slope form using principal components analysis. *Journal of Soil Science* 26:131–43.

Anderson, M. G., and Burt, T. P. 1990. Subsurface runoff. In M. G. Anderson and T. P. Burt (eds.), *Process Studies in Hillslope Hydrology.* Chichester, UK: Wiley, 365–400.

Aspie, J. M. 1989. *Influence of Groundwater on Streambank Soil Moisture Content, Storm Runoff Production, and Sediment Transport in a Semi-arid Watershed.* M.S. Thesis, Montana State University.

Aspinall, R., and Veitch, N. 1993. Habitat mapping from satellite imagery and wildlife survey data using a Bayesian modeling procedure in GIS. *Photogrammetric Engineering and Remote Sensing* 59:537–43.

Auerbach, S., and Schaeben, H. 1990. Surface representation reproducing given digitized contour lines. *Mathematical Geology* 22:723–42.

Aull, G. H. 1980. *Cropland, Buffer, and Stream: A Case Study in Agricultural Nonpoint Source Water Pollution.* Ph.D. Thesis, Michigan State University.

Aumann, G., Ebner, H., and Tang, L. 1992. Automatic derivation of skeleton lines from digitized contours. *ISPRS Journal of Photogrammetry and Remote Sensing* 46:259–68.

Aurenhammer, F. 1991. Voronoi diagrams: a survey of a fundamental geometric data structure. *ACM Computing Surveys* 23:345–405.

429

Austin, M. P. 1985. Continuum concept, ordination methods, and niche theory. *Annual Review of Ecology and Systematics* 16:39–61.

Austin, M. P., and Heyligers, P. C. 1989. Vegetation survey design for conservation: gradsect sampling of forests in northeastern New South Wales. *Biological Conservation* 50:13–32.

Austin, M. P., and McKenzie, N. J. 1988. Data analysis. In R. H. Gunn, J. A. Beattie, R. E. Reid, and R. H. M. van de Graff (eds.), *Australian Soil and Land Survey Handbook: Guidelines for Conducting Surveys*. Melbourne: Inkata, 210–32.

Austin, M. P., and Smith, T. M. 1989. A new model for the continuum concept. *Vegetatio* 83:35–47.

Austin, M. P., Cunningham, R. B., and Good, R. B. 1983. Altitudinal distribution of several eucalypt species in relation to other environmental factors in southern New South Wales. *Australian Journal of Ecology* 8:169–80.

Austin, M. P., Cunningham, R. B., and Fleming, P. M. 1984. New approaches to direct gradient analysis using environmental scalars and statistical curve-fitting procedures. *Vegetatio* 55:11–27.

Austin, M. P., Nicholls, A. O., and Margules, C. R. 1990. Measurement of the realized quantitative niche: environmental niches of five *Eucalyptus* species. *Ecological Monographs* 60:161–77.

Austin, M. P., Nicholls, A. O., Doherty, M. D., and Meyers, J. A. 1994. Determining species response functions to an environmental gradient by means of a beta-function. *Journal of Vegetation Science* 5:215–28.

Aziz, F. P. 1989. *Soil Survey of Golden Valley County, North Dakota*. Washington, DC: United States Department of Agriculture, Soil Conservation Service.

Baker, F. S. 1944. Mountain climates of the western United States. *Ecological Monographs* 14:225–54.

Baker, W. L. 1989. Macro- and micro-scale influences on riparian vegetation in Western Colorado. *Annals of the Association of American Geographers* 79:65–78.

Band, L. E. 1986. Topographic partition of watersheds with digital elevation models. *Water Resources Research* 22:15–24.

Band, L. E. 1993a. Distributed parameterization of complex terrain. *Surveys in Geophysics* 12:249–70.

Band, L. E. 1993b. Extraction of channel networks and topographic parameters from digital elevation data. In K. J. Beven and M. J. Kirkby (eds.), *Channel Network Hydrology*. New York: Wiley, 17–42.

Band, L. E., Peterson, D. L., Running, S. W., Coughlan, J., Lammers, R., Dungan, J., and Nemani, R. 1991. Forest ecosystem processes at the watershed scale: basis for distributed simulation. *Ecological Modelling* 56:171–96.

Barker, R. D. 1981. *Soil Survey of Latah County Area*. Washington, DC: United States Department of Agriculture, Soil Conservation Service

Barling, R. D. 1992. *Saturation Zones and Ephemeral Gullies on Arable Land in Southeastern Australia*. Ph.D. Thesis, University of Melbourne.

Barling, R. D., Moore, I. D., and Grayson, R. B. 1994. A quasi-dynamic wetness index for characterizing the spatial distribution of zones of surface saturation and soil water content. *Water Resources Research* 30:1029–44.

Bates, P. D., Anderson, M. G., and Horrit, M. 1998. Terrain information in geomorphological models: stability, resolution, and sensitivity. In S. N. Lane, K. S. Richards, and J. H. Chandler (eds.), *Landform Monitoring, Modelling, and Analysis.* New York: Wiley, 279–310.

Bathurst, J. C. 1986. Physically based distributed modeling of an upland catchment using the Systeme Hydrologique European. *Journal of Hydrology* 87:79–102.

Bauer, M. E., Burk, T. E., Ek, A. R., Coppin, P. R., Lime, S. D., Walsh, T. A., Walters, D. K., Befort, W., and Heinzen, D. F. 1994. Satellite inventory of Minnesota forest resources. *Photogrammetric Engineering and Remote Sensing* 60:287–98.

Bayes, A. J., and Mackey, B. G. 1991. Algorithms for monotonic functions and their application to ecological studies in vegetation science. *Ecological Modelling* 56:135–59.

Beasley, D. B., and Huggins, L. F. 1982. *ANSWERS (Areal Nonpoint Source Watershed Environmental Response Simulation) User's Manual.* Chicago, IL: United States Environmental Protection Agency Report No. 905/9-82-001.

Beasley, D. B., Huggins, L. F., and Monke, E. J. 1980. A model for watershed planning. *Transactions for the American Society of Agricultural Engineers* 24:938–44.

Beattie, J. A. 1972. *Ground Surfaces of the Wagga Wagga Region, New South Wales.* Canberra: CSIRO Australian Soil Publication No. 28.

Beauchamp, R. M. 1986. *A Flora of San Diego County.* National City, CA: Sweetwater Press.

Beckett, P. H. T., and Webster, R. 1971. Soil variability: a review. *Soils and Fertilizers* 34:1–15.

Bell, J. C., Cunningham, R. L., and Havens, M. W. 1992. Calibration and validation of a soil-landscape model for predicting soil drainage class. *Soil Science Society of America Journal* 56:1860–6.

Bell, J. C., Cunningham, R. L., and Havens, M. W. 1994a. Soil drainage class probability mapping using a soil-landscape model. *Soil Science Society of America Journal* 58:464–70.

Bell, J. C., Thompson, J. A., Butler, C. A., and McSweeney, K. 1994b. Modeling soil genesis from a landscape perspective. In *Transactions of the Fifteenth World Congress of Soil Science, Acapulco, Mexico,* Volume 6A. Madison, WI: International Society of Soil Science, 179–90.

Bendix, J. 1994. Scale, direction, and pattern in riparian vegetation-environment relationships. *Annals of the Association of American Geographers* 84:652–65.

Bergeron, Y., and Dansereau, P. 1993. Predicting the composition of Canadian southern boreal forest in different fire cycles. *Journal of Vegetation Science* 4:827–32.

Betson, R. P., and Marius, J. B. 1969. Source areas of storm runoff. *Water Resources Research* 5:574–82.

Beven, K. J., 1981. Kinematic subsurface stormflow. *Water Resources Research* 17:1419–24.

Beven, K. J. 1987. Towards the use of catchment geomorphology in flood frequency predictions. *Earth Surface Processes and Landforms* 12:69–82.

Beven, K. J. 1989. Changing ideas in hydrology: the case of physically based models. *Journal of Hydrology* 105:157–72.

Beven, K. J., and Kirkby, M. J. 1979. A physically-based, variable contributing area model of basin hydrology. *Hydrological Sciences Bulletin* 24:43–69.

Beven, K. J., Calver, A., and Morris, E. M. 1987. *The Institute of Hydrology Distributed Model.* Wallingford, UK: Institute of Hydrology Report No. 98.

Bezdek, J. C., Trivedi, M., Ehrlich, R., and Full, W. 1981. Fuzzy clustering: a new approach for geostatistical analysis. *International Journal of Systems, Measurements, and Decisions* 2:13–23.

Bezdek, J. C., Ehrlich, R., and Full, W. 1984. FCM: the fuzzy *c*-means clustering algorithm. *Computers and Geosciences* 10:191–203.

Binford, M. W., and Buchenau, M. J. 1993. Riparian greenways and water resources. In D. S. Smith and P. C. Hellmund (eds.), *Ecology of Greenways*. Minneapolis, MN: University of Minnesota Press, 69–104.

Birkeland, P. W. 1984. *Soils and Geomorphology*. New York: Oxford University Press.

Birkeland, P. W. 1990. Soil-geomorphic research: a selective overview. *Geomorphology* 3:207–24.

Black, S. C., and Townsend, Y. E. 1997. Nevada Test Site Annual Site Environmental Report for Calendar Year 1996. Washington, DC: U.S. Department of Energy Report No. DOE/NV/11718-137.

Blaszczynski, J. S. 1997. Landform characterization with geographic information systems. *Photogrammetric Engineering and Remote Sensing* 63:183–91.

Blöschl, G., and Sivapalan, M. 1995. Scale issues in hydrological modelling: a review. *Hydrological Processes* 9:313–30.

Bluemle, J. P. 1975. *Guide to the Geology of Southwest North Dakota*. Bismarck, ND: North Dakota Geological Survey Educational Series No. 8.

Boast, R. R., and Shelito, R. G. 1989. *Soil Survey of Madison County Area, Montana*. Washington, DC: U.S. Department of Agriculture, Soil Conservation Service.

Boer, M., Del Barrio, G., and Puigdefábregas, J. 1996. Mapping soil depth classes in dry Mediterranean areas using terrain attributes derived from a digital elevation model. *Geoderma* 72:99–118.

Bolstad, P. V., and Lillesand, T. M. 1992. Improved classification of forest vegetation in northern Wisconsin through a rule-based combination of soils, terrain, and Landsat Thematic Mapper data. *Forest Science* 38:5–20.

Bolstad, P. V., and Stowe, T. 1994. An evaluation of DEM accuracy: elevation, slope, and aspect. *Photogrammetric Engineering and Remote Sensing* 60:1327–32.

Bonan, G. B., and Shugart, H. H. 1989. Environmental factors and ecological processes in boreal forests. *Annual Review of Ecology and Systematics* 20:1–28.

Botkin, D. B., Janak, J. F., and Wallis, J. R. 1972. Some ecological consequences of a computer model of forest growth. *Journal of Ecology* 60:849–73.

Bourgeron, P. S., Engelking, L. D., Humphries, H. C., Muldavin, E., and Moir, W. H. 1995. Assessing the conservation value of the Gray Ranch: rarity, diversity, and representativeness. *Desert Plants* 11:5–51.

Boyer, D. G., Wright, R. J., Winant, W. M., and Perry, H. D. 1990. Soil water relations on a hilltop cornfield in central Appalachia. *Soil Science* 149:383–92.

Brabyn, L. 1997. Classification of macro landforms using GIS. *ITC Journal* 14:26–40.

Brakensiek, D. 1967. Kinematic flood routing. *Transactions of the American Society of Agricultural Engineers* 11:340–3.

Brakke, T. W., and Kanemasu, E. T. 1981. Insolation estimation from satellite measurements. *Remote Sensing of Environment* 11:157–67.

Breimen, L., Friedman, J. H., Olshen, R. A., and Stone, C. J. 1984. *Classification and Regression Trees.* Belmont, CA: Wadsworth.

Briggs, I. C. 1974. Machine contouring using minimum curvature. *Geophysics* 39:39–48.

Bristow, K. L., and Campbell, G. S. 1984. On the relationship between incoming solar radiation and daily maximum and minimum temperature. *Agricultural and Forest Meteorology* 31:159–66.

Brown, D. A. 1993. Early nineteenth-century grasslands of the midcontinent plains. *Annals of the American Association of Geographers* 84:589–612.

Brown, D. B., and Bara, T. J. 1994. Recognition and reduction of systematic error in elevation and derivative surfaces from 7½-minute DEMs. *Photogrammetric Engineering and Remote Sensing* 60:189–94.

Brubaker, S. C., Jones, A. J., Frank, K., and Lewis, D. T. 1994. Regression models for estimating soil properties by landscape position. *Soil Science Society of America Journal* 58:1763–7.

Bruneau, P., Gascuel-Odoux, C., Robin, P., Merot, P., and Beven, K. 1995. Sensitivity to space and time resolution of a hydrological model using digital elevation data. *Hydrological Processes* 9:69–81.

Brutsaert, W. 1986. Catchment-scale evaporation and the atmospheric boundary layer. *Water Resources Research* 22:39–45.

Budyko, M. I. 1974. *Climate and Life.* New York, Academic Press.

Burrough, P. A. 1989. Fuzzy mathematical methods for soil survey and land evaluation. *Journal of Soil Science* 40:477–92.

Burrough, P. A. 1993. Soil variability: a late 20th century view. *Soils and Fertilizers* 56:529–62.

Burrough, P. A. 1996. Opportunities and limitations of GIS-based modeling of solute transport at the regional scale. In D. L. Corwin and K. Loague (eds.), *Application of GIS to the Modeling of Non-point Source Pollution in the Vadose Zone.* Madison, WI: Soil Science Society of America, 19–38.

Burrough, P. A. 1998. Dynamic modeling and geocomputation. In Longley, P. A., Brooks, S. M., McDonnell, R. A., and MacMillan, B. (eds.) *Geocomputation: A Primer.* New York: Wiley, 165–91.

Burrough, P. A., and McDonnell, R. A. 1998. *Principles of Geographical Information Systems.* Oxford, UK: Oxford University Press.

Burrough, P. A., van Gaans, P. F. M., and MacMillan, R. A. 2000a. High-resolution landform classification using fuzzy k-means. *Journal of Fuzzy Sets and Systems* 113:37–52.

Burrough, P. A., Wilson, J. P., van Gaans, P. F. M., and Hansen, A. J. 2000b. Fuzzy *k*-means classification of digital elevation models as an aid to forest mapping in the Greater Yellowstone Area, USA. *Landscape Ecology* (submitted).

Burt, T. P., and Butcher, D. P. 1985. Topographic controls of soil moisture distributions. *Journal of Soil Science* 36:469–86.

Busacca, A. J., Cook, C. A., and Mulla, D. J. 1993. Comparing landscape-scale estimation of soil erosion in the Palouse using Cs-137 and RUSLE. *Journal of Soil and Water Conservation* 48:361–7.

Butler, B. E. 1964. Assessing the soil factor in agricultural production. *Journal of Australian Institute of Agricultural Science* 30:232–40.

Butler, B. E., and Churchward, H. M. 1983. Aeolian processes. In *Soils: an Australian Viewpoint*. London: Academic, 91–100.

Calder, I. R. 1990. *Evaporation in the Uplands*. Chichester, UK: Wiley.

Calver, A. 1988. Calibration, sensitivity and validation of a physically-based rainfall-runoff model. *Journal of Hydrology* 103:103–15.

Calver, A., and Cammeraat, L. H. 1993. Testing a physically-based runoff model against field observations on a Luxembourg hillslope. *Catena* 20:273–88.

Calver, A. and Wood, W. L. 1989. On the discretization and cost-effectiveness of a finite element solution for hillslope subsurface flow. *Journal of Hydrology* 110:165–79.

Campbell, G. S. 1974. A simple method for determining unsaturated conductivity from moisture retention data. *Soil Science* 117:311–4.

Carrara, A., Bitelli, G., and Carla, R. 1997. Comparison of techniques for generating digital terrain models from contour lines. *International Journal of Geographical Information Science* 11:451–73.

Carsel, R. F., and Parrish, R. S. 1988. Developing joint probability distributions of soil water retention characteristics. *Water Resources Research* 24:755–69.

Carter, J. R. 1988. Digital representations of topographic surfaces. *Photogrammetric Engineering and Remote Sensing* 54:1577–80.

Carter, J. R. 1992. The effect of data precision on the calculation of slope and aspect using gridded DEMs. *Cartographica* 29:22–34.

Cary, G. J. 1998. *Predicting Fire Regimes and Their Ecological Effects in Spatially Complex Landscapes*. Ph.D. Thesis, Australian National University.

Chairat, S., and Delleur, J. W. 1993. Effects of the topographic index distribution on predicted runoff using GRASS. *Water Resources Bulletin* 29:1029–34.

Chang, K.-T., and Tsai, B.-W. 1991. The effect of DEM resolution on slope and aspect mapping. *Cartography and Geographical Information Systems* 18:69–77.

Cherubini, C., Cotechia, V., Renna, G., and Schiraldi, B. 1983. The use of bivariate probability density functions in Monte Carlo simulation of slope stability in soils. In *Proceedings of ICASP 4: The Fourth International Conference on Applications of Statistics and Probability in Soil and Structural Engineering, Firenze, Italy, Bologna, Italy, 13–17 June 1983*. Bologna: Pitagora Editrice, 1401–11.

Chesters, G., and Schierow, L. J. 1985. A primer on non-point pollution. *Journal of Soil and Water Conservation* 40:9–13.

Churchward, H. M., and Gunn, R. H. 1983. Stripping of deep weathered mantles and its significance to soil patterns. In *Soils: an Australian Viewpoint*. London: Academic, 73–82.

Clapp, R. B., and Hornberger, G. M. 1978. Empirical equations for some soil hydraulic properties. *Water Resources Research* 14:601–4.

Clark, L. A., and Pregibon, D. 1992. Tree-based models. In J. M. Chambers and T. J. Hastie (eds.), *Statistical Models in S*. New York: Chapman & Hall, 377–419.

Clarke, A. L., Grün, A., and Loon, J. C. 1982. The application of contour data for generating high fidelity grid digital elevation models. In *Proceedings of AutoCarto V*. Falls Church, VA: American Society of Photogrammetry and Remote Sensing, 213–22.

Clarke, K. C. 1990. *Analytical and Computer Cartography*. Englewood Cliffs, NJ: Prentice Hall.

Cohen, S. J. (ed.). 1994. *Proceedings of the Sixth Biennial AES/DIAND Meeting of Northern Climate and Mid Study Workshop of the Mackenzie Basin Impact Study.* Downsview: Environment Canada.

Cole, G., MacInnes, S., and Miller, J. 1990. Conversion of contoured topography to digital terrain data. *Computers and Geosciences* 16:101–9.

Collins, S. H., and Moon, G. C. 1981. Algorithms for dense digital terrain models. *Photogrammetric Engineering and Remote Sensing* 47:71–6.

Collins, S. L., Glenn, S. M., and Roberts, D. W. 1993. The hierarchical continuum concept. *Journal of Vegetation Science* 4:149–56.

Conacher, A. J., and Dalrymple, J. B. 1977. The nine unit landsurface model: an approach to pedogeomorphic research. *Geoderma* 18:1–154.

Congalton, R. G. 1991. A review of assessing the accuracy of classifications of remotely sensed data. *Remote Sensing of the Environment* 37:35–46.

Cook, S. E., Corner, R. J., Grealish, G. J., Gessler, P. E., and Chartres, C. J. 1996a. A rule based system to map soil properties. *Soil Science Society of America Journal* 60:1893–1900.

Cook, S. E., Corner, R. J., Groves, P. R., and Grealish, G. J. 1996b. Use of airborne gamma radiometric data for soil mapping. *Australian Journal of Soil Research* 34:183–94.

Cooper, J. R., Gilliam, J. W., Daniels, R. B., and Robarge, W. P. 1987. Riparian areas as filters for agricultural sediment. *Soil Science Society of America Journal* 51:416–20.

Costa-Cabral, M. C., and Burges, S. J. 1994. Digital elevation model networks (DEMON): a model of flow over hillslopes for computation of contributing and dispersal areas. *Water Resources Research* 30:1681–92.

Cramer, W., and Prentice, I. C. 1988. Simulation of regional soil moisture deficits on a European scale. *Norsk Geografisk Tidsskrift* 42:149–51.

Crist, P., and Cicone, R. C. 1984. A physically based transformation of Thematic Mapper data: the TM tassled cap. *IEEE Transactions on Geoscience and Remote Sensing* 22:256–63.

Crocker, R. L. 1952. Soil genesis and the pedogenic factors. *Biological Review* 27:139–68.

Crosson, L. S., and Protz, R. 1973. Prediction of soil properties from stereorthophoto measurements of landform properties. *Canadian Journal of Soil Science* 53:259–62.

Curran, P. 1980. Multispectral remote sensing of vegetation amount. *Progress in Physical Geography* 4:315–41.

Daly, C., Neilson, R. P., and Phillips, D. L. 1994. A statistical-topographic model for mapping climatological precipitation maps over mountainous terrain. *Journal of Applied Meteorology* 33:140–58.

David, M. B. 1988. Use of loss-on-ignition to assess soil orgainic carbon in forest soils. *Communications in Soil Science and Plant Analysis* 19:1593–9.

Davis, C. E., and Shovic, H. C. 1984. *Soil Survey of Gallatin National Forest.* Washington, DC: U.S. Department of Agriculture, Forest Service.

Davis, F. W., and Dozier, J. 1990. Information analysis of a spatial database for ecological land classification. *Photogrammetric Engineering and Remote Sensing* 56:605–13.

Davis, F. W., and Goetz, S. 1990. Modeling vegetation pattern using digital terrain data. *Landscape Ecology* 4:69–80.

Dawes, W. R., and Short, D. 1994. The significance of topology for modeling the surface hydrology of fluvial landscapes. *Water Resources Research* 30:1045–55.

Day, T., and Muller, J.-P. 1988. Quality assessment of digital elevation models produced by automatic stereo matchers from SPOT image pairs. *International Archives of Photogrammetry and Remote Sensing* 27:148–59.

De Grandpre, L., Gagnon, L., and Bergeron, Y. 1993. Changes in the understory of Canadian southern boreal forest after fire. *Journal of Vegetation Science* 4:803–10.

Denning, P. J. 1990. Modeling reality. *American Scientist* 76:495–8.

Dent, D., and Young, A. 1981. *Soil Survey and Land Evaluation.* London: Allen & Unwin.

Desmet, P. J. J., and Govers, G. 1996a. Comparison of routing algorithms for digital elevation models and their implications for predicting ephemeral gullies. *International Journal of Geographical Information Systems* 10:311–31.

Desmet, P. J. J., and Govers, G. 1996b. A GIS procedure for automatically calculating the USLE LS factor on topographically complex landscape units. *Journal of Soil and Water Conservation* 51:427–33

Desmet, P. J. J., and Govers, G. 1997. Comment on 'Modelling topographic potential for erosion and deposition using GIS.' *International Journal of Geographical Information Science* 11:603–10.

Deumlich, D. 1993. Beitrag zur Erarbeitung einer Isoerodentkarte Deutschlands. *Archive fur Acker- und Pflanzenbau und Bodenkunde* 37:17–24.

Deumlich, D., and Frielinghaus, M. 1994. Eintragspfade Bodenerosion und Oberflächenabfluß im Lockergesteinsbereich. In W. Werner and W.-P. Wodsak (eds.), *Stickstoff- und Phosphateintrag in Fließgewässern Deutschlands unter besonderer Berücksichtigung des Eintragsgeschehens im Lockergesteinsbereich der ehemaligen DDR.* Frankfurt: Verlagsunion AGRAR, 48–84.

Dickinson, R. E. 1984. Modelling evapotranspiration for three-dimensional global climate models. In J. E. Hanson and T. Takahashi (eds.), *Climate Processes and Climate Sensitivity.* Falls Church, VA: American Geophysical Union, 58–72.

Dicks, S. E., and Lo, T. H. C. 1990. Evaluation of thematic map accuracy in a land-use and land-cover mapping program. *Photogrammetric Engineering and Remote Sensing* 56:1247–52.

Dietrich, W. E., Wilson, C. J., and Reneau, S. L. 1986. Hollows, colluvium and landslides in soil-mantled landscapes. In A. D. Abrahams (ed.), *Hillslope Processes.* London: Allen & Unwin, 361–88.

Dietrich, W. E., Wilson, C. J., Montgomery, D. R., McKean, J., and Bauer, R. 1992. Erosion thresholds and land surface morphology. *Geology* 20:675–9.

Dietrich, W. E., Wilson, C. J., Montgomery, D. R., and McKean, J. 1993. Analysis of erosion thresholds, channel networks and landscape morphology using a digital terrain model. *Journal of Geology* 101:259–78.

Dietrich, W. E., Reiss, R., Hsu, M. L., and Montgomery, D. R. 1995. A process-based model for colluvial soil depth and shallow landsliding using digital elevation data. *Hydrological Processes* 9:383–400.

Dikau, R. 1989. The application of a digital relief model to landform analysis in geomorphology. In J. Raper (ed.), *Three Dimensional Applications of Geographic Information Systems.* London: Taylor & Francis: 55–77.

Dillaha, T. A., Sherrard, J. H., and Lee, D. 1989. Long-term effectiveness of vegetated filter strips. *Water Environment and Technology* 1:417–21.

Dixon, L. F. J., Barker, R., Bray, M., Farres, P., Hooke, J., Inkpen, R., Merel, A., Payne, D., and Shelford, A. 1998. Analytical photogrammetry for geomorphological research. In S. N.

Lane, K. S. Richards, and J. H. Chandler (eds.), *Landform Monitoring, Modelling, and Analysis.* New York: Wiley, 63–94.

Dixon, T. H. 1991. An introduction to the global positioning system and some geological applications. *Reviews of Geophysics* 29:249–76.

Dixon, T. H. (ed.). 1995. *SAR Interferometry and Surface Change Detection.* Miami, FL: University of Miami RSMAS Technical Report TR 95-003.

Douglas, D. H. 1986. Experiments to locate ridges and channels to create a new type of digital elevation model. *Cartographica* 23:29–61.

Dozier, J., Bruno, J., and Downey, P. 1981. A faster solution to the horizon problem. *Computers and Geosciences* 7:145–51.

Dubayah, R. 1992. Estimating net solar radiation using Landsat Thematic Mapper and digital elevation data. *Water Resources Research* 28:2469–84.

Dubayah, R. 1994. A solar radiation topoclimatology for the Rio Grande River Basin. *Journal of Vegetation Science* 5:627–40.

Dubayah, R., and Rich, P. M. 1995. Topographic solar radiation models for GIS. *International Journal of Geographical Information Systems* 9:405–19.

Dubayah, R., and van Katwijk, V. 1992. The topographic distribution of annual incoming solar radiation in the Rio Grande River basin. *Geophysical Letters* 19:2231–4.

Duda, A. M., and Johnson, R. J. 1985. Cost effective targeting of agricultural non-point source pollution controls. *Journal of Soil and Water Conservation* 40:108–11.

Duda, R. O., and Hart, P. E. 1973. *Pattern Recognition and Scene Analysis.* New York: Wiley.

Dunne, T. 1978. Field studies of hillslope processes. In M. J. Kirkby (ed.), *Hillslope Hydrology.* New York: Wiley, 227–93.

Dunne, T., and Black, R. D. 1970. Partial area contributions to storm runoff in a small New England watershed. *Water Resources Research* 6:1296–1311.

Dunne, T., and Leopold, L. B. 1978. *Water in Environmental Planning.* New York: W. H. Freeman.

Dymond, J. R., Derose, R. C., and Harmsworth, G. R. 1995. Automated mapping of land components from digital elevation data. *Earth Surface Processes and Landforms* 20:131–7.

Dyrness, C. T. 1969. *Hydrologic Properties of Soils on Three Small Watersheds in the Western Cascades of Oregon.* Portland, OR: United States Department of Agriculture, Forest Service, Pacific Northwest Forest and Range Experiment Station Research Note No. PNW-111.

Eberhardt, K. E., and Woodward, P. M. 1987. Distribution of residual vegetation associated with large fires in Alberta. *Canadian Journal of Forest Research* 17:1207–12.

Ebner, H., Reinhardt, W., and Hössler, R. 1988. Generation, management and utilization of high fidelity digital terrain models. *International Archives of Photogrammetry and Remote Sensing* 27:556–65.

Ecological Stratification Working Group. 1995. *A National Ecological Framework for Canada.* Ottawa/Hull: Agricultural and Agri-Food Canada, Research Branch, Centre for Land and Biological Resources Research and Environment Canada, State of the Environment Directorate, Ecozone Analysis Branch (report and national map at scale of 1:17.5 million).

Edwards, N. T. 1975. Effects of temperature and moisture on carbon dioxide evolution in a mixed deciduous forest floor. *Soil Science Society of America Journal* 39:361–5.

Eklundh, L., and Martensson, U. 1995. Rapid generation of digital elevation models from topographic maps. *International Journal of Geographical Information Systems* 9:329–40.

Engel, R., McFarlane, D. J., and Street, G. 1987. The influence of dolerite dykes on saline seeps in southwestern Australia. *Australian Journal of Soil Research* 25:125–36.

Environment Canada. 1982a. *Canadian Climate Normals,* Volume 2: *Temperature.* Ottawa: Atmospheric Environment Service, Environment Canada.

Environment Canada. 1982b. *Canadian Climate Normals,* Volume 3: *Precipitation.* Ottawa: Atmospheric Environment Service, Environment Canada.

Eswaran, H., van den Berg, E., and Reich, P. 1993. Organic carbon in soils of the world. *Soil Science Society of America Journal* 57:192–4.

Evans, I. S. 1980. An integrated system of terrain analysis and slope mapping. *Zeitschrift für Geomorphologie,* Supplement Band 36:274–95.

Fairfield, J., and Leymarie, P. 1991. Drainage networks from grid digital elevation models. *Water Resources Research* 27:709–717, 2809.

Farrar, J. L. 1995. *Trees in Canada.* Markham: Fitzhenry & Whiteside.

Federal Geographic Data Committee Vegetation Subcommittee. 1996. *FGDC Vegetation Classification and Information Standards.* Reston, VA: Federal Geographic Data Committee.

Feddes, R. A., Kowalik, P., Kolinska-Malinka, K., and Zaradny, H. 1976. Simulation of field water uptake by plants using a soil water dependent root extraction function. *Journal of Hydrology* 31:13–26.

Felicisimo, A. M. 1994. Parametric statistical method for error detection in digital elevation models. *ISPRS Journal of Photogrammetry and Remote Sensing* 49:29–33.

Fels, J. E. 1994. *Modeling and Mapping Potential Vegetation Using Digital Terrain Data.* Ph.D. Thesis, North Carolina State University.

Fenske, P., and Carnahan, C. 1975. *Water Table and Related Maps for the Nevada Test Site and Central Test Area.* Reno, NV: University of Nevada, Desert Research Institute Publication No. 45009.

Ferren, W. R., Jr., and Fiedler, P. L. 1993. Rare and threatened wetlands of central and southern California. In J. E. Keeley (ed.), *Interface Between Ecology and Land Development in California.* Fairfield, WA: International Association of Wildland Fires, 119–32.

Fix, R. E., and Burt, T. P. 1995. Global Positioning System: an effective way to map a small area or catchment. *Earth Surface Processes and Landforms* 20:817–27.

Flannigan, M. D., and van Wagner, C. E. 1991. Climate change and wildfire in Canada. *Canadian Journal of Forest Research* 21:66–72.

Fleming, P. M. 1987. *Notes on a Radiation Index for Use in Studies of Aspect Effects on Radiation Climate.* Canberra: Commonwealth Scientific and Industrial Research Organization, Division of Water Resources Research, Institute of Biological Resources Technical Memorandum.

Florinsky, I. V. 1998. Accuracy of local topographic variables derived from digital elevation models. *International Journal of Geographical Information Science* 12:47–62.

Forest Service. 1981. *CALVEG: a Classification of California Vegetation.* San Francisco: Regional Ecology Group.

Forest Service. 1994. *Forest Inventory Analysis User's Guide.* Sacramento, CA: United States Department of Agriculture, Forest Service Region 5 Remote Sensing Laboratory.

Forman, R. T. T. 1964. Growth under controlled conditions to explain the hierarchical distributions of a moss *Teraphis pellucidia. Ecological Monographs* 34:1–25.

Foster, D. R. 1983. The history and pattern of fire in the boreal forest of southeastern Labrador. *Canadian Journal of Botany* 61:2459–71.

Foster, G. R. 1990. Process-based modeling of soil erosion by water on agricultural land. In J. R. Boardman, I. D. L. Foster, and J. A. Dearing (eds.), *Soil Erosion on Agricultural Land.* New York: Wiley, 429–45.

Foster, G. R. 1994. Comment on "Length-slope factors for the Revised Universal Soil Loss Equation: Simplified method of estimation." *Journal of Soil and Water Conservation* 49:171–3.

Foth, H. 1984. *Fundamentals of Soil Science,* 7th ed. New York: Wiley.

Franchini, M., Wendling, J., Obled, C., and Todini, E. 1996. Physical interpretation and sensitivity analysis of TOPMODEL. *Journal of Hydrology* 175:293–338.

Franke, R. 1982. Smooth interpolation of scattered data by local thin plate splines. *Computers and Mathematics with Applications* 8:273–81.

Franklin, J. 1991. Land cover stratification using Landsat Thematic Mapper data in Sahelian and Sudanian woodlands and wooded grassland. *Journal of Arid Environments* 20:141–63.

Franklin, J. 1995. Predictive vegetation mapping: geographic modeling of biospatial patterns in relation to environmental gradients. *Progress in Physical Geography* 19:474–99.

Franklin, J. 1998. Predicting the distributions of shrub species in California chaparral and coastal sage communities from climate and terrain-derived variables. *Journal of Vegetation Science* 9:733–48.

Franklin, J., and Stephenson, J. 1996. Integrating GIS and remote sensing to produce regional vegetation databases: attributes related to environmental modeling. In NCGIA (ed.), *Proceedings of the Third International Conference on Integrating GIS and Environmental Modeling, Santa Fe, New Mexico, 21–25 January 1996.* Santa Barbara, CA: National Center for Geographic Information and Analysis, University of California: CD-ROM and WWW.

Franklin, J., and Woodcock, C. E. 1997. Multiscale vegetation data for the mountains of southern California: spatial and categorical resolution. In D. A. Quattrochi and M. F. Goodchild (eds.), *Scaling of Remote Sensing Data for GIS.* Boca Raton, FL: CRC/Lewis, 141–68.

Franklin, J., Logan, T., Woodcock, C. E., and Strahler, A. H. 1986. Coniferous forest classification and inventory using Landsat and digital terrain data. *IEEE Transactions on Geoscience and Remote Sensing* 24:139–49.

Franzreb, K. E. 1990. An analysis of options for reintroducing a migratory, native passerine, the endangered Least Bell's Vireo (*Vireo bellii pusillus*) in the Central Valley in California. *Biological Conservation* 53:105–12.

Frederiksen, P., Jacobi, O., and Kubik, K. 1985. A review of current trends in terrain modelling. *ITC Journal* 2:101–6.

Freeman, G. T. 1991. Calculating catchment area with divergent flow based on a regular grid. *Computers and Geosciences* 17:413–22.

Freeze, R. A., and Cherry, J. A. 1979. *Groundwater.* Englewood Cliffs, NJ: Prentice-Hall.

Friedl, M. A., Davis, F. W., Michaelsen, J., and Moritz, M. A. 1995. Scaling and uncertainty in the relationship between the NDVI and land surface biophysical variables: an analysis using a scene simulation model and data from FIFE. *Remote Sensing of the Environment* 54:233–46.

Fritz, S., and MacDonald, T. H. 1949. Average solar radiation in the United States. *Air Conditioning, Heating and Ventilating* 46:61–4.

Fryer, J. G., Chandler, J. H., and Cooper, M. A. H. 1994. On the accuracy of heighting from aerial photographs and maps: implications to process modellers. *Earth Surface Processes and Landforms* 19:577–583.

Fuller, R. M., Groom, G. B., and Jones, A. R. 1994. The land cover map of Great Britain: an automated classification of Landsat TM data. *Photogrammetric Engineering and Remote Sensing* 60:553–62.

Furley, P. A. 1968. Soil formation and slope development, II: the relationship between soil formation and gradient angle in the Oxford area. *Zeitschrift für Geomorphologie* 12:25–42.

Furley, P. A. 1974a. Soil–slope–plant relationships in the northern Maya Mountains, Belize, Central America, I: the sequence over metmorphic sandstones and shales. *Journal of Biogeography* 1:171–86.

Furley, P. A. 1974b. Soil–slope–plant relationships in the northern Maya Mountains, Belize, Central America, II: the sequence over phyllites and granites. *Journal of Biogeography* 1:263–79.

Furley, P. A. 1976. Soil–slope–plant relationships in the northern Maya Mountains, Belize, Central America, III: variations in the properties of soil profiles. *Journal of Biogeography* 3:303–19.

Gallant, J. C. 1996. Topographic analysis using elevation residuals. Centre for Resource and Environmental Studies, The Australian National University, Canberra. *http://cres20.anu .edu.au/~johng/elevresid.html.*

Gallant, J. C. 1997. Modeling solar radiation in the forests of southeastern Australia. Unpublished paper presented at the International Congress on Modeling and Simulation (MODSIM97), Hobart, Tasmania.

Gallant, J. C., and Hutchinson, M. F. 1996. Towards an understanding of landscape scale and structure. In NCGIA (ed.), *Proceedings of the Third International Conference Integrating GIS and Environmental Modeling.* Santa Barbara, CA: University of California, National Center for Geographic Information and Analysis: CD-ROM and WWW.

Gallant, J. C., and Hutchinson, M. F. In preparation. Specific catchment area and its relationship to plan curvature.

Gallant, J. C., and Wilson, J. P. 1996. TAPES-G: a grid-based terrain analysis program for the environmental sciences. *Computers and Geosciences* 22:713–22.

Gao, J. 1994. A C program for detecting slope forms from grid DEMs. *Earth Surface Processes and Landforms* 19:827–37.

Garbrecht, J., and Martz, L. W. 1994. Grid size dependency of parameters extracted from digital elevation models. *Computers and Geosciences* 20:85–7.

Garbrecht, J., and Starks, P. 1995. Note on the use of USGS level 1 7.5–minute DEM coverages for landscape drainage analyses. *Photogrammetric Engineering and Remote Sensing* 61:519–22.

Gaskin, J. W., Dowd, J. F., Nutter, W. L., and Swank, W. T. 1989. Vertical and lateral components of soil nutrient flux in a hillslope. *Journal of Environmental Quality* 18:403–10.

Gates, D. M. 1980. *Biophysical Ecology.* New York: Springer.

Gates, D. M., Keegan, H. J., Schleter, J. C., and Weidner, V. R. 1965. Spectral properties of plants. *Applied Optics* 4:11–20.

Gauthier, S., Gagnon, J., and Bergeron, Y. 1993. Population age structure of *Pinus Banksiana* at the southern edge of the Canadian boreal forest. *Journal of Vegetation Science* 4:783–90.

Gerrard, A. J. 1981. *Soils and Landforms: An Integration of Geomorphology and Pedology.* London: Allen & Unwin.

Gerrard, A. J. 1990. Soil variations on hillslopes in humid temperate climates. *Geomorphology* 3:225–44.

Gerrard, A. J. 1992. *Soil Geomorphology.* London: Chapman & Hall.

Gerrard, R. A., Church, R. L., Stoms, D. M., and Davis, F. W. 1997. Selecting conservation reserves using species covering models: adapting the ARC/INFO GIS. *Transactions in GIS* 2:45–60.

Gessler, P. E. 1996. *Statistical Soil–Landscape Modeling for Environmental Management.* Ph.D. Thesis, Australian National University.

Gessler, P. E., Moore, I. D., McKenzie, N. J., and Ryan, P. J. 1995. Soil–landscape modeling and spatial prediction of soil attributes. *International Journal of Geographic Information Systems* 9:421–32.

Gessler, P. E., McKenzie, N. J., and Hutchinson, M. F. 1996. Progress in soil–landscape modelling and spatial prediction of soil attributes for environmental models. In NCGIA (ed.) *Proceedings of the Third International Conference Integrating GIS and Environmental Modeling.* Santa Barbara, CA: University of California, National Center for Geographic Information and Analysis: CD-ROM and WWW.

Giles, P. T. 1998. Geomorphological signatures: classification of aggregated slope unit objects from digital elevation and remote sensing data. *Earth Surface Processes and Landforms* 20:581–94.

Giles, P. T., and Franklin, S. E. 1996. Comparison of derivative topographic surfaces of a DEM generated from stereoscopic SPOT images with field measurements. *Photogrammetric Engineering and Remote Sensing* 62:1165–71.

Gill, A. M. 1975. Fire and the Australian flora: a review. *Australian Forestry* 38:4–25.

Glassy, J. M., and Running, S. W. 1994. Validating diurnal climatology logic of the MT-CLIM model across a climatic gradient in Oregon. *Ecological Applications* 4:248–57.

Goodchild, M. F. 1994. Integrating GIS and remote sensing for vegetation analysis and modeling: methodological issues. *Journal of Vegetation Science* 5:615–26.

Goodchild, M. F., and Mark, D. M. 1987. The fractal nature of geographic phenomena. *Annals of the Association of American Geographers* 77:265–78.

Goodrich, D. C. 1990. *Geometric Simplification of a Distributed Rainfall–Runoff Model Over a Large Range of Basin Scales.* Ph.D. Thesis, University of Arizona.

Goovaerts, P. 1997. *Geostatistics for Natural Resources Evaluation.* Oxford, Oxford University Press.

Gordon, H., and White, T. C. 1994. *Ecological Guide to Southern California Chaparral Plant Series.* San Diego, CA: USDA Forest Service, Pacific Southwest Region.

Gorham, E. 1991. Northern peatlands: role in the carbon cycle and probable responses to global warming. *Ecological Applications* 1:182–95.

Graf, W. L. 1982. Tamarisk and river channel management. *Environmental Management* 6:283–96.

Gray, C. A. 1995. *Predicting the Location of Riparian Vegetation using Landsat TM and Digital Terrain Data in the Cleveland National Forest.* M.A. Thesis, San Diego State University.

Gray, S. L. 1995. *A Descriptive Forest Inventory of Canada's Forest Regions.* Chalk River, Natural Resources Canada, Canadian Forest Service, Petawawa National Forestry Institute Informal Report No PI-X-122

Grayson, R. B., and Moore, I. D. 1991. Effect of land surface configuration on catchment hydrology. In A. J. Parsons and A. D. Abrahams (eds.), *Hydraulics and Erosion Mechanics of Overland Flow.* London: UCL, 147–75.

Grayson, R. B., Moore, I. D., and McMahon, T. A. 1992a. Physically based hydrologic modeling, 1: a terrain-based model for investigative purposes. *Water Resources Research* 28:2639–58.

Grayson, R. B., Moore, I. D., and McMahon, T. A. 1992b. Physically based hydrologic modeling, 2: is the concept realistic? *Water Resources Research* 28:2659–66.

Grayson, R. B., Blöschl, G., Barling, R. D., and Moore, I. D. 1993. Process, scale, and constraints to hydrological modelling in GIS. In K. Kovar and H. P. Nachtnebel (eds.), *Application of Geographic Information Systems in Hydrology and Water Resources: Proceedings of the HydroGIS 93 Conference held in Vienna, April 1993*. Wallingford, UK: International Association of Hydrological Sciences Publication No. 211:83–92.

Grayson, R. B., Blöschl, G., and Moore, I. D. 1994. Distributed parameter hydrologic modelling using vector elevation data: THALES and TAPES-C. In V. P. Singh (ed.), *Computer Models of Watershed Hydrology*. Boca Raton, FL: CRC, 669–96.

Griffin, M. L., Beasley, D. B., Fletcher, J. J., and Foster, G. R. 1988. Estimating soil loss on topographically non-uniform field and farm units. *Journal of Soil and Water Conservation* 43:326–31.

Griffith, D. A., and Amrhein, C. G. 1991. *Statistical Analysis for Geographers*. Englewood Cliffs, NJ: Prentice-Hall.

Grigal, D. F., and Ohmann, L. F. 1992. Carbon storage in upland forests of the lake states. *Soil Science Society America Journal* 56:935–43.

Grigal, D. F., Brovold, S. L., Nord, W. S., and Ohmann, L. F. 1989. Bulk density of surface soils and peat in the North Central United States. *Canadian Journal Soil Science* 69:895–900.

Grigal, D. F., Chamberlain, L. M., Finney, H. R., Wroblewski, D. V., and Gross, E. R. 1974. *Soils of the Cedar Creek Natural History Area*. St. Paul, MN: Minnesota Agricultural Experiment Station Miscellaneous Report No. 123.

Guercio, R., and Soccodato, F. M. 1996. GIS procedure for automatic extraction of geomorphological attributes from TIN-DTMs. In K. Kovar and H. P. Nachtnebel (eds.) *Application of Geographic Information Systems in Hydrology and Water Resources: Proceedings of the HydroGIS 96 Conference held in Vienna, April 1996*. Wallingford, UK: International Association of Hydrological Sciences Publication No. 235:175–82.

Hack, J. T., and Goodlett, J. C. 1960. *Geomorphology and Forest Ecology of a Mountain Region in the Central Appalachians*. Washington, DC, United States Geological Survey Paper 347.

Hairsine, P. B., and Rose, C. W. 1991. Rainfall detachment and deposition: sediment transport in the absence of flow-driven processes. *Soil Science Society of America Journal* 55:320–4.

Hairsine, P. B., and Rose, C. W. 1992a. Modelling water erosion due to overland flow using physical principles, I: sheet flow. *Water Resources Research* 28:237–43.

Hairsine, P. B., and Rose, C. W. 1992b. Modelling water erosion due to overland flow using physical principles, II: rill flow. *Water Resources Research* 28:245–50.

Hall, G. F., and Olson, C. G., 1991. Predicting variability of soils from landscape models. In M. J. Mausbach and L. P. Wilding (eds.), *Spatial Variability of Soils and Landforms*. Madison, WI: Soil Science Society of America, 9–24.

Halladin, S., Grip, H., and Perttu, K. 1979. Model for energy exchange of a pine forest canopy. In S. Halladin (ed.), *Comparison of Forest Water and Energy Exchange Models*. Copenhagen: International Society of Ecological Modelling, 59–75.

Hammer, R. D., Young, F. J., Wollenhaupt, N. C., Barney, T. L., and Haithcoate, T. W. 1995. Slope class maps from soil survey and digital elevation models. *Soil Science Society of America Journal* 59:509–19.

Hammond, C. D., Hall, D., Miller, S., and Swetik, P. 1992. *Level I Stability Analysis (LISA): Documentation for Version 2.0.* Ogden, UT: United States Department of Agriculture, Forest Service, Intermountain Research Station General Technical Report No. 285.

Hanes, T. J. 1977. Chaparral. In M. G. Barbour and J. Major (eds.), *Terrestrial Vegetation of California.* New York: Wiley, 417–69.

Hannah, M. J. 1981. Error detection and correction in digital elevation models. *Photogrammetric Engineering and Remote Sensing* 47:63–9.

Harding, D. J., Bufton, J. L., and Frawley, J. 1994. Satellite laser altimetry of terrestrial topography: vertical accuracy as a function of surface slope, roughness and cloud cover. *IEEE Transactions on Geoscience and Remote Sensing* 32:329–39.

Harris, R. R. 1987. Occurrence of vegetation on geomorphic surfaces in the active floodplain of a California alluvial stream. *American Midland Naturalist* 118:393–405.

Hartmann, K. 1988. *Untersuchungen zur Erosivität der Niederschläge unter Berücksichtigung bodenphysikalischer Kenngrößen auf Jungmoränen-Standorten der DDR.* Thesis, AdL d DDR Berlin.

Heckert, P. 1982. Color image quantization for frame buffer display. *Computer Graphics* 16:297–307.

Heller, M. 1990. Triangulation algorithms for adaptive terrain modelling. In *Proceedings of the Fourth International Symposium on Spatial Data Handling.* Columbus, OH: International Geographical Union, 163–74.

Helvey, J. D., and Patric, J. H. 1965. Canopy and litter interception of rainfall by hardwoods of Eastern United States. *Water Resources Research* 1:193–206.

Helvey, J. D., and Patric, J. H. 1988. Research on interception losses and soil moisture relationships. In W. T. Swank and D. A. Crossley (eds.), *Forest Hydrology and Ecology at Coweeta.* New York: Springer, 129–37.

Helvey, J. D., Hewlett, J. D., and Douglass, J. E. 1972. Predicting soil moisture in the southern Appalachians. *Soil Science Society of America Proceedings* 36:954–9.

Henderson-Sellers, A. 1990. Evaluation for the continent of Australia of the simulation of the surface climate using the Biosphere-Atmosphere Transfer Scheme (BATS) coupled into a global climate model. *Climate Research* 1:43–62.

Henderson-Sellers, A., and Robinson, P. J. 1986. *Contemporary Climatology.* Harlow, Longman.

Hession, W. C., and Shanholtz, V. O. 1988. A geographic information system for targeting non-point source agricultural pollution. *Journal of Soil and Water Conservation* 43:264–6.

Hetrick, W. A., Rich, P. M., Barnes, F. J., and Weiss, S. B. 1993a. GIS-based solar radiation flux models. In *Proceedings of the ASPRS-ACSM Annual Convention,* Vol. 3. Bethesda, MD: American Society for Photogrammetry and Remote Sensing, 132–43.

Hetrick, W. A., Rich, P. M., and Weiss, S. B. 1993b. Modeling insolation on complex surfaces. In *Proceedings of the Thirteenth Annual ESRI Users Conference,* Vol. 2. Redlands, CA: Environmental Systems Research Institute: 447–58.

Hewitt, A. E. 1993. Predictive modeling in soil survey. *Soils and Fertilizers* 3:305–15.

Hewlett, J. D. 1962 *Internal Water Balance of Forest Trees on the Coweeta Watershed.* Ph.D. Thesis, Duke University.

Hewlett, J. D., and Hibbert, A. R. 1963. Moisture and energy conditions within a sloping soil mass during drainage. *Journal of Geophysical Research* 68:1080–7.

Hewlett, J. D., and Hibbert, A. R. 1967. Factors affecting the response of small watersheds to precipitation in humid areas. In W. E. Soper and H. W. Lull (eds.), *International Symposium on Forest Hydrology.* New York: Pergamon, 275–90.

Hill, M. O. 1979. *TWINSPAN: A FORTRAN Program for Arranging Multivariate Data in an Ordered Two-way Table by Classification of the Individuals and Attributes.* Ithaca, NY: Cornell University.

Hinton, M. J., Schiff, S. L., and English, M. C. 1993. Physical properties governing groundwater flow in a glacial till catchment. *Journal of Hydrology* 142:229–49.

Hobbs, F. 1995. The rendering of relief images from digital contour data. *The Cartographic Journal* 32:111–6.

Hodgson, M. E., 1995. What cell size does the computed slope/aspect angle represent? *Photogrammetric Engineering and Remote Sensing* 61:513–7.

Holland, R. F. 1986. *Preliminary Descriptions of the Terrestrial Natural Communities of California.* Sacramento, CA: The Resource Agency, Nongame Heritage Program, California Department of Fish and Game.

Holmgren, P. 1994. Multiple flow direction algorithms for runoff modeling in grid-based elevation models: an empirical evaluation. *Hydrological Processes* 8:327–34.

Hoosbeek, M. R., and Bryant, R. B. 1992. Towards the quantitative modeling of pedogenesis: a review. *Geoderma* 55:183–210.

Hornberger, G. M., and Cosby, B. J. 1985. Selection of parameter values in environmental models using sparse data: a case study. *Applied Mathematics and Computation* 17:335–55.

Hornberger, G. M., Mills, A. L., and Herman, J. S. 1992. Bacterial transport in porous media: evaluation of a model using laboratory observation. *Water Resources Research* 28:915–23.

Houghton, H. G. 1954. On the annual heat balance of the Northern Hemisphere. *Journal of Meteorology* 11:1–9.

Houghton, R. A. 1995. Changes in the storage of terrestrial carbon since 1850. In R. Lal, J. Kimble, E. Levine, and B. Stuart (eds.), *Soils and Global Change.* London: Lewis/CRC Press: 45–66.

Hubbard, J. P. 1977. Importance of riparian ecosystems: biotic considerations. In R. R. Johnson and D. A. Jones (eds.) *Importance, Preservation and Management of Riparian Habitat: A Symposium.* Fort Collins, CO: USDA Forest Service: 14–8.

Hubble, G. D., and Isbell, R. F. 1983. Eastern Highlands. In *Soils: an Australian Viewpoint.* London: Academic, 219–30.

Hudson, B. D. 1990. Concepts of soil mapping and interpretation. *Soil Survey Horizons* 31:63–72.

Hudson, B. D. 1992. The soil survey as a paradigm-based science. *Soil Science Society of America Journal* 56:836–41.

Huggett, R. J. 1975. Soil landscape systems: a model of soil genesis. *Geoderma* 13:1–22.

Hungerford, R. D., Nemani, R. R., Running, S. W., and Coughlan, J. C. 1989. *MTCLIM: A Mountain Microclimate Simulation Model.* Ogden, UT: United States Department of Agriculture, Forest Service, Intermountain Research Station Research Paper No. INT-414.

Hunsaker, C. T., and Levine, D. A. 1995. Hierarchical approaches to the study of water quality in rivers. *Bioscience* 45:193–203.

Hunter, G. J., and Goodchild, M. F. 1995. Dealing with error in spatial databases: a simple case study. *Photogrammetric Engineering and Remote Sensing* 61:529–37.

Hunter, G. J., and Goodchild, M. F. 1996. Communicating uncertainty in spatial databases. *Transactions in GIS* 1:13–24.

Hunter, G. J., Caetano, M., and Goodchild, M. F. 1995. A methodology for reporting uncertainty in spatial database products. *URISA Journal* 7:11–21.

Hunter, M. L., 1993. Natural fire regimes as spatial models for managing boreal forests. *Biological Conservation* 65:115–20.

Hutchins, R. B., Bevins, R. L., Hill, J. D., and White, E. H. 1976. The influence of soils and microclimate on vegetation of forested slopes in eastern Kentucky. *Soil Science* 121:234–41.

Hutchinson, G. E. 1957. Concluding remarks: Cold Spring Harbor Symposium. *Quaternary Biology* 22:415–27.

Hutchinson, M. F. 1988. Calculation of hydrologically sound digital elevation models. In *Proceedings of the Third International Symposium on Spatial Data Handling, Sydney, 17–19 August 1988.* Columbus, OH: International Geographical Union, 117–33.

Hutchinson, M. F. 1989a. A New Method for Spatial Interpolation of Meteorological Variables From Irregular Networks Applied to the Estimation of Monthly Mean Solar Radiation, Temperature, Precipitation and Windrun. *CSIRO Division of Water Resources Technical Memorandum* 89/5:95–104.

Hutchinson, M. F. 1989b. A new procedure for gridding elevation and stream line data with automatic removal of pits. *Journal of Hydrology* 106:211–32.

Hutchinson, M. F. 1995. Interpolating mean rainfall using thin plate smoothing splines. *International Journal of Geographical Information Systems* 9:385–403.

Hutchinson, M. F. 1996. A locally adaptive approach to the interpolation of digital elevation models. In NCGIA (ed.), *Proceedings of the Third International Conference Integrating GIS and Environmental Modeling,* Santa Fe, New Mexico, 21–25 January, 1996. Santa Barbara, CA: University of California, National Center for Geographic Information and Analysis: CD-ROM and WWW.

Hutchinson, M. F. 1997. *ANUDEM* Version 4.6. Centre for Resource and Environmental Studies, Australian National University, Canberra. *http://cres.anu.edu.au/software/anudem.html.*

Hutchinson, M. F. 1998. Interpolation of rainfall using thin plate smoothing splines, II: analysis of topographic dependence. *Journal of Geographic Information and Decision Making* 2:168–85.

Hutchinson, M. F., and Bischof, R. J. 1993. A new method for estimating the spatial distribution of mean seasonal and annual rainfall applied to the Hunter Valley, New South Wales. *Australian Meteorological Magazine* 31:179–84.

Hutchinson, M. F., and Dowling, T. I. 1991. A continental hydrological assessment of a new grid-based digital elevation model of Australia. *Hydrological Processes* 5:45–58.

Hutchinson, M. F., and Gallant, J. C. 1999. Representation of terrain. In P. A. Longley, M. F. Goodchild, D. J. Maguire, and D. W. Rhind (eds.), *Geographical Information Systems: Principles and Technical Issues,* Vol. 1. New York: Wiley, 105–24.

Hutchinson, M. F., Booth, T. H., McMahon, J. P., and Nix, H. A. 1984. Estimating monthly mean values of daily total solar radiation for Australia. *Solar Energy* 32:277–90.

Hutchinson, M. F., Nix, H. A., McMahon, J. P., and Ord, K. D. 1996. The development of a topographic and climate database for Africa. In NCGIA (ed.), *Proceedings of the Third*

International Conference on Integrating GIS and Environmental Modelling, Sante Fe, New Mexico, 21–25 January, 1996. Santa Barbara, CA: National Center for Geographic Information and Analysis, University of California: CD-ROM and WWW.

Hutchinson, M. F., Gessler, P. E., Xu, T., and Gallant, J. C. 1997. Filtering Wagga TOPSAR data to improve drainage accuracy. In A. K. Milne (ed.), *Proceedings of the International Workshop on Radar Image Processing and Applications.* Canberra: Commonwealth Scientific and Industrial Research Organization, 80–83.

Hydrosphere. 1993. *Climatedata West 2.* Boulder, CO: Hydrosphere Inc. (CD-ROM).

Idso, S. B. 1969. Atmospheric attenuation of solar radiation. *Journal of the Atmospheric Sciences* 26:1088–95.

Ida, T. 1984. A hydrological method of estimation of topographic effect on saturated throughflow. *Transactions of Japan Geomorphology Union* 5:1–12.

Iqbal, M. 1983. *An Introduction to Solar Radiation.* Toronto: Academic.

Isard, S. A. 1986. Evaluation of models for predicting insolation on slopes within the Colorado alpine tundra. *Solar Energy* 36:559–64.

Issacson, D. L., and Ripple, W. J. 1991. Comparison of 7.5 minute and 1 degree digital elevation models. *Photogrammetric Engineering and Remote Sensing* 56:1523–7.

James, D. E., and Hewitt, M. J. 1995. To save a river: building a resource decision support system for the Blackfoot River drainage. In J. G. Lyon and J. McCarthy (eds.), *Wetland and Environmental Applications of GIS.* Boca Raton, FL: CRC, 313–29.

Jenny, H. 1941. *Factors of Soil Formation: A System of Quantitative Pedology.* New York: McGraw-Hill.

Jenny, H. 1980. *The Soil Resource: Origin and Behaviour.* New York: Springer.

Jenson, J. R. 1983. Biophysical remote sensing. *Annals of the Association of the American Geographers* 73:111–32.

Jenson, J. R. 1996. *Introductory Digital Image Processing: A Remote Sensing Perspective,* 2nd ed. Upper Saddle River, NJ: Prentice Hall.

Jenson, S. K. 1991. Applications of hydrologic information automatically extracted from digital elevation models. *Hydrological Processes* 5:31–44.

Jenson, S. K., and Domingue, J. O. 1988. Extracting topographic structure from digital elevation model data for geographic information system analysis. *Photogrammetric Engineering and Remote Sensing* 54:1593–1600.

Jersey, J. K. 1993. *Assessing Vegetation Patterns and Hydrologic Characteristics of a Semiarid Environment Using a Geographic Information System and Terrain-based Models.* M.S. Thesis, Montana State University.

João, E. M., and Walsh, S. J. 1992. GIS implications for hydrologic modeling: simulation of non-point pollution generated as a consequence of watershed development scenarios. *Computers, Environment and Urban Systems* 16:43–63.

Johnson, D. L., and Hole, F. D. 1994. Soil formation theory: a summary of its principal impacts on geography, geomorphology, soil-geomorphology, quaternary geology and palaeopedology. In R. Amundson, J. Harden, and M. Singer (eds.), *Factors of Soil Formation: A Fiftieth Anniversary Retrospective.* Madison, WI: Soil Science Society of America, 111–26.

Johnson, D. L., Keller, E. A., and Rockwell, T. K. 1990. Dynamic pedogenesis: new views on some key soil concepts, and a model for interpreting quaternary soils. *Quaternary Research* 33:306–19.

Johnson, N. L., and Kotz, S. 1970. *Distributions in Statistics: Continuous Univariate Distributions*. Boston, MA: Houghton Mifflin.

Johnson, W. C. 1994. Woodland expansion in the Platte River, Nebraska: patterns and causes. *Ecological Monographs* 64:45–84.

Jones, J. A. A. 1986. Some limitations to the a/s index for predicting basin-wide patterns of soil water drainage. *Zeitschrift für Geomorphologie* 60:7–20.

Jones, J. A. A. 1987. The initiation of natural drainage networks. *Progress in Physical Geography* 11:205–45.

Jones, K. H. 1996. A comparison of eight algorithms used to compute slopes as a local property of the DEM. In *Proceedings of GISRUK96, 10–12 April 1996,* University of Kent.

Jones, N. L., Wright, S. G., and Maidment, D. R. 1990. Watershed delineation with triangle-based terrain models. *Journal of Hydraulic Engineering* 116:1232–51.

Joria, P., and Jorgenson, J. 1996. Comparison of three methods for mapping tundra with Landsat digital data. *Photogrammetric Engineering and Remote Sensing* 62:163–9.

Kachanoski, R. G., Rolston, E. E., and de Jong, E. 1985. Spatial and spectral relationships of soil properties and microtopography, I: density and thickness of A horizon. *Soil Science Society of America Journal* 49:804–12.

Kasischke, E. S., Christensen, N. L., and Stocks, B. J. 1995. Fire, global warming, and the carbon balance of boreal forests. *Ecological Applications* 5:437–51.

Katibah, E. F. 1989. A brief history of riparian forests in the Central Valley of California. In R. E. Warner and K. M. Hendrix (eds.), *California Riparian Systems: Ecology, Conservation and Productive Management*. Berkeley, CA: University of California Press, 23–9.

Keeley, J. E., and Keeley, S. C. 1988. Chaparral. In M. G. Barbour and W. D. Billings (eds.), *North American Terrestrial Vegetation*. Cambridge, UK: Cambridge University Press, 165–207.

Kelly, R. E., McConnell, P. R. H., and Mildenberger, S. J. 1977. The Gestalt photomapping system. *Photogrammetric Engineering and Remote Sensing* 43:1407–17.

Kemp, K. K. 1997a. Fields as a framework for integrating GIS and environmental process models, 1: representing spatial contiguity. *Transactions in GIS* 1:219–34.

Kemp, K. K. 1997b. Fields as a framework for integrating GIS and environmental process models, 2: specifying field variables. *Transactions in GIS* 1:235–46.

Kessel, S. R. 1996. The integration of empirical modeling, dynamic process modeling, visualization and GIS for bushfire decision support in Australia. In M. F. Goodchild, L. T. Steyaert, B. O. Parks, M. P. Crane, C. A. Johnston, D. R. Maidment, and S. Glendenning (eds.), *GIS and Environmental Modeling: Progress and Research Issues*. Fort Collins, CO: GIS World Books, 367–71.

Kidner, D. B., and Smith, D. H. 1992. Compression of digital elevation models by Huffman coding. *Computers and Geosciences* 18:1013–34.

Kimerling, A. J., and Moellering, H. 1989. The development of digital slope-aspect displays. In *Proceedings of AutoCarto 9, Baltimore, MD, April 2–7, 1989*. Little Falls, VA: American Society for Photogrammetry and Remote Sensing and American Congress on Surveying and Mapping, 241–44.

Kirkby, M. J., and Chorley, R. J. 1967. Throughflow, overland flow, and erosion. *Bulletin of the International Association of Scientific Hydrology* 12:5–12.

Kirkby, M. J., Imeson, A. C., Bergkamp, G., and Cammeraat, L. H. 1996. Scaling up processes and models from the field plot to the watershed and regional scales. *Journal of Soil and Water Conservation* 51:391–6.

Kirkpatrick, J. B., and Hutchinson, C. F. 1980. The environmental relationships of California coastal sage scrub and some of its component communities. *Journal of Biogeography* 7:23–8.

Kirkpatrick, J. B., and Nunez, M. 1980. Vegetation-radiation relationships in mountainous terrain: eucalypt-dominated vegetation in the Risdon Hills, Tasmania. *Journal of Biogeography* 7:197–208.

Klingebiel, A. A., Horvath, E. H., Moore, D. G., and Reybold, W. U. 1987. Use of slope, aspect and elevation maps derived from digital elevation model data in making soil surveys. In W. U. Reybold and G. W. Peterson (eds.), *Soil Survey Techniques*. Madison, WI: Soil Science Society of America: 77–90.

Kollias, V. J., and Voliotis, A. 1991. Fuzzy reasoning in the development of geographical information systems. *International Journal of Geographical Information Systems* 5:209–23.

Kondratyev, K. Y. 1969. *Radiation Regime of Inclined Surfaces*. Geneva: World Meterological Organization Technical Note No. 152.

Konecny, G., Lohmann, P., Engel, H., and Kruck, E. 1987. Evaluation of SPOT imagery on analytical instruments. *Photogrammetric Engineering and Remote Sensing* 53:1223–30.

Krajewski, S. A., and Gibbs, B. L. 1994. Computer contouring generates artifacts. *Geotimes* 39:15–9.

Kraus, K. 1994. Visualization of the quality of surfaces and their derivatives. *Photogrammetric Engineering and Remote Sensing* 60:457–62.

Kreznor, W. R., Olson, K. R., Banwart, W. L., and Johnson, D. L. 1989. Soil, landscape, and erosion relationships in a northwest Illinois watershed. *Soil Science of America Journal* 53:1763–71.

Kristensen, K. J., and Jensen, S. E. 1975. A model for estimating actual evapotranspiration from potential evapotranspiration. *Nordic Hydrology* 6:170–88.

Krysanova, V., Müller-Wohlfeil, D. I., and Lahmer, W. 1995. Integrated modelling of nonpoint source pollution for mesoscale watersheds using GIS. In M. J. Xern (ed.), *Proceedings of the Second International Association for Water Quality (IAWQ) Specialized Conference and Symposia on Diffuse Pollution*. Prague: Czech Geological Survey, 351–6.

Kuchment, L. S. 1992. The construction of continental scale models of the terrestrial hydrological cycle: an analysis of the state-of-the-art and future prospects. In J. P. O'Kane (ed.), *Advances in Theoretical Hydrology*. Amsterdam: Elsevier, 113–28.

Kulshreshtha, S. N. 1993. *World Water Resources and Regional Vulnerability: Impact of Future Changes*. Laxenburg, Austria: International Institute for Applied System Analysis Research Report No. 93-10.

Kumar, L., Skidmore, A. K., and Knowles, E. 1997. Modelling topographic variation in solar radiation in a GIS environment. *International Journal of Geographical Information Science* 11:475–97.

Kumler, M. P. 1994. An intensive comparison of triangulated irregular networks (TINs) and digital elevation models. *Cartographica* 31:1–99.

Laflen, J. M., Lane, L. J., and Foster, G. R. 1991a. WEPP: a new generation of erosion prediction technology. *Journal of Soil and Water Conservation* 46:34–38.

Laflen, J. M., Elliot, W. J., Simanton, J. R., Holzhey, C. S., and Kohl, K. D. 1991b. WEPP: soil erodibility experiments for rangeland and cropland soils. *Journal of Soil and Water Conservation* 46:39–44.

Lagacherie, P., Moussa, R., Cormary, D., and Molenat, J. 1996. Effects of DEM data source and sampling pattern on topographical parameters and on a topography-based hydrological model. In K. Kovar and H. P. Nachtnebel (eds.) *Application of Geographic Information Systems in Hydrology and Water Resources: Proceedings of the HydroGIS 96 Conference held in Vienna, April 1996*. Wallingford, UK: International Association of Hydrological Sciences Publication No. 235:191–9.

Lanari, R., Fornaro, G., Riccio, D., Migliaccio, M., Papathanassiou, K., Moreira, J., Scwabisch, M., Dutra, L., Puglisi, G., Franceschetti, G., and Coltelli, M. 1997. Generation of digital elevation models by using SIR-C/X-SAR multifrequency two-pass interferometry: the Etna case study. *IEEE Transactions on Geoscience and Remote Sensing* 34:1097–114.

Langbein, W., Hains, C., and Culler, R. 1951. *Hydrology of Stock-water Reservoirs in Arizona*. Washington, DC: United States Geological Survey Circular No. 110.

Lange, A. F., and Gilbert, C. 1999. Using GPS for GIS data capture. In P. A. Longley, M. F. Goodchild, D. J. Maguire, and D. W. Rhind (eds.), *Geographical Information Systems: Principles, Techniques, Applications and Management Issues*. New York: Wiley, 467–76.

Lauver, C. L., and Whistler, J. L. 1993. A hierarchical classification of *Landsat* TM imagery to identify natural grassland areas and rare species habitat. *Photogrammetric Engineering and Remote Sensing* 59:629–34.

Lea, N. J. 1992. An aspect-driven kinematic routing algorithm. In A. J. Parsons and A. D. Abrahams (eds.), *Overland Flow: Hydraulics and Erosion Mechanics*. London: UCL Press.

Lee, J. 1991. Comparison of existing methods for building triangular irregular network models of terrain from grid digital elevation models. *International Journal of Geographical Information Systems* 5:267–85.

Lee, J., Snyder, P. K., and Fisher, P. F. 1992. Modeling the effect of data errors on feature extraction from digital elevation models. *Photogrammetric Engineering and Remote Sensing* 58:1461–7.

Lee, R. 1978. *Forest Microclimatology*. New York: Columbia University Press.

Lees, B. G., and Ritman, K. 1991. Decision-tree and rule-induction approach to integration of remotely sensed and GIS data in mapping vegetation in disturbed or hilly environments. *Environmental Management* 15:823–31.

Legates, D. R., and Willmott, C. J. 1986. Interpolation of point values from isoline maps. *The American Cartographer* 13:308–23.

Lemmens, M. J. P. M. 1988. A survey on stereo matching techniques. *International Archives of Photogrammetry and Remote Sensing* 27:V11–V23.

Li, Z. 1991. Effects of check points on the reliability of DTM accuracy estimates obtained from experimental tests. *Photogrammetric Engineering and Remote Sensing* 57:1333–40.

Li, Z. 1994. A comparative study of the accuracy of digital terrain models (DTMs) based on various data models. *ISPRS Journal of Photogrammetry and Remote Sensing* 49:2–11.

Likens, G. E., Bormann, F. H., Pierce, R. S., Eaton, J. S., and Johnson, N. M. 1977. *Biogeochemistry of a Forested Ecosystem*. New York: Springer.

Linacre, E. 1992. *Climate Data and Resources: A Reference and Guide*. London: Routledge.

List, R. J. 1968. *Smithsonian Meteorological Tables*. Washington DC: Smithsonian Miscellaneous Collections No. 114.

Little, E. L. 1971. *Atlas of United States Trees.* Washington DC: United States Department of Agriculture.

Liu, B. Y. H., and Jordan, R. C. 1960. The interrelationships and characteristic distribution of direct, diffuse, and total solar radiation. *Solar Energy* 4:1–19.

Lopez, C. 1997. Locating some types of random errors in digital terrain models. *International Journal of Geographical Information Science* 11:677–98.

Lowell, K. 1994. An uncertainty-based spatial representation for natural resources phenomena. In T. C. Waugh and R. G. Healey (eds.) *Advances in GIS Research: Proceedings of the Sixth International Symposium on Spatial Data Handling.* London: Taylor & Francis, 933–44.

Lowrance, R. S., Leonard, R., and Sheridan, J. 1985. Managing riparian ecosystems to control nonpoint pollution. *Journal of Soil and Water Conservation* 40:87–91.

Lowrance, R. S., McIntyre, S., and Lance, C. 1988. Sediment deposition in a field/forest system estimated using cesium-137 activity. *Journal of Soil and Water Conservation* 43:195–9.

Ma, Z., and Redmond, R. L. 1993. *Using Landsat TM Data and a GIS to Classify and Map Existing Vegetation.* Missoula, MT: University of Montana, Wildlife Spatial Analysis Laboratory.

Ma, Z., and Redmond, R. L. 1996. Building attribute tables for raster GIS files with ARC/INFO. In S. Morain and S. L. Baros (eds.), *Raster Imagery in Geographic Information Systems.* Santa Fe, NM: Onward Press, 198–201.

Ma, Z., Hart, M. M., and Redmond, R.L. 2000. Mapping vegetation across large geographic areas: integration of remote sensing and GIS to classify multi-source data. *Photogrammetric Engineering and Remote Sensing* 66: in press.

Maas, R. P., Smolen, M. D., and Dressing, S. A. 1985. Selecting critical areas for nonpoint source pollution control. *Journal of Soil and Water Conservation* 40:68–71.

Mackey, B. G. 1994a. Predicting the potential distribution of rain-forest structural characteristics. *Journal of Vegetation Science* 5:43–54.

Mackey, B. G. 1994b. A spatial analysis of the environmental relations of rainforest structural types. *Journal of Biogeography* 20:303–36.

Mackey, B. G. 1996. The role of GIS and environmental modeling in the conservation of biodiversity. In NCGIA (ed.), *Proceedings of the Third International Conference on Integrating GIS and Environmental Modelling, Sante Fe, New Mexico, 21–25 January, 1996.* Santa Barbara, CA: National Center for Geographic Information and Analysis, University of California: CD-ROM and WWW.

Mackey, B. G., McKenney, D. W., Widdifield, C. A., Sims, R. A., Lawrence, K. M., and Szcyrek, N. 1994. *A New Digital Elevation Model of Ontario.* Sault Ste Marie, Ontario: Natural Resources Canada and Canadian Forest Service–Ontario NODA/NFP Technical Report No. 6.

Mackey, B. G., Sims, R. A., Baldwin, K. A., and Moore, I. D. 1996. Spatial analysis of boreal forest ecosystems: results from the Rinker Lake case study. In M. F. Goodchild, L. T. Steyaert, B. O. Parks, C. Johnston, D. R. Maidment, M. Crane, and S. Glendinning (eds.), *GIS and Environmental Modelling: Progress and Research Issues.* Fort Collins, CO: GIS World Books, 187–90.

Maidment, D. R. 1993. GIS and hydrological modeling. In M. F. Goodchild, B. O. Parks, and L. T. Steyaert (eds.). *Environmental Modeling with GIS.* New York: Oxford University Press, 147–67.

Malanson, G. P., and Armstrong, M. P. 1997. Issues of spatial representations: effects of number of cells and between-cell step size on models of environmental processes. *Geographical and Environmental Modelling* 1:47–64.

Malanson, G. P., Westman, W. E., and Yan, Y.-L. 1992. Realized versus fundamental niche functions in a model of chaparral response to climate change. *Ecological Modeling* 64:261–77.

Malo, D. D., Worcester, B. K., Cassel, D. K., and Matzdorf, K. D. 1974. Soil–landscape relationships in a closed drainage system. *Soil Science Society of America Proceedings* 38:813–18.

Mancini, M., and Rosso, R. 1989. Using GIS to assess spatial variability of SCS curve number at the basin scale. In M. L. Kavvas (ed.), *New Directions in Surface Water Hydrology.* Wallingford, UK: Institute of Applied Hydrological Sciences Publication No. 181:435–44.

March, W. J., Wallace, J. R., and Swift, L. W. 1979. An investigation into the effect of storm type on precipitation in a small mountain watershed. *Water Resources Research* 15:298–304.

Mark, D. M. 1986. Knowledge-based approaches for contour-to-grid interpolation on desert pediments and similar surfaces of low relief. In *Second International Symposium on Spatial Data Handling.* Columbus, OH: International Geographical Union, 225–34.

Markarovic, B. 1984. Structures for geo-information and their application in selective sampling for digital terrain models. *ITC Journal* 1:285–95.

Martz, L. W., and De Jong, E. 1987. Using cesium-137 to assess the variability of net soil erosion and its association with topography in a Canadian prairie landscape. *Catena* 14:439–51.

Martz, L. W., and De Jong, E. 1988. CATCH: a FORTRAN program for measuring catchment area from digital elevation models. *Computers and Geosciences* 14:627–40.

Martz, L. W., and Garbrecht, J. 1998. The treatment of flat areas and depressions in automated drainage analysis of raster digital elevation models. *Hydrological Processes* 12:843–56.

McArthur, D. S. 1992. Building the region: The geomorphology of San Diego County. In P. R. Pryde (ed.), *San Diego: An Introduction to the Region,* 3d ed. Dubuque, IA: Kendall/Hunt, 13–30.

McBratney, A. B., and de Gruijter, J. J. 1992. A continuum approach to soil classification by modified fuzzy *k*-means with extragrades. *Journal of Soil Science* 43:159–75.

McBratney, A. B., and Moore, A. W. 1985. Application of fuzzy sets to climatic classification. *Agricultural and Forest Meteorology* 35:165–85.

McCarthy, M. A., Gill, A. M., and Bradstock, R. A. In press. Theoretical fire-interval distribution. *International Journal of Wildland Fire.*

McCool, D. K., Brown, L. C., Foster, G. R., Mutchler, C. K., and Meyer, L. D. 1987. Revised slope steepness factor for the Universal Soil Loss Equation. *Transactions of the American Society of Agricultural Engineers* 30:1387–96.

McCool, D. K., Foster, G. R., and Mutchler, C. K. 1989. Revised slope length factor for the Universal Soil Loss Equation. *Transactions of the American Society of Agricultural Engineers* 32:1571–6.

McCullagh, M. J. 1988. Terrain and surface modelling systems: theory and practice. *Photogrammetric Record* 12:747–79.

McCullagh, P., and Nelder, J. A. 1989. *Generalized Linear Models,* 2d ed. London: Chapman & Hall.

McCullough, P. E. 1995. *Predictive Vegetation Mapping Using Classification Tree Analysis.* M.A. Thesis, San Diego State University.

McGinty, D. T. 1976. *Comparative Root and Soil Dynamics on a White Pine Watershed and in the Hardwood Forest in the Coweeta Basin.* Ph.D. Thesis, University of Georgia.

McKenney, D. W., Mackey, B. G., and Zavitz, B. L. 1999. Calibration and sensitivity analysis of a spatially distributed solar radiation model. *International Journal of Geographical Information Science* 13:49–65.

McKenzie, N. J. 1991. *A Strategy for Soil Survey and Land Evaluation in Australia.* Canberra: CSIRO Division of Soils Divisional Report No. 114.

McKenzie, N. J., and Austin, M. P. 1993. A quantitative Australian approach to medium and small scale surveys based on soil stratigraphy and environmental correlation. *Geoderma* 57:329–55.

McKenzie, N. J. and Ryan, P. J. 1999. Spatial prediction of soil properties using environmental correlation. *Geoderma* 89:67–94.

McNab, W. H. 1989. Terrain shape index: Quantifying effect of minor landforms on tree height. *Forest Science* 35:91–104.

McNaughton, K. G., and Jarvis, P. G. 1983. Predicting effects of vegetation changes on transpiration and evaporation. In T. T. Kozlowski (ed.), *Water Deficits and Plant Growth.* New York: Academic, 1–47.

McSweeney, K., Gessler, P. E., Slater, B. K., Hammer, R. D., Bell, J. C., and Petersen, G. W. 1994. Toward a new framework for modeling the soil-landscape continuum. In R. Amundson, J. Harden, and M. Singer (eds.), *Factors of Soil Formation: A Fiftieth Anniversary Retrospective.* Madison, WI: Soil Science Society of America, 127–45.

Michaelsen, J., Davis, F. W., and Borchert, M. 1987. A non-parametric method for analyzing hierarchical relationships in ecological data. *Coenoses* 2:39–48.

Michaelsen, J., Schimel, D., Friedl, M., Davis, F. W., and Dubayah, R. C. 1994. Regression tree analysis of satellite and terrain data to guide vegetation sampling and surveys. *Journal of Vegetation Science* 5:673–86.

Miller, P. C., Hajek, E., and Miller, P. M. 1983. The influence of annual precipitation, topography and vegetative cover on soil moisture and summer drought in southern California. *Oecologia* 56:385–91.

Milne, G. 1935. Some suggested units of classification and mapping particularly for East African soils. *Soil Research* 4:183–98.

Milnes, A. R., Bourman, R. P., and Northcote, K. H. 1985. Field relationships of ferricretes and weathered zones in southern South Australia: a contribution to "laterite" studies in Australia. *Australian Journal of Soil Research* 23:441–65.

Minty, B. R. S. 1997. Fundamentals of airborne gamma-ray spectometry. *AGSO Journal of Australian Geology & Geophysics* 17:39–50.

Mitas, L., Brown, W. M., and Mitasova, H. 1997. Role of dynamic cartography in simulations of landscape processes based on multivariate fields. *Computers and Geosciences* 23:437–46.

Mitas, L., Mitasova, H., Brown, W. M., and Astley, M. 1996. Interacting fields approach for evolving spatial phenomenon: application to erosion simulation for optimized land use. In NCGIA (ed.), *Proceedings of the Third International Conference Integrating GIS and Environmental Modeling, Santa Fe, New Mexico, 21–25 January 1996.* Santa Barbara, CA: National Center for Geographic Information and Analysis, University of California: CD-ROM and WWW.

Mitasova, H., and Hofierka, J. 1993. Interpolation by regularized spline with tension, II: application to terrain modeling and surface geometry analysis. *Mathematical Geology* 25:657–69.

Mitasova, H., and Mitas, L. 1993. Interpolation by regularised spline with tension, I: theory and implementation. *Mathematical Geology* 25:641–55.

Mitasova, H., Mitas, L., Brown, W. M., Gerdes, D. P., Kosinovsky, I., and Baker, T. 1995. Modeling spatially and temporally distributed phenomena: new methods and tools for the GRASS GIS. *International Journal of Geographical Information Systems* 9:433–56.

Mitasova, H., Hofierka, J., Zlocha, M., and Iverson, L. R. 1996. Modeling topographic potential for erosion and deposition using GIS. *International Journal of Geographical Information Systems* 10:629–41.

Mitasova, H., Hofierka, J., Zlocha, M., and Iverson, L. R. 1997. Reply to Comment by Desmet and Govers. *International Journal of Geographical Information Science* 11:611–8.

Montgomery, D. R., and Dietrich, W. E. 1989. Source areas, drainage density and channel initiation. *Water Resources Research* 25:1907–18.

Montgomery, D. R., and Dietrich, W. E. 1992. Channel initiation and the problem of landscape scale. *Science* 255:826–30.

Montgomery, D. R., and Dietrich, W. E. 1994. A physically based model for the topographic control on the shallow landsliding. *Water Resources Research* 30:1153–71.

Montgomery, D. R., and Dietrich, W. E. 1995. Hydrologic processes in a low-gradient source area. *Water Resources Research* 31:1–10.

Montgomery, D. R., and Foufoula-Georgiou, E. 1993. Channel network source representation using digital elevation models. *Water Resources Research* 29:3925–34.

Mooney, H. A. 1977. Southern coastal scrub. In M. G. Barbour and J. Major (eds.), *Terrestrial Vegetation of California.* New York: Wiley, 471–89.

Moore, D. M., Lees, B. G., and Davey, S. M. 1991. A new method for predicting vegetation distributions using decision tree analysis in a geographic information system. *Environmental Management* 15:59–71.

Moore, I. D. 1992. Terrain analysis programs for the environmental sciences. *Agricultural Systems and Information Technology* 2:37–9.

Moore, I. D. 1996. Hydrologic modelling and GIS. In M. F. Goodchild, L. T. Steyaert, B. O. Parks, M. P. Crane, C. A. Johnston, D. R. Maidment, and S. Glendinning (eds.) *GIS and Environmental Modeling: Progress and Research Issues.* Fort Collins, CO: GIS World Books: 143–8.

Moore, I. D., and Burch, G. J. 1986a. Sediment transport capacity of sheet and rill flow: application of unit stream power theory. *Water Resources Research* 22:1350–60.

Moore, I. D., and Burch, G. J. 1986b. Physical basis of the length-slope factor in the Universal Soil Loss Equation. *Soil Science Society of America Journal* 50:1294–8.

Moore, I. D., and Burch, G. J. 1986c. Modelling erosion and deposition: topographic effects. *Transactions of the American Society of Agricultural Engineers* 29:1624–30, 1640.

Moore, I. D., and Foster, G. R. 1990. Hydraulics and overland flow. In M. G. Anderson and T. P. Burt (eds.), *Process Studies in Hillslope Hydrology.* New York: Wiley, 215–52.

Moore, I. D., and Grayson, R. B. 1991. Terrain-based catchment partitioning and runoff prediction using vector elevation data. *Water Resources Research* 27:1177–91.

Moore, I. D., and Hutchinson, M. F. 1991. Spatial extension of hydrologic process modelling. In *Proceedings of the International Hydrology and Water Resources Symposium, Perth, 2–4 October 1991.* Canberra: Institute of Australian Engineers, 803–8.

Moore, I. D., and Nieber, J. L. 1989. Landscape assessment of soil erosion and non-point source pollution. *Journal of the Minnesota Academy of Science* 55:18–25.

Moore, I. D., and Wilson, J. P. 1992. Length–slope factors for the Revised Universal Soil Loss Equation: simplified method of estimation. *Journal of Soil and Water Conservation* 47:423–8.

Moore, I. D., and Wilson, J. P. 1994. Reply to comments by Foster on "Length–slope factors for the Revised Universal Soil Loss Equation: Simplified method of estimation." *Journal of Soil and Water Conservation* 49:174–80.

Moore, I. D., Mackey, S. M., Wallbrink, P. J., Burch, G. J., and O'Loughlin, E. M. 1986. Hydrologic characteristics and modelling of a small forested catchment in southeastern New South Wales: Prelogging condition. *Journal of Hydrology* 83:307–335.

Moore, I. D., Burch, G. J., and MacKenzie, D. H. 1988a. Topographic effects on the distribution of surface soil water and the location of ephemeral gullies. *Transactions of the American Society of Agricultural Engineers* 31:1098–107.

Moore, I. D., O'Loughlin, E. M., and Burch, G. J. 1988b. A contour-based topographic model for hydrological and ecological applications. *Earth Surface Processes and Landforms* 13:305–20.

Moore, I. D., Grayson, R. B., and Ladson, A. R. 1991. Digital terrain modeling: a review of hydrological, geomorphological, and biological applications. *Hydrological Processes* 5:3–30.

Moore, I. D., Wilson, J. P., and Ciesiolka, C. A. 1992. Soil erosion prediction and GIS: linking theory and practice. In S. H. Luk and J. Whitney (eds.), *Proceedings of the International Conference on the Application of Geographic Information Systems to Soil Erosion Management.* Toronto: University of Toronto Press, 31–48.

Moore, I. D., Gallant, J. C., Guerra, L., and Kalma, J. D. 1993a. Modelling the spatial variability of hydrologic processes using GIS. In K. Kovar and H. P. Nachtnebel (eds.), *Application of Geographic Information Systems in Hydrology and Water Resources: Proceedings of the HydroGIS 93 Conference held in Vienna, April 1993.* Wallingford, UK: International Association of Hydrological Sciences Publication No. 211:83–92.

Moore, I. D., Gessler, P. E., Nielsen, G. A., and Peterson, G. A. 1993b. Soil attribute prediction using terrain analysis. *Soil Science Society of America Journal* 57:443–52.

Moore, I. D., Gessler, P. E., Nielsen, G. A., and Peterson, G. A. 1993c. Terrain analysis for soil specific crop management. In P. C. Robert, R. H. Rust, and W. E. Larson (eds.) *Soil Specific Crop Management.* Madison, WI: Soil Science Society of America, 27–55.

Moore, I. D., Lewis, A., and Gallant, J. C. 1993d. Terrain attributes: estimation methods and scale effects. In A. J. Jakeman, M. B. Beck, and M. J. McAleer (eds.), *Modelling Change in Environmental Systems.* New York: Wiley, 189–214.

Moore, I. D., Norton, T. W., and Williams, J. E. 1993e. Modeling environmental heterogeneity in forested landscapes. *Journal of Hydrology* 150:717–47.

Moore, I. D., Turner, A. K., Wilson, J. P., Jenson, S. K., and Band, L. E. 1993f. GIS and land surface–subsurface modeling. In M. F. Goodchild, B. O. Parks, and L. T. Steyaert (eds.), *Environmental Modeling With GIS.* New York: Oxford University Press, 196–230.

Morrison, M. L., Marcot, B. G., and Mannan, R. W. 1992. *Wildlife–Habitat Relationships.* Madison, WI: University of Wisconsin Press.

Mullen, I. C. 1995. *Environmental Processes and Landscape Pattern: Rainforests of the South Coast of New South Wales.* B.Sc. Hons. Thesis, Australian National University.

Müller-Wohlfeil, D. I., Lahmer, W., and Krysanova, V. 1996. Topography-based hydrological modeling in the Elbe drainage basin. In NCGIA (ed.), *Proceedings of the Third International Conference Integrating GIS and Environmental Modeling, Santa Fe, New Mexico, 21–25 January 1996.* Santa Barbara, CA: National Center for Geographic Information and Analysis, University of California: CD-ROM and WWW.

Munn, L. C., Buchanan, B. A., and Nielsen, G. A. 1978. Soil temperatures in adjacent high elevation forests and meadows in Montana. *Soil Society of America Journal* 42:982–4.

Munn, L. C., Nielsen, G. A., Caprio, J. M., and Lossing, L. 1981. *Long-term soil climate observations in Montana.* Bozeman, MT: Montana Agricultural Experiment Station Miscellaneous Report No. 18.

Munnik, M. C., Verster, E., and Rooyen, T. H. 1984. Pedogeomorphic aspects of the Roodepoort area, Transvaal: soil depth–slope relationships. *South African Journal of Plant and Soil* 1:61–6.

Nash, J. E., and Sutcliffe, J. V. 1979. River flow forecasting through conceptual models, 1: a discussion of principles. *Journal of Hydrology* 10:282–90.

National Oceanic and Atmospheric Administration. 1993a. *Climatological Data, North Dakota.* Asheville, NC: National Oceanic and Atmospheric Administration.

National Oceanic and Atmospheric Administration. 1993b. *Local Climatological Data Annual Summary, With Comparative Data, Bismarck, North Dakota.* Asheville, NC: National Oceanic and Atmospheric Administration.

National Renewable Energy Laboratory. 1992. *User's Manual for the National Solar Radiation Database (1961–90).* Asheville, NC: National Oceanic and Atmospheric Administration, National Climatic Data Center.

National Research Council. 1993. *Soil and Water Quality: an Agenda for Agriculture.* Washington, DC: National Academy Press.

Neuman, S. P. 1975. Galerkin approach to saturated-unsaturated flow in porous media. In R. H. Gallagher, J. T. Oden, C. Taylor, and O. C. Zienkiewicz (eds.) *Finite Elements in Fluids: Viscous Flow and Hydrodynamics.* London: Wiley, 201–17.

Niemann, O. 1993. Automated forest cover mapping using Thematic Mapper and ancillary data. *Applied Geography* 13:86–95.

Nix, H. A. 1982. Environmental determinants of evolution in Terra Australis. In W. R. Barker and P. J. M. Greenslade (eds.) *Evolution of the Flora and Fauna of Arid Australia.* Adelaide: Peacock, 47–66.

Nix, H. A. 1986. A biogeographic analysis of Australian elapid snakes. In R. Longmore (ed.) *Atlas of Elapid Snakes of Australia.* Canberra: Australian Government Printing Service, 4–15.

Nizeyimana, E., and Bicki, T. J. 1992. Soil and soil–landscape relationships in the north central region of Rwanda, East-Central Africa. *Soil Science* 153:225–36.

Noble, I. R., and Slatyer, R. O. 1980. The use of vital attributes to predict successional changes in plant community subject to recurrent disturbances. *Vegetatio* 43:5–21.

Noguchi, Y. 1992a. Vegetation asymmetry in Hawaii under the trade wind regime. *Journal of Vegetation Science* 3:223–30.

Noguchi, Y. 1992b. Hydrometeorological differences between opposite valley slopes and vegetation asymmetry in Hawaii. *Journal of Vegetation Science* 3:231–8.

Novotny, V., and Chesters, G. 1989. Delivery of sediment and pollutants from nonpoint sources: a water quality perspective. *Journal of Soil and Water Conservation* 44:568–76.

Novotny, V., and Olem, H. 1994. *Water Quality: Prevention, Identification, and Management of Diffuse Pollution.* New York: Van Nostrand Reinhold.

O'Callaghan, J. F., and Mark, D. M. 1984. The extraction of drainage networks from digital elevation data. *Computer Vision, Graphics and Image Processing* 28:323–44.

O'Connell, D. A. 1997. *Predicting soil and hydraulic properties in small forested catchments.* Ph.D. thesis, Australian National University, Canberra.

Odeh, I. O. A., McBratney, A. B., and Chittleborough, D. J. 1992. Soil pattern recognition with fuzzy *c*-means: application to classification and soil–landform interrelationships. *Soil Science Society of America Journal* 56:506–16.

Odeh, I. O. A., McBratney, A. B., and Chittleborough, D. J. 1994. Spatial prediction of soil properties from landform attributes derived from a digital elevation model. *Geoderma* 63:197–214.

Oke, T. R. 1987. *Boundary Layer Climates,* 2d ed. London, Methuen.

Okimura, T., and Ichikawa, R. 1985. A prediction method for surface failure by movements of infiltrated water in a surface soil layer. *Natural Disaster Science* 7:41–51.

Okimura, T., and Iida, T. 1983. Evolution of hillslopes including landslides. *Transactions of Japan Geomorphology Union* 4:191–200.

O'Leary, J. F., Hope, A. S., and Wright, R. D. 1994. *Vegetation and Land Cover Types: Mirimar Naval Air Station Technical Report.* San Diego, CA: San Diego State University, Department of Geography, CESAR Technical Report.

Oleson, K. W., Sarlin, S., Garrison, J., Smith, S., Privette, J., and Emery, W. 1995. Unmixing multiple land-cover type reflectances from coarse spatial resolution satellite data. *Remote Sensing of Environment* 54:98–112.

Ollier, C. D. 1982. The Great Escarpment of eastern Australia: tectonic and geomorphic significance. *Journal of the Geological Society of Australia* 29:13–23.

Ollier, C. D. 1991. *Ancient Landforms.* London: Belhaven.

Ollier, C. D., Chan, R. A., Craig, M. A., and Gibson, D. L. 1988. Aspects of landscape evolution and regolith in the Kalgoorlie region. *Bureau of Mineral Resources Journal of Geology and Geophysics* 10:309–21.

O'Loughlin, E. M. 1986. Prediction of surface saturation zones in natural catchments by topographic analysis. *Water Resources Research* 22:794–804.

Olson, C. G., and Hupp, C. R. 1986. Coincidence and spatial variability of geology, soils and vegetation, Mill Run Watershed, Virginia. *Earth Surface Processes and Landforms* 11:619–29.

O'Neill, R. V., DeAngelis, D. L., Waide, J. B., and Allen, T. F. H. 1986. *A Hierarchical Concept of Ecosystems.* Princeton, NJ: Princeton University Press.

Onstad, C. A., and Brakensiek, D. L. 1968. Watershed simulation by the stream path analogy. *Water Resources Research* 4:965–71.

Ordinance Survey. 1993. *Digital Map Data Catalogue.* Southampton, UK: Ordinance Survey.

Osborne, L. L., and Kovacic, D. A. 1993. Riparian vegetated buffer strips in water quality restoration and stream management. *Freshwater Biology* 29:243–58.

Oswald, H., and Raetzsch, H. 1984. A system for generation and display of digital elevation models. *GeoProcessing* 2:197–218.

Owenby, J. R., and Ezell, D. S. 1992. *Monthly Station Normals of Temperature, Precipitation, and Heating and Cooling Degree Days, 1961–90, North Dakota.* Asheville, NC: National Oceanic and Atmospheric Administration.

Paetzold, R. F. 1988. *Soil Climate Occasional Notes.* Lincoln, NE: United States Department of Agriculture, Soil Conservation Service, National Soil Survey Center.

Paniconi, C., and Wood, E. 1993. A detailed model for simulation of catchment scale subsurface hydrologic processes. *Water Resources Research* 29:1601–20.

Panuska, J. C., Moore, I. D., and Kramer, L. A. 1991. Terrain analysis: integration into the agriculture non-point source (AGNPS) pollution model. *Journal of Soil and Water Conservation* 46:59–64.

Parlange, J., Lisle, I., Braddock, R. D., and Smith, R. E. 1982. The three-parameter infiltration equation. *Soil Science* 133:337–41.

Parrett, C., and Hull, J. A. 1985. *Streamflow Characteristics of Mountain Streams in Western Montana.* Washington DC: United States Geological Survey Professional Paper No. 2260.

Parton, W. J., and Logan, J. A. 1981. A model for diurnal variation in soil and air temperature. *Agricultural Meteorology* 23:205–16.

Paruelo, J. M., and Lauenroth, W. K. 1995. Regional patterns of normalized difference vegetation index in North American shrublands and grasslands. *Ecology* 76:1888–98.

Paton, T. R., Humphreys, G. S., and Mitchell, P. B. 1995. *Soils: A New Global View.* London: UCL Press.

Peltzer, G. R. 1992. *A Comprehensive Unsupervised Clustering Technique for the Classification of Remotely Sensed Data.* M.S. Thesis, University of Wisconsin–Madison.

Pennock, D. J., van Kessel, C., Farrell, R. E., and Sutherland, R. A. 1992. Landscape–scale variations in denitrification. *Soil Science Society of America Journal* 56:770–6.

Pennock, D. J., Zebarth, B. J., and De Jong, E. 1987. Landform classification and soil distribution in hummocky terrain, Saskatchewan, Canada. *Geoderma* 40:297–315.

Phillips, J. D. 1986. Sediment storage, sediment yield, and time scales in sediment denudation studies. *Geographical Analysis* 18:161–7.

Phillips, J. D. 1988. The role of spatial scale in geomorphic systems. *Geographical Analysis* 20:308–17.

Phillips, J. D. 1989. An evaluation of the factors determining the effectiveness of water quality buffer strips. *Journal of Hydrology* 107:133–45.

Phillips, J. D. 1990. A saturation-based model of relative wetness for wetland identification. *Water Resources Bulletin* 26:333–42.

Pickup, G., Chewings, V. H., and Nelson, D. J. 1993. Estimating changes in vegetation cover over time in arid rangelands using *Landsat* MSS data. *Remote Sensing of Environment* 43:243–63.

Pickup, G., Bastin, G. N., and Chewings, V. H. 1994. Remote sensing-based condition assessment for non-equilibrium rangelands under large-scale commercial grazing. *Ecological Applications* 4:497–517.

Pierce, R. L. 1954. *Vegetation Cover Types and Land Use History of the Cedar Creek Natural History Reservation, Anoka and Isanti Counties, Minnesota.* M.S. Thesis, University of Minnesota.

Pilgrim, D. H., and Cordery, I. 1993. Flood runoff. In D. R. Maidment (ed.), *Handbook of Hydrology.* New York: McGraw-Hill, 9.1–9.42.

Pinder, G. F., and Gray, W. G. 1977. *Finite Element Simulation of Flow in Surface and Subsurface Hydrology.* New York: Academic.

Pogacnik, T. M. 1985. *Time of Grazing Effects on Stream Channel Stability and Instream Sediment Loads.* M.S. Thesis, Montana State University.

Pohll, G. M., Warwick, J. J., and Tyler, S. W. 1996. Coupled surface/subsurface hydrologic model of a nuclear subsidence crater at the Nevada Test Site. *Journal of Hydrology* 186:43–62.

Polidori, L., Chorowicz, J., and Guillande, R. 1991. Description of terrain as a fractal surface, and application to digital elevation model quality assessment. *Photogrammetric Engineering and Remote Sensing* 57:1329–32.

Ponce-Hernandez, R., Marriott, F. H. C., and Beckett, P. H. T. 1986. An improved method for reconstructing a soil profile from analyses of a small number of samples. *Journal of Soil Science* 37:455–67.

Poole, D. K., and Miller, P. C. 1981. The distribution of plant water stress and vegetation characteristics in southern California chaparral. *American Midland Naturalist* 105:32–43.

Prescott, J. A. 1948. A climatic index for the leaching factor in soil formation. *Journal of Soil Science* 1:9–19.

Pries, R. A. 1995. A system for large-scale image mapping and GIS data collection. *Photogrammetric Engineering and Remote Sensing* 61:503–11.

Priestly, C. H. B., and Taylor, R. J. 1972. On the assessment of surface heat flux and evaporation using large-scale parameters. *Monthly Weather Review* 100:81–92.

Qian, J., Ehrich, R. W., and Campbell, J. B. 1990. DNESYS: an expert system for automatic extraction of drainage networks from digital elevation data. *IEEE Transactions on Geoscience and Remote Sensing* 28:29–45.

Quine, T. A., and Walling, D. E. 1993. Use of caesium-137 measurements to investigate relationships between erosion rates and topography. In D. S. Thomas and R. J. Allison (eds.), *Landscape Sensitivity.* New York: Wiley, 31–48.

Quinn, P. F., Beven, K. J., Chevallier, P., and Planchon, O. 1991. The prediction of hillslope flow paths for distributed hydrological modelling using digital terrain models. *Hydrological Processes* 5:59–79.

Quinn, P. F., Beven, K. J., and Lamb, R. 1995. The ln(a/tan β) index: how to calculate it and how to use it within the TOPMODEL framework. *Hydrological Processes* 9:161–82.

Ratliff, L. F., Ritchie, J. T., and Cassel, D. K. 1983. Field-measured limits of soil water availability as related to laboratory-measured properties. *Soil Science Society of America Journal* 47:770–5.

Rawls, W. J., and Brakensiek, D. L. 1989. Estimation of soil water retention and hydraulic properties. In H. J. Morel-Seytoux (ed.), *Unsaturated Flow in Hydrologic Modeling: Theory and Practice.* Amsterdam: Kluwer, 275–300.

Rawls, W. J., Brakensiek, D. L., and Sexton, K. E. 1982. Estimation of soil water properties. *Transactions of the American Society of Agricultural Engineers* 25:1316–20, 1328.

Rawls, W. J., Ahuja, L. R., Brakensiek, D. L., and Shirmohammadi, A. 1993. Infiltration and soil water movement. In D. R. Maidment (ed.) *Handbook of Hydrology.* New York: McGraw-Hill, 5.1–5.51.

Reiger, W. 1998. A phenomenon-based approach to upslope contributing areas and depressions in DEMs. *Hydrological Processes* 12:857–72.

Renard, K. G., Foster, G. R., Weesies, G. A., and Porter, J. P. 1991. RUSLE: Revised Universal Soil Loss Equation. *Journal of Soil and Water Conservation* 41:30–3.

Rich, P. M., Hetrick, W. A., and Saving, S. C. 1994. *Modeling Topographic Influences on Solar Radiation: a Manual for the SOLARFLUX Model.* Lawrence, KS: University of Kansas.

Ripple, W. J. 1985. Asymptotic reflectance characteristics of grass vegetation. *Photogrammetric Engineering and Remote Sensing* 51:1915–21.

Ritchie, J. C., and McHenry, J. R. 1990. Application of radioactive fallout cesium-137 for measuring soil erosion and sediment accumulation rates and patterns: a review. *Journal of Environmental Quality* 19:215–33.

Roberts, D. W., Dowling, T. I., and Walker, J. 1997. *FLAG: a Fuzzy Landscape Analysis GIS Method for Dryland Salinity Assessment.* Canberra: CSIRO Land and Water Technical Report No. 8.

Robinson, J. S., Sivapalan, M., and Snell, J. D. 1995. On the relative roles of hillslope processes, channel routing, and network geomorphology in the hydrologic response of natural catchments. *Water Resources Research* 31:3089–101.

Robinson, N. (ed.). 1966. *Solar Radiation.* Amsterdam: Elsevier.

Rogers, C. C. M., Beven, K. J., Morris, E. M., and Anderson, M. G. 1985. Sensitivity analysis, calibration and predictive uncertainty of the Institute of Hydrology Distributed Model. *Journal of Hydrology* 81:179–91.

Rogowski, A. S., and Hoover, J. R. 1996. Catchment infiltration, Part 1: distribution of variables. *Transactions in GIS* 1:95–110.

Rohdenburg, H. 1989. Methods for the analysis of agro-ecosystems in central Europe, with emphasis on geoecological aspects. *Catena* 16:1–57.

Rosenberg, N. J. 1993. Towards an integrated impact assessment of climate change: the MINK study. *Climatic Change* 24:1–173.

Rothacher, J., Dyrness, C. T., and Fredriksen, R. 1967. *Hydrologic and Related Characteristics of Three Small Watersheds in the Oregon Cascades.* Portland, OR: United States Department of Agriculture, Forest Service, Pacific Northwest Forest and Range Experiment Station.

Rowe, J. S. 1972. *Forest Regions of Canada.* Ottawa: Department of Environment and Canadian Forest Service Publication No. 1300.

Roy, A. G., Jarvis, R. S., and Arnett, R. R. 1980. Soil–slope relationships within a drainage basin. *Annals of the Association of American Geographers* 70:397–413.

Rudd, V. E. 1951. Geographical affinities of the flora of North Dakota. *American Midland Naturalist* 45:722–39.

Ruhe, R. V. 1969. *Quaternary Landscapes in Iowa.* Ames, IA: Iowa University Press.

Ruhe, R. V., and Walker, P. H. 1960. Elements of the soil landscape. In *Transactions of the Seventh Congress of the International Soil Science Society, Madison, Wisconsin,* 165–70.

Ruhe, R. V., and Walker, P. H. 1968. Hillslope models and soil formation, II: open systems. In *Transactions of the Ninth Congress of the International Soil Science Society, Adelaide, Australia,* 551–60.

Ruiz, M. 1997. A causal analysis of error in viewsheds from USGS digital elevation models. *Transactions in GIS* 2:85–94.

Runge, E. C. A. 1973. Soil development sequences and energy models. *Soil Science* 115:183–93.

Running, S. W. 1991. Computer simulation of regional evapotranspiration by integrating landscape biophysical attributes with satellite data. In T. J. Schmugge and J. Andre (eds.) *Land Surface Evaporation: Measurement and Parameterization.* London: Springer, 359–69.

Running, S. W., and Thornton, P. E. 1996. Generating daily surfaces of temperature and precipitation over complex topography. In M. F. Goodchild, L. T. Steyaert, B. O. Parks, C. Johnston, D. R. Maidment, M. Crane, and S. Glendinning (eds.) *GIS and Environmental Modeling: Progress and Research Issues.* Fort Collins, CO: GIS World Books, 93–8.

Running, S. W., Nemani, R. R., and Hungerford, R. D. 1987. Extrapolation of synoptic meteorological data in mountainous terrain and its use for simulating forest evapotranspiration and photosynthesis. *Canadian Journal of Forest Research* 17:472–83.

Ruspini, E. H. 1969. A new approach to clustering. *Information and Control* 15:22–32.

Rutter, A. J., Kershaw, K. A., Robins, P. C., and Morton, A. J. 1971. A predictive model of rainfall interception in forests, I: derivation of the model from observations in a plantation of Corsican pine. *Agricultural Meteorology* 9:367–84.

Rutter, A. J., Morton, A. J., and Robins, P. C. 1975. A predictive model of rainfall interception in forests, II: generalization of the model and comparison with observations in some coniferous and hardwood stands. *Journal of Applied Ecology* 12:367–80.

Sambridge, M., Braun, J., and McQueen, H. 1995. Geophysical parameterization and interpolation of irregular data using natural neighbours. *Geophysical Journal International* 122:837–57.

Samson, S. A. 1993. Two indices to characterize temporal patterns of spectral response of vegetation. *Photogrammetric Engineering and Remote Sensing* 59:511–7.

Sasowsky, K. C., Petersen, G. W., and Evans, B. M. 1992. Accuracy of SPOT digital elevation model and derivatives: utility for Alaska's North Slope. *Photogrammetric Engineering and Remote Sensing* 58:815–24.

Sauerborn, P. 1993. Eine Gefahrenstufenkarte zur potentiellen Erosionsgefährdung durch Niederschläge in Deutschland. *Mitteilungen Deutschen Bodenkundlichen Gesellschaft* 72:1237–40.

Saupe, G. 1985. Die Erosivität der Niederschläge im Süden der DDR ein Beitrag zur quantitativen Prognose der Bodenerosion. *Archiv für Naturschutz und Landschaftsforschung* 25:155–69.

Savory, D. J. 1992. *Digital Terrain Classification via Scale-Sensitive Edge Detection: Hillslope Characterization for Soil–Landscape Analysis.* M.S. Thesis, University of Wisconsin–Madison.

Sawyer, J. O., and Keeler-Wolf, T. 1995. *A Manual of California Vegetation.* Sacramento, CA: California Native Plant Society.

Schaffer, W. M. 1981. Ecological abstraction: the consequences of reduced dimensionality in ecological models. *Ecological Monographs* 5:383–401.

Schuster, R. L., and Krizek, R. J. 1978. Landslides: analysis and control. *Transportation Research Board Special Report* 176:1–10.

Shafi, M. I., and Yarranton, G. A. 1973. Vegetational heterogeneity during a secondary postfire succession. *Canadian Journal of Botany* 51:73–90.

Shandley, J. 1993. *Mapping Vegetation Stands Using Image Segmentation in the Pine Creek Watershed.* M.A. Thesis, San Diego State University.

Shary, P. A. 1991. The second derivative topographic method. In I. N. Stepanov (ed.) *The Geometry of the Earth Surface Structures.* Pushchino: Pushchino Research Centre Press, 30–60.

Shary, P. A. 1995. Land surface in gravity points classification by complete system of curvatures. *Mathematical Geology* 27:373–90.

Shugart, H. H. 1984. *A Theory of Forest Dynamics.* New York: Springer.

Shuttleworth, W. J. 1993. Evaporation. In D. R. Maidment (ed.), *Handbook of Hydrology.* New York: McGraw-Hill, 4.1–4.53.

Sibson, R. 1981. A brief description of natural neighbour interpolation. In V. Barnett (ed.), *Interpreting Multivariate Data.* Chichester, UK: Wiley, 21–36.

Sidle, R. C. 1987. A dynamic model of slope stability in zero-order basins. In R. L. Beschta, T. Blinn, G. E. Grant, G. G. Ice, and F. J. Swanson (eds.), *Proceedings of the Erosion and Sedimentation in the Pacific Rim Symposium, Corvallis, Oregon, 3–7 August 1987.* Wallingford, UK: International Association of Hydrological Sciences Publication No. 165:101–10.

Sidle, R. C. 1991. A conceptual model of changes in root cohesion in response to vegetation management. *Journal of Environmental Quality* 20:43–52.

Sidle, R. C. 1992. A theoretical model of the effects of timber harvesting on slope stability. *Water Resources Research* 28:1897–1910.

Sidle, R. C., Pearce, A. J., and O'Loughlin, C. L. 1985. *Hillslope Stability and Landuse.* Washington, DC: American Geophysical Union, Water Resource Monograph No. 11.

Simpson, J. J., Gobat, J. I., and Frouin, R. 1995. Improved destriping of GOES images using finite impulse response filters. *Remote Sensing of Environment* 52:15–35.

Sims, R. A., and Addison, P. A. 1994. Biodiversity and Canadian forests. In D. W. McKenney, R. A. Sims, M. E. Soule, B. G. Mackey, and K. L. Campbell (eds.), *Towards a Set of Biodiversity Indicators for Canadian Forests: Proceedings of a Forest Biodiversity Indicator Workshop.* Sault Ste Marie: Canadian Forest Service and Natural Resources Canada.

Sims, R. A., Towill, W. D., Baldwin, K. A., and Wickware, G. M. 1989. *Field Guide to the Forest Classification for Northwestern Ontario.* Thunder Bay: Ontario Ministry of Natural Resources.

Sims, R. A., Kershaw, M. K., and Wickware, G. M. 1990. *The Autoecology of Major Tree Species in the North Central Region of Ontario.* Thunder Bay: Forestry Canada, Ontario Region COFRDA Report No. 3302.

Sims, R. A., Baldwin, K. A., Walsh, S. A., Lawrence, K. M., McKenney, D. W., Ford, M. J., and Mackey, B. G. 1997. *The Derivation of Spatially Referenced Ecological Databases for Ecosystem Mapping and Modeling in the Rinker Lake Research Area, Northwestern Ontario.* Sault Ste Marie: Canadian Forest Service, Natural Resources Canada, and Ontario Ministry of Natural Resources Open File Report No. 33.

Simunek, J., Vogel, T., and van Genuchten, M. T. 1994. *The SWMS_2D Code for Simulating Water Flow and Solute Transport in Two-Dimensional Variably Saturated Media (Version 1.2).* Riverside, CA: United States Salinity Laboratory.

Sivapalan, M., and Wood, E. F. 1986. Spatial heterogeneity and scale in the infiltration response of catchments. In V. K. Gupta, I. Rodriguez-Itrube, and E. F. Wood (eds.), *Scale Problems in Hydrology.* Dordrecht: Reidel, 81–106.

Sivapalan, M., Beven, K. J., and Wood, E. F. 1987. On hydrologic similarity, 2: a scaled model of storm runoff production. *Water Resources Research* 23:2266–78.

Skidmore, A. K. 1989. A comparison of techniques for the calculation of gradient and aspect from a gridded digital elevation model. *International Journal of Geographic Information Systems* 3:323–34.

Skidmore, A. K. 1990. Terrain position as mapped from a gridded digital elevation model. *International Journal of Geographic Information Systems* 4:33–49.

Skidmore, A. K., and Turner, B. J. 1988. Forest mapping accuracies are improved using a supervised nonparametric classifier with SPOT data. *Photogrammetric Engineering and Remote Sensing* 54:1415–21.

Slater, B. K. 1994. *Continuous Classification and Visualization of Soil Layers.* Ph.D. Thesis, University of Wisconsin–Madison.

Slater, B. K., McSweeney, K., McBratney, A. B., Ventura, S. J., and Irvin, B. J. 1994. A spatial framework for integrating soil–landscape and pedogenic models. In R. B. Bryant and R. W. Arnold (eds.), *Quantitative Modeling of Soil Forming Processes*. Madison, WI: Soil Science Society of America, 169–85.

Slatyer, R. O., and McIlroy, I. C. 1961. *Practical Microclimatology With Special Reference to the Water Factor in Soil–Plant–Atmosphere Relationships*. Canberra: UNESCO/CSIRO.

Sloan, P. G., and Moore, I. D. 1984. Modeling subsurface stormflow on steeply sloping forested watersheds. *Water Resources Research* 20:1815–22.

Smith, D. H., and Lewis, M. 1994. Optimal predictors for compression of digital elevation models. *Computers and Geosciences* 20:1137–41.

Smith, G. D., Newhall, F., Robinson, L. H., and Swanson, D. 1964. *Soil Temperature Regimes: Their Characteristics and Predictability*. Washington, DC: United States Department of Agriculture, Soil Conservation Service Report No. TP-144.

Smith, R., Corradini, C., and Melone, F. 1993. Modeling infiltration for multi-storm runoff events. *Water Resources Research* 29:133–44.

Smith, S. D., Wellington, B. A., and Nachlinger, J. L. 1991. Functional responses of riparian vegetation to streamflow diversion in the Eastern Sierra Nevada. *Ecological Applications* 1:89–97.

Smith, T. R., Zhan, X., and Gao, P. 1990. A knowledge based, two-step procedure for extracting channel networks from noisy DEM data. *Computers and Geosciences* 16:777–86.

Smith, W. H. F., and Wessel, P. 1990. Gridding with continuous curvature. *Geophysics* 55:293–305.

Soil Survey Staff. 1994. *Keys to Soil Taxonomy,* 6th ed. Washington, DC: United States Department of Agriculture, Soil Conservation Service.

Speight, J. G. 1974. A parametric approach to landform regions. In E. H. Brown and R. S. Waters (eds.), *Progress in Geomorphology*. London: Alden, 213–30.

Speight, J. G. 1980. The role of topography in controlling throughflow generation: a discussion. *Earth Surface Processes* 5:187–91.

Speight, J. G. 1990. Landform. In R. C. McDonald, R. F. Isbell, J. G. Speight, J. Walker, and M. S. Hopkins (eds.), *Australian Soil and Land Survey Field Handbook*. Melbourne: Inkata, 9–57.

Srinivasan, R., and Engel, B. A. 1991. Effect of slope prediction methods on slope and erosion estimates. *Applied Engineering in Agriculture* 7:779–83.

Srivastava, K. P., and Moore, I. D. 1989. Application of terrain analysis to land resource investigations of small catchments in the Caribbean. In *Proceedings of the Twentieth International Conference of the Erosion Control Association, Steamboat Springs, Colorado*, 229–42.

Steele, B. M., Winne, J. C., and Redmond, R. L. 1998. Estimation and mapping of misclassification probabilities for thematic land cover maps. *Remote Sensing of Environment* 66:192–202.

Strahler, A. H. 1981. Stratification of natural vegetation for forest and rangeland inventory using *Landsat* digital imagery and collateral data. *International Journal of Remote Sensing* 2:15–41.

Stromberg, J. C., and Patten, D. T. 1990. Riparian vegetation instream flow requirements: a case study for a diverted stream in the eastern Sierra Nevada, California, USA. *Environmental Management* 14:185–94.

Su, W., and Mackey, B. G., 1997. A spatially explicit and temporal dynamic simulation model of forested landscape ecosystems. In A. D. McDonald and M. McAleer (eds.), *MODSIM 97: International Congress on Modeling and Simulation Proceedings, Hobart, 8–11 December 1997.* Hobart: University of Tasmania, 1635–40.

Suppnick, J. D. 1992. A nonpoint source pollution load allocation for Sycamore Creek in Ingham County, Michigan. In *Proceedings of the Surface Water Quality and Ecology Symposium.* New Orleans, LA: Water Environment Federation, 293–302.

Sutherland, R. A. 1991. Caesium-137 and sediment budgeting within a partially closed drainage basin. *Zeitschrift für Geomorphologie* 35:47–63.

Swank, W. T., and Douglass, J. E. 1975. Nutrient flux in undisturbed and manipulated forest ecosystems in the southern Appalachian Mountains. In *Proceedings of the Tokyo Symposium on the Hydrological Characteristics of River Basins.* Wallingford, UK: International Association of Hydrological Sciences, 445–56.

Swanson, D. K., and Grigal, D. F. 1989. Vegetation indicators of organic soil properties in Minnesota. *Soil Science Society of America Journal* 53:491–5.

Swanson, F. J., and Dyrness, C. T. 1975. Impact of clearcutting and road construction on soil erosion by landslides in the western Cascade Range, Oregon. *Geology* 3:393–6.

Swanson, F. J., Benda, L. E., Duncan, S. H., Grant, G. E., Megahan, W. F., Reid, L. M., and Ziemer, R. R. 1987. Mass failures and other processes of sediment production in Pacific Northwest forest landscapes. In E. O. Salo and T. W. Cundy (eds.) *Proceedings of the Streamside Management, Forestry, and Fishery Interactions Symposium, Seattle, Washington, 12–14 February, 1986.* Seattle, WA: University of Washington, Institute of Forest Resources Report No. 57:9–38.

Swift, L. W. 1976. Algorithm for solar radiation on mountain slopes. *Water Resources Research* 12:108–12.

Swift, L. W., Swank, W. T., Mankin, J. B., Luxmoore, R. J., and Goldstein, R. A. 1975. Simulation of evapotranspiration and drainage from mature and clear-cut deciduous forest and young pine plantation. *Water Resources Research* 11:667–73.

Swift, L. W., Cunningham, G. B., and Douglass, J. E. 1988. Climatology and hydrology. In W. T. Swank and D. A. Crossley (eds.) *Forest Hydrology and Ecology at Coweeta.* New York: Springer, 35–55.

Tajchman, S. J. 1981. On computing topographic characteristics of a mountainous catchment. *Canadian Journal of Forest Research* 11:768–74.

Tajchman, S. J., and Lacey, C. J. 1986. Bioclimatic factors in forest site potential. *Forest Ecology and Management* 14:211–8.

Tansley, A. G., and Chipp, T. F. 1926. *Aims and Methods in the Study of Vegetation.* London: British Empire Vegetation Committee.

Thomas, D. J. 1996. *Soil Survey of Macon County, North Carolina.* Franklin, NC: United States Department of Agriculture, Natural Resource Conservation Service.

Thompson, J. A. and Bell, J. C. 1996. A color index for identifying hydric conditions for seasonally-saturated prairie soils in Minnesota. *Soil Science Society of America Journal* 60:1979–88.

Thompson, J. A., Bell, J. C., and Butler, C. A. 1997. Quantitative soil–landscape modeling for estimating the areal extent of hydromorphic soils. *Soil Science Society of America Journal* 61:971–80.

Thompson, K. W. 1978. *Soil Survey of Slope County, North Dakota.* Washington, DC: United States Department of Agriculture, Soil Conservation Service.

Thorne, C. R., Zevenbergen, L. W., Grissinger, F. H., and Murphey, J. B. 1986. Ephemeral gullies as sources of sediment. In *Proceedings of the Fourth Federal Interagency Sedimentation Conference, Las Vegas, Nevada, 24–27 March (Volume 1)*, 3.152–61.

Thornthwaite, C. W., and Mather, J. R. 1957. *Instructions and Tables for Computing Potential Evapotranspiration and the Water Balance.* Centerton, PA: Drexel Institute of Technology Publications in Climatology No. 10-3.

Tim, U. S., Motaghimi, S., Shanholtz, V. O., and Zhang, N. 1992. *Identification of Critical Nonpoint Pollution Source Area Using Geographic Information Systems and Simulation Modeling.* Falls Church, VA: American Society of Agricultural Engineers Presentation Paper No. 912114.

Tompson, A. F., Ababou, R., and Gelhar, L. W. 1989. Implementation of the three-dimensional turning bands random field generator. *Water Resources Research* 25:2227–43.

Topp, G. C., Davis, J. L., and Annan, A. P. 1985. Measurement of soil water content using time-domain reflectometry. *Soil Science Society of America Journal* 49:19–24.

Tou, J. T., and Gonzalez, R. C. 1974. *Pattern Recognition Principles.* Reading, MA: Addison-Wesley.

Toy, T. J., Kuhaida, A. J., and Munson, B. E. 1978. The prediction of mean monthly soil temperature from mean monthly air temperature. *Soil Science* 126:181–9.

Troch, P. A., Mancini, M., Paniconi, C., and Wood, E. F. 1993. Evaluation of a distributed catchment scale water balance model. *Water Resources Research* 29:1805–17.

Troeh, F. R. 1964. Landform parameters correlated to soil drainage. *Soil Science Society of America Proceedings* 28:808–12.

Tsai, V. 1993. Delauney triangulations in TIN creation: an overview and linear-time algorithm. *International Journal of Geographical Information Systems* 7:501–24.

Tueller, P. T. 1989. Remote sensing technology for rangeland management applications. *Journal of Range Management* 42:442–53.

Twigg, D. R. 1998. The Global Positioning System and its use for terrain mapping and monitoring. In S. N. Lane, K. S. Richards, and J. H. Chandler (eds.), *Landform Monitoring, Modelling, and Analysis.* New York: Wiley, 37–62.

Tyler, S. W., McKay, W. A., and Mihevc, T. M. 1992. Assessment of soil moisture movement in nuclear subsidence craters. *Journal of Hydrology* 139:159–81.

United States Army Corps of Engineers. 1987. *GRASS Reference Manual.* Champaign, IL: Construction Engineering Research Laboratory.

United States Department of Agriculture, Soil Conservation Service. 1993. *State Soil Geographic Database User's Guide.* Washington, DC: United States Department of Agriculture, Soil Conservation Service.

United States Department of Agriculture, Soil Conservation Service, Cooperative Extension Service, and Agricultural Stabilization and Conservation Service. 1990. *Sycamore Creek Watershed Quality Plan, Ingham County.* East Lansing, MI: United States Department of Agriculture, Soil Conservation Service, Michigan State Office Unpublished Report.

United States Geological Survey. 1993. *Digital Elevation Models: Data Users Guide.* Reston, VA: United States Geological Survey.

United States Geological Survey. 1999. *USGS Digital Elevation Model Data.* WWW document, *http://edcwww.cr.usgs.gov/glis/hyper/guide/usgs_dem.*

United States Soil Conservation Service. 1985. *National Engineering Handbook: Section 4, Hydrology.* Washington, DC: United States Department of Agriculture.

United States Soil Conservation Service. 1986. *Urban Hydrology for Small Watersheds.* Washington, DC: United States Department of Agriculture Technical Release No. 55.

van Genuchten, M. T. 1980. A closed-form equation for predicting the hydraulic conductivity of unsaturated soils. *Soil Science Society of America Journal* 44:892–98.

Venables, W. M., and Ripley, B. D. 1994. *Modern Applied Statistics With S-Plus.* New York: Springer.

Ventura, S. J., Crisman, N. R., Connors, K., Gurda, R. F., and Martin, R. W. 1988. A land information system for soil erosion control planning. *Journal of Soil and Water Conservation* 43:230–3.

Vertessey, R. A., Hatton, T. J., O'Shaughnessy, P. J., and Jayasuriya, M. D. A. 1993. Predicting water yield from a mountain ash forest catchment using a terrain analysis based catchment model. *Journal of Hydrology* 150:665–700.

Vieux, B. E. 1993. DEM aggregation and smoothing effects on surface runoff modeling. *Journal of Computing in Civil Engineering* 7:310–38.

Vieux, B. E., and Needham, S. 1993. Nonpoint-pollution model sensitivity to grid-cell size. *Journal of Water Resources Planning and Management* 119:141–57.

Vieux, B. E., Farajalla, N. S., and Gaur, N. 1996. Integrated GIS and distributed storm water runoff modeling. In M. F. Goodchild, L. T. Steyaert, B. O. Parks, C. Johnston, D. R. Maidment, M. F. Crane, and S. Glendinning (eds.), *GIS and Environmental Modeling: Progress and Research Issues.* Fort Collins, CO: GIS World Books, 199–204.

Vörösmarty, C. J., Moore, B., Grace, A., Gildea, M. P., Melillo, J. M., Peterson, B. J., Rastetter, E. B., and Steudler, P. A. 1989. Continental scale models of water balance and fluvial transport: an application to South America. *Global Biogeochemical Cycles* 3:241–65.

Vose, J. M., and Swank, W. T. 1991. Water balances. In D. W. Johnson and S. E. Lindberg (eds.) *Atmospheric Deposition and Forest Nutrient Cycling: A Synthesis of the Integrated Forest Study.* New York: Springer, 29–49.

Walker, P. H., Hall, G. F., and Protz, R. 1968. Relation between landform parameters and soil properties. *Soil Science Society of America Proceedings* 32:101–4.

Walling, D. E. 1983. The sediment delivery problem. *Journal of Hydrology* 65:209–37.

Wang, F., Hall, G. B., and Subaryono. 1990. Fuzzy information representation and processing in conventional GIS software: database design and application. *International Journal of Geographical Information Systems* 4:261–83.

Ward, A. W., Ward, W. T., McBratney, A. B., and de Gruijter, J. J. 1992. *MacFuzzy: A Program for Data Analysis by Fuzzy k-Means.* Glen Osmond, South Australia: CSIRO Division of Soils.

Ward, T. J. 1976. *Factor of Safety Approach to Landslide Potential Delineation.* Ph.D. Thesis, Colorado State University.

Ward, T. J. 1985. Computer-based landslide delineation and risk assessment procedures for management planning. In *Proceedings of a Workshop on Slope Stability: Problems and Solutions in Forest Management,* 51–7.

Ward, T. J., Li, R., and Simons, D. B. 1978. Landslide potential and probability considering randomness of controlling factors. In *Proceedings of the International Symposium on Risk and Reliability in Water Resource, Waterloo, Canada, 26–28 June 1978.* Waterloo: University of Waterloo, 592–608.

Ward, T. J., Li, R., and Simons, D. B. 1982. Mapping landslide hazards in forest watersheds. *Transactions of the American Society of Civil Engineers* 108:319–24.

Waring, R. H., Rogers, J. J., and Swank, W. T. 1980. Water relations and hydrologic cycles. In D. E. Reichle (ed.), *Dynamic Properties of Forest Ecosystems.* Cambridge, UK: Cambridge University Press, 205–64.

Watson, D. F., and Philip, G. M. 1984. Triangle based interpolation. *Mathematical Geology* 16:779–95.

Webster, R. 1977. *Quantitative and Numerical Methods in Soil Classification and Survey.* Oxford, UK: Clarendon.

Webster, R., and Oliver, M. A. 1990. *Statistical Methods in Soil and Land Resource Survey.* Oxford, UK: Oxford University Press.

Weibel, R., and Heller, M. 1991. Digital terrain modelling. In D. J. Maguire, M. F. Goodchild, and D. W. Rhind (eds.), *Geographic Information Systems,* Vol. 1: *Principles.* Harlow: Longman: 269–97.

Weih, R. C., and Smith, J. L. 1997. The influence of cell slope computation algorithms on a common forest management decision. In M. J. Kraak and M. Molenaar (eds.), *Advances in GIS Research II: Proceedings of the Seventh International Symposium on Spatial Data Handling.* London: Taylor & Francis, 857–75.

Wells, P. V. 1962. Vegetation in relation to geological substratum and fire in the San Luis Obispo Quadrangle, California. *Ecological Monographs* 32:79–103.

Wendland, F., Albert, H., Bach, M., and Schmidt, R. 1993. *Atlas zum Nitratstrom in der Bundesrepublik Deutschland.* Berlin: Springer.

Westman, W. E. 1981a. Diversity relationships and succession in Californian coastal sage scrub. *Ecology* 62:170–84.

Westman, W. E. 1981b. Factors influencing the distribution of species of California coastal sage scrub. *Ecology* 62:439–55.

Westman, W. E. 1991. Measuring realized niche spaces: climatic response of chaparral and coastal sage scrub. *Ecology* 72:1678–84.

White, J. D., Kroh, G. C., and Pinder, J. E., III. 1995. Forest mapping at Lassen Volcanic National Park, California using *Landsat* TM data and a geographical information system. *Photogrammetric Engineering and Remote Sensing* 61:299–305.

Whittaker, R. H. 1951. A criticism of the plant association and climax community concepts. *Northwest Scientist* 25:17–31.

Wigmosta, M. S., Vail, L. W., and Lettenmaier, D. P. 1994. A distributed hydrology-vegetation model for complex terrain. *Water Resources Research* 30:1665–79.

Wilkin, D. C., and Hebel, S. J. 1982. Erosion, redeposition, and delivery of sediment to midwestern streams. *Water Resources Research* 18:1278–82.

Wilkinson, G. G. 1996. A review of current issues in the integration of GIS and remote sensing. *International Journal of Geographical Information Systems* 10:85–101.

Willgoose, G., Bras, R. L., and Rodriguez-Iturbe, I. 1991. A coupled channel network growth and hillslope evolution model, 1: theory. *Water Resources Research* 27:1671–84.

Wilson, J. P. 1996. GIS-based land surface/subsurface models: new potential for new models. In NCGIA (ed.), *Proceedings of the Third International Conference Integrating GIS and Environmental Modeling, Santa Fe, New Mexico, 21–25 January 1996.* Santa Barbara, CA: National Center for Geographic Information and Analysis, University of California: CD-ROM and WWW.

Wilson, J. P. 1999a. Local, regional, and global applications of GIS in agriculture. In P. A. Longley, M. F. Goodchild, D. J. Maguire, and D. W. Rhind (eds.), *Geographical Information Systems,* Vol. 2: *Management Issues and Applications.* New York: Wiley, 981–98.

Wilson, J. P. 1999b. Current and future trends in the development of integrated methodologies for assessing non-point source pollutants. In D. L. Corwin, K. Loague, and T. R. Ellsworth (eds.), *Assessment of Non-Point Source Pollution in the Vadose Zone.* Washington, DC: American Geophysical Union, 343–61.

Wilson, J. P., and Burrough, P. A. 1999. Dynamic modeling, geostatistics, and fuzzy classification: new sneakers for a new geography? *Annals of the Association of American Geographers* 89:736–46.

Wilson, J. P., and Gallant, J. C. 1996. EROS: a grid-based program for estimating spatially distributed erosion indices. *Computers and Geosciences* 22:707–12.

Wilson, J. P., and Gallant, J. C. 1998. Terrain-based approaches to environmental resource evaluation. In S. N. Lane, K. S. Richards, and J. H. Chandler (eds.), *Landform Monitoring, Modelling, and Analysis.* New York: Wiley, 219–40.

Wilson, J. P., and Gallant, J. C. 2001. SRAD: a program for estimating radiation and temperature in complex terrain. *Transactions in GIS* 5: in press.

Wilson, J. P., and Lorang, M. S. 1999. Spatial models of soil erosion and GIS. In A. S. Fotheringham and M. Wegener (eds.), *Spatial Models and GIS: New Potential and New Models.* London: Taylor & Francis, 83–108.

Wilson, J. P., Inskeep, W. P., Wraith, J. M., and Snyder, R. D. 1996. GIS-based solute transport modeling applications: scale effects of soil and climate data input. *Journal of Environmental Quality* 25:445–53.

Wilson, J. P., Spangrud, D. J., Landon, M. A., Jacobsen, J. S., and Nielsen, G. A. 1994. Mapping soil attributes for site-specific management of a Montana field. In G. F. Meyer and J. A. DeShazer (eds.) *Optics in Agriculture, Forestry, and Biological Processing.* Bellingham, WA, International Society for Optical Engineering Publication No. 2345:324–35.

Wilson, J. P., Spangrud, D. S., Nielsen, G. A., Jacobsen, J. S., and Tyler, D. A. 1998. GPS sampling intensity and pattern effects on computed terrain attributes. *Soil Science Society of America Journal* 62:1410–7.

Wischmeier, W. H., and Smith, D. D. 1978. *Predicting Rainfall Erosion Losses to Conservation Planning.* Washington, DC: United States Department of Agriculture Handbook No. 537.

Wise, S. M. 1998. The effect of GIS interpolation errors on the use of digital elevation models in geomorphology. In S. N. Lane, K. S. Richards, and J. H. Chandler (eds.), *Landform Monitoring, Modelling, and Analysis.* New York: Wiley, 139–64.

Wolock, D. M., and McCabe, G. J. 1995. Comparison of single and multiple flow direction algorithms for computing topographic parameters in TOPMODEL. *Water Resources Research* 31:1315–24.

Wolock, D. M., and Price, C. V. 1994. Effects of digital elevation model and map scale and data resolution on a topography-based watershed model. *Water Resources Research* 30:3041–52.

Wood, E. F., Sivapalan, M., and Beven, K. J. 1990. Similarity and scale in catchment storm response. *Reviews in Geophysics* 28:1–18.

Wood, R., Sivapalan, M., and Robinson, J. 1997. Modeling the spatial variability of surface runoff using a topographic index. *Water Resources Research* 33:1061–73.

Wood, W. L., and Calver, A. 1992. Initial conditions for hillslope hydrology modelling. *Journal of Hydrology* 130:379–97.

Woodcock, C. D., Collins, J., Gopal, S., Jakabhazy, V. D., Li, X., Macomber, S., Ryherd, S., Harward, V. J., Levitan, J., Wu, Y., and Warbington, R. 1994. Mapping forest vegetation using *Landsat* TM imagery and a canopy reflectance model. *Remote Sensing of the Environment* 50:240–54.

Wu, W. and Sidle, R. C. 1995. A distributed slope stability model for steep forested basins. *Water Resources Research* 31:2097–2110.

Xiang, W. N. 1993. Application of a GIS-based stream buffer generation model to environmental policy evaluation. *Environmental Management* 17:817–27.

Yaalon, D. H. 1975. Conceptual models in pedogenesis: can soil-forming functions be solved? *Geoderma* 14:189–205.

Yeakley, J. A., Meyer, J. L., and Swank, W. T. 1994. Hillslope nutrient flux during near-stream vegetation removal, I: a multi-scaled modeling design. *Water, Air and Soil Pollution* 77:229–46.

Yeakley, J. A., Swank, W. T., Swift, L. W., Hornberger, G. M., and Shugart, H. H. 1998. Soil moisture gradients and controls on a southern Appalachian hillslope from drought through recharge. *Hydrology and Earth System Sciences* 2:31–9.

Young, R. A., Onstad, C. A., Bosch, D. D., and Anderson, W. P. 1989. AGNPS: a nonpoint source pollution model for evaluating agricultural watersheds. *Journal of Soil and Water Conservation* 44:168–73.

Yu, S., van Kreveld, M., and Snoeyink, J. 1997. Drainage queries in TINs: from local to global and back again. In M. J. Kraak and M. Molenaar (eds.), *Advances in GIS Research II: Proceedings of the Seventh International Symposium on Spatial Data Handling*. London: Taylor & Francis, 829–42.

Zadeh, L. A. 1965. Fuzzy sets. *Information and Control* 8:338–53.

Zar, Z. H. 1984. *Biostatistical Analysis,* 2d ed. Englewood Cliffs, NJ: Prentice-Hall.

Zebker, H. A., Werner, C., Rosen, P. A., and Hensley, S. 1994. Accuracy of topographic maps derived from ERS-1 interferometric radar. *IEEE Transactions on Geoscience and Remote Sensing* 32:823–36.

Zevenbergen, L. W., and Thorne, C. R. 1987. Quantitative analysis of land surface topography. *Earth Surface Processes and Landforms* 12:47–56.

Zhang, W. H., and Montgomery, D. R. 1994. Digital elevation model grid size, landscape representation, and hydrologic simulations. *Water Resources Research* 30:1019–28.

Zheng, D., Hunt, E. R., and Running, S. W. 1995. Comparison of available water capacity estimated from topography and soil series information. *Landscape Ecology* 11:4–13.

Zhuang, X., Engel, B. A., Xiong, X., and Johannsen, C. J. 1995. Analysis of classification results of remotely sensed data and evaluation of classification algorithms. *Photogrammetric Engineering and Remote Sensing* 61:427–33.

Zienkiewicz, O. C. 1977. *The Finite Element Method.* London: McGraw-Hill.

Zoltai, S. C. 1965. Glacial features of the Quetico–Nipigon area, Ontario. *Canadian Journal of Earth Sciences* 2:247–69.

remote sensing data, 400
slope, 403
topographic wetness index, 398, 403,
    405, 410, 412–414
trembling aspen:
    autecology of, 399
    distribution of, 414, 415, 417
upslope contributing area, 398, 403, 406
Riparian areas. *See* Investigative riparian
    buffers; Stream buffers
Runoff curve number method, 90, 111–113
RUSLE. *See* Revised Universal Soil Loss
    Equation

Scale. *See* Spatial scale; Temporal scale
Secondary topographic attributes, 6–15,
    87–131, 424. *See also* Solar radia-
    tion indices; Stream power index;
    Temperature indices; Topographic
    wetness index; DYNWET-C pro-
    gram; DYNWET-G program;
    EROS program; SRAD program;
    WET programs
simplifying assumptions, 20–21, 87
Sediment transport capacity indices,
    11–12, 88–89, 119–120, 122,
    142–148, 150, 154–156. *See also*
    Stream power index
correlation with Cesium-137 soil redis-
    tribution rates, 156
correlation with selected primary and
    secondary topographic attributes,
    155–156
differences from RUSLE length-slope
    factor, 89, 142–144, 146
effect of DEM source and resolution on,
    138–141
geomorphological applications, 11–12
SHE hydrological model, 226
Slope, 6–7, 53–54, 73, 83–84, 138,
    155–156, 196, 254, 277–278,
    368
contour-based methods, 83–84
correlation with Cesium-137 soil redis-
    tribution rates, 156
correlation with selected primary and
    secondary topographic attributes,
    155–156
D8 method, 53–54

effect of DEM source and resolution on,
    137–138, 141
finite difference method, 53–54, 196
Soil erosion. *See also* Nonpoint source
    pollution; Water quality
Cesium-137 soil redistribution rates,
    153–154, 186
detachment-limited erosion, 159–161,
    187. *See also* Universal Soil Loss
    Equation; Revised Universal Soil
    Loss Equation
sediment delivery ratio, 186
transport-limited erosion, 159–161, 187.
    See also Sediment transport capac-
    ity indices
Soil formation. *See* Pedogenesis
Soil-landscape models, 296–297, 301–302,
    304–305, 307, 309. *See also* Soil
    survey; Spatial prediction methods
Soil moisture, 6, 8, 10–11, 14–15, 192,
    200–201, 205, 331
field observation of wet spots, 192,
    200–201
map evidence of high flows, 192,
    200
modeling in humid mountainous land-
    scapes, 215–216, 218–224. *See
    also* IHDM4 hydrological model
topographic controls, 10, 205
Soil properties. *See* Soil survey
Soil survey, 245–246, 265, 267. *See also*
    Landform classification; Pedogene-
    sis
sampling issues, 256
soil organic carbon, 295–296
spatial distribution of soil properties,
    245–246, 251–253, 261, 296–297,
    305–309. *See also* Spatial predic-
    tion methods
soil map units, 245–246
SOLARFLUX model, 14, 276, 280–281,
    337
Solar radiation indices:
biological applications, 12
incident short-wave solar radiation. *See*
    Short-wave radiation
long-wave radiation, 9, 13–14, 99
net radiation, 9, 13–14, 99–100,
    123–125